Real Estate Marketing

Real Estate Marketing is specifically designed to educate real estate students in the art and science of the real estate marketing profession.

The ideal textbook for undergraduate- and graduate-level classes in business schools as well as professional and continuing education programs in real estate, this book will also be of interest to professional real estate entrepreneurs looking to boost their knowledge and marketing techniques.

The book is divided into five major parts. Part 1 focuses on introducing students to fundamental concepts of marketing as a business philosophy and strategy. Concepts discussed include strategic analysis, target marketing, and the four elements of the marketing mix: property planning, site selection, pricing of properties, and promotion of properties.

Part 2 looks at personal selling in real estate. Students will learn the exact process and steps involved in representing real estate buyers and sellers.

Part 3 centers on negotiations in real estate. How do effective real estate professionals use various negotiation approaches such as collaboration, competition, accommodation, and compromise as a direct function of the situation and personalities involved in either buying or selling real estate properties?

Part 4 concentrates on human resource management issues such as recruiting and training real estate agents; performance evaluation, motivation, and compensation; and leadership.

Finally, Part 5 focuses on legal and ethical issues in the real estate industry. Students will learn how to address difficult situations and legal/ethical dilemmas by understanding and applying a variety of legal/ethical tests. Students will also become intimately familiar with the industry's code of ethics.

M. Joseph Sirgy is Professor of Marketing and Virginia Real Estate Research Fellow at Virginia Polytechnic Institute and State University, U.S.A.

Real Estate Marketing

Strategy, Personal Selling, Negotiation, Management, and Ethics

M. Joseph Sirgy

Routledge
Taylor & Francis Group

LONDON AND NEW YORK

First published 2014
by Routledge
2 Park Square, Milton Park, Abingdon, Oxon OX14 4RN

and by Routledge
711 Third Avenue, New York, NY 10017

Routledge is an imprint of the Taylor & Francis Group, an informa business

© 2014 M. Joseph Sirgy

British Library Cataloguing in Publication Data
A catalogue record for this book is available from the British Library

Library of Congress Cataloging-in-Publication Data
Sirgy, M. Joseph.
    Real estate marketing : strategy, personal selling, negotiation, management, and ethics / M. Joseph Sirgy.
        pages cm
    Includes bibliographical references and index.
    1. Real estate business.   2. Real estate development.   3. Residential
real estate—Marketing.   4. Commercial real estate—Marketing.   I. Title.
    HD1375.S374 2015
    333.33068′8—dc23
    2013046506

ISBN: 978-0-415-72394-7 (hbk)
ISBN: 978-0-415-72401-2 (pbk)
ISBN: 978-1-315-77596-8 (ebk)

Typeset in Goudy
by Apex CoVantage, LLC

Printed and bound in Great Britain by
TJ International Ltd, Padstow, Cornwall

Contents

Figures

Tables

About the Author

M. JOSEPH SIRGY is a management psychologist (Ph.D., U/Massachusetts, 1979), Professor of Marketing, and Virginia Real Estate Research Fellow at Virginia Polytechnic Institute and State University (Virginia Tech). He has published extensively in the area of marketing, business ethics, and quality of life (QOL) as both an author and an editor. His particular research interests lie in housing well-being, neighborhood well-being, and community well-being.

Sirgy co-founded the International Society for Quality-of-Life Studies (ISQOLS) in 1995, serving as its Executive Director/Treasurer from 1995 to 2011 and its Development Director from 2011 to 2012. ISQOLS has honored him with the Distinguished Fellow Award in 1998 and as the Distinguished QOL Researcher, for research excellence and a record of lifetime achievement in QOL research, in 2003. He has also served as President of the Academy of Marketing Science, from which he received the Distinguished Fellow Award in the early 1990s and the Harold Berkman Service Award, a lifetime achievement award for serving the marketing professoriate, in 2007.

In 2008, Sirgy received the Virginia Tech's Pamplin Teaching Excellence Award/Holtzman Outstanding Educator Award and University Certificate of Teaching Excellence. In 2010, ISQOLS honored him for excellence and lifetime service to the society. He has won the 2010 Best Paper Award in the *Journal of Happiness Studies* for his theory on the balanced life and the 2011 Best Paper Award in the *Journal of Travel Research* for his goal theory of leisure travel satisfaction. In 2012, he was awarded the EuroMed Management Research Award for outstanding achievements and groundbreaking contributions to well-being and quality-of-life research.

In the early 2000s, Sirgy helped co-found the Macromarketing Society and the Community Indicators Consortium, and he has served as a board member of these two professional associations. He co-founded the journal, *Applied Research in Quality of Life*, the official journal of the International Society for Quality-of-Life Studies, in 2005, and has served as editor since then. He serves as editor of the QOL section in the *Journal of Macromarketing* (1995–present) and is the current editor of ISQOLS/Springer book series on handbooks in QOL research and the community QOL indicators best practices.

Preface

This book is about real estate marketing. It is written for college students taking the basic *real estate marketing course*—usually at the undergraduate level, although the same course is customarily offered in real estate graduate-level programs at many universities. The focus is mostly on the U.S., although the vast variety of the marketing concepts and illustrations are generalizable to many other countries, specifically the English-speaking countries such as the Great Britain, Canada, Australia, and New Zealand. For example, in Part 5 of the book we discuss real estate marketing laws mostly from a U.S. perspective. Nevertheless, we also discuss variations of real estate marketing laws across countries.

The book is divided into five major sections reflecting five major themes: *strategy, personal selling, negotiation, management,* and *ethics*. These themes are reflected in the book's subtitle. Part 1 (*Strategy in Real Estate Development Firms*) focuses on strategy issues related to real estate development firms. This part has three chapters. Chapter 1 (*Marketing Strategy*) discusses marketing strategy by marketing executives in real estate development firms. We make the distinction between selling and marketing and provide a formal definition of real estate marketing. We then discuss important concepts in marketing strategy such as strategic analysis and target marketing (or what we call prospect strategy). Chapters 2 and 3 tackle what marketing scholars refer to as the "four Ps"—product, place, price, and promotion. Specifically, Chapter 2 (*Product and Place Strategy*) focuses on product and place strategy in real estate development firms, while Chapter 3 (*Price and Promotion Strategy*) focuses on price and promotion decision-making.

The focus of much of Part 1 is on the marketing manager in real estate development firm. In contrast, Part 2 (*Personal Selling—the Real Estate Agent*) focuses on the real estate agent and the business of buying and selling in the context of real estate brokerage firms. Specifically, Chapter 4 (*The Seller Representative*) describes the process of personal selling by the seller representative (or the listing agent). Chapter 5 (*The Buyer Representative*) focuses on the process of personal selling by the buyer representative.

Part 3 (*How Real Estate Agents Negotiate*) transitions to negotiation concepts and applications. The focus remains on the real estate agent who is assists the buyer or seller clients to negotiate in a manner most likely to result in a mutually satisfactory deal—mutually satisfactory for both seller and buyer parties. Here we emphasize win-win negotiations, not zero-sum (or win-lose) negotiation. Thus, Chapter 6 (*The Social Psychology of Real Estate Negotiation*) discusses many the social psychological principles of negotiation that have been demonstrated through good science to lead to effective (win-win) outcome. Chapter 7 (*Negotiation Strategies and Tactics*) takes many of the social psychological principles and translates them into negotiation strategies and tactics that are readily used by the real estate agent.

Part 4 (*Sales Management—the Real Estate Broker*) focuses on the real estate broker or manager—the real estate professional who manages the brokerage firm. This part has three chapters focusing on sales management issues. Specifically, Chapter 8 (*Recruitment and Training of Real Estate Salespeople*) discusses many of the principles related to effective recruitment and training of real estate agents in a brokerage firm. Chapter 9 (*Motivation and Compensation Issues in Real Estate Marketing*) discusses principles related to motivating the sales force and effective compensation. Chapter 10 (*Leadership Issues in Real Estate Firms*) focuses on leadership principles as applied to real estate brokers.

Part 5 (*Law and Ethics in Real Estate Marketing*) is the final section of the book; and the title suggests, the focus here is on laws and ethics pertinent to real estate marketing. Chapter 11 (*Real Estate Marketing Laws*) focuses on the many marketing laws designed to regulate the real estate industry. Although the primary focus is on the U.S. we discuss real estate marketing laws in the English-speaking countries such as Great Britain, Canada, Australia, and New Zealand. Chapter 12 (*Ethics in Real Estate Marketing*) covers many ethics principles and applies them to situations most relevant to real estate marketers. Finally, Chapter 13 (*A Code of Ethics for Real Estate Marketing Professionals*) describes in some detail the U.S. National Association of Realtors' (NAR) code of ethics and its 17 main articles. We provide many examples that bring those articles to life. Although the NAR code is created from a U.S. perspective, the vast majority of its articles readily apply globally, particularly the English-speaking countries.

We hope that students in real estate marketing will find this book useful in helping pave a career in real estate marketing. Happy reading and fruitful learning!

Part 1

Strategy in Real Estate Development Firms

Chapter 1

Marketing Strategy

LEARNING OBJECTIVES

This chapter covers many topics related to the marketing strategy of real estate development firms. It is designed to help students of real estate marketing learn:

■ what the marketing concept is and how it is different from the selling concept in the context of real estate development firms
■ a formal definition of real estate marketing
■ how the marketing manager performs a sales analysis
■ how the marketing manager performs a customer analysis
■ how the marketing manager performs a market analysis
■ how the marketing manager performs a competitive analysis
■ how the marketing manager makes target marketing decisions

THE MARKETING CONCEPT

What is the marketing concept? How is it different from the selling concept? How do we formally define the real estate marketing discipline? Here are some answers.[1]

Let's start with the selling concept. This concept is captured in Figure 1.1. Many traditional real estate firms practice marketing with a selling concept in mind. A real estate development firm develops a parcel of land (e.g., development of a residential subdivision involving 30 single-family homes). Once the units are available for sales, the marketing staff becomes involved in advertising and promotion of these units to prospective buyers. In other words, the marketing function is essentially a selling function—nothing more, nothing less. This is very different from a real estate development firm that takes the marketing concept to heart. Selling is only one small element of the marketing function. Marketing involves marketing research that guides the formulation of an integrated marketing plan. The goal is to achieve customer satisfaction, which ultimately leads to profitability (see Figure 1.2). This means that the real estate development firms starts out with good marketing research to understand the housing needs of a customer group. Suppose a real estate development firm is planning to build a retirement community with a variety of living options:

■ *active adult*—adult community for those 55 to 60 years old in which residents enjoy the benefits of home ownership in single-family, low-maintenance homes
■ *independent living*—single-level homes designed for the active, independent senior (ages 60 and older)
■ *assisted-living level 1*—for residents who want a catered lifestyle in a safe and secure environment
■ *assisted-living level 2*—for residents who require assistance with daily living activities but who are not quite ready for long-term nursing care

- *memory care*—for residents with symptoms of dementia
- *long-term nursing care*—for residents requiring assistance with all or most activities of daily living and 24-hour nursing care

To do so, the firm has to do thorough marketing research to uncover the exact housing needs and preferences of six different market segments (active adult, independent living, assisted living level 1, assisted living level 2, memory care, and long-term nursing care). The product (different housing structures and amenities), the price (pricing of the different elements of the product line), the place (the location of these different elements of the product line), and the promotion (the messages and media placement of these different elements of the product line) have to be guided by marketing research related to the aforementioned six market segments. Customer need assessment (i.e., marketing research) is paramount to marketing effectiveness, and it is essentially the first stage in the marketing cycle (see Figure 1.2). The information unearthed from marketing research should pave the way for formulating an integrated marketing plan and implementing this plan (integrated marketing effort), which is essentially the second stage in the marketing cycle. This means that the real estate developer uses marketing research to plan the product mix (e.g., the various types of residential structures suited for the six different market segments), to price the various housing structures (as a function of cost, competition, and customers' willingness to pay), to find optimal location

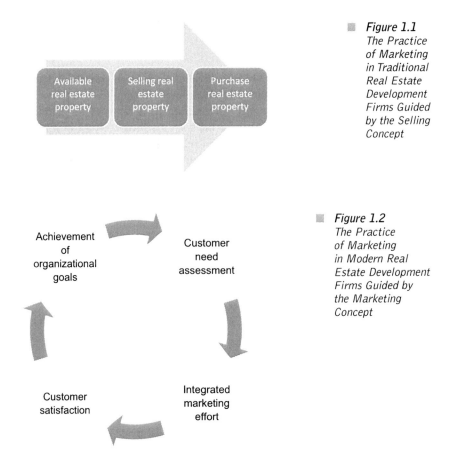

Figure 1.1
The Practice of Marketing in Traditional Real Estate Development Firms Guided by the Selling Concept

Figure 1.2
The Practice of Marketing in Modern Real Estate Development Firms Guided by the Marketing Concept

sites for the planned structures (as a function of customers' location preferences and other structural, environmental, and legal conditions), and to promote these housing structures to the various market segments (as a function of understanding the housing needs of the various customer groups and their media habits). The third stage of the marketing cycle is customer satisfaction. This means that customer satisfaction is a very important objective. The marketing effort of the firm is evaluated as a direct function of customer satisfaction. Marketing success is acknowledged only if customer satisfaction is achieved. How do real estate development firms recognize whether customer satisfaction is achieved? Through customer satisfaction research! That is, the firm conducts periodic surveys of the various customer groups to capture the degree of satisfaction they experience with the various elements of the marketing mix (product, price, place, and promotion). The result of customer satisfaction is profitability—high levels of customer satisfaction should translate to high levels of profitability. The feedback loop is reflected in situations when the firm does not achieve its stated goals and objectives. This causes stress in the system, prompting the real estate developer to conduct more marketing research to uncover problems in the planning of the marketing mix, the implementation of the mix, and the method of performance evaluation (i.e., customer satisfaction research). Once the problems are identified, corrective action follows to ensure that the system meets the stated organizational goals and objectives.

A FORMAL DEFINITION OF REAL ESTATE MARKETING

Here we present a formal definition of real estate marketing guided by the marketing concept. Real estate marketing involves anticipating, managing, and satisfying demand via the exchange process between buyer and seller of a property. As such, marketing encompasses all facets of real estate buyer/seller relationships. Specific marketing activities include strategic analysis, target marketing, property planning, site selection, pricing of the property, promotion planning, and marketing management.[2]

This definition of real estate marketing can be depicted as another process or cycle as shown in Figure 1.3. These are strategic analysis, prospect strategy (target marketing),

Figure 1.3
The Marketing Process in Real Estate Development Firms

product strategy (property planning), place strategy (site selection), price strategy (pricing of the property), and promotion strategy (promotion of the property). We will discuss strategic analysis and target marketing in some detail in the remaining parts of this chapter. Chapter 2 will cover product and place strategy, while Chapter 3 will cover price and promotion strategy.

STRATEGIC ANALYSIS

Strategic analysis involves an assessment of the internal and external environments (see Figure 1.4). An assessment of the internal environment involves an analysis of sales and customers, whereas an assessment of the external environment involves an analysis of the market and the competition.[3]

Note that strategic analysis is not conducted in a vacuum. In other words, the marketing manager does not simply start out with strategic analysis, and then all prospect, product, place, price, and promotion decisions are based on the information reflected in that strategic analysis. Strategic analysis is a continuous process of information gathering, data collection, and analysis. The marketing manager starts out with an assessment of the internal and external environments, but this assessment is further guided by the many decisions the manager has to make concerning prospect, product, place, price, and promotion strategies. This point is accentuated by the double arrows between strategic analysis and the five Ps (prospect strategy, product strategy, place strategy, price strategy, and promotion strategy) as shown in Figure 1.3.

Internal Analysis

As previously mentioned, internal analysis (or an assessment of the internal environment) involves a sales analysis and a customer analysis.

Sales Analysis

A *sales analysis* focuses on plotting sales trends over the last several years (or as far back as possible, depending on data availability) and making an attempt to explain sales fluctuations.[4] That is, the marketing manager makes an attempt to explain factors that may have contributed to high levels of sales as well as low sales. Examine Figure 1.5. The figure shows a sales trend of residential homes. The x-axis shows the time scale: 1994–2014. The y-axis

Figure 1.4
The Elements Involved in Strategic Analysis Performed by the Marketing Staff in Real Estate Development Firms

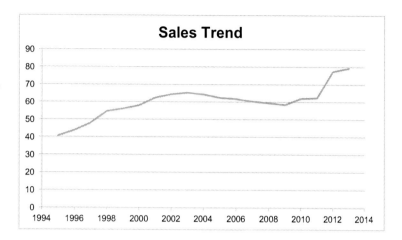

Figure 1.5
Sales Analysis
of Residential
Homes for the
Retired Sold
by Real Estate
Development
Firm XYZ

shows dollar sales in millions. The sales trend for residential homes for retired people seems to have continuously increased up to 2003, then decreased slightly all the way to 2009, and then shot up in 2012 and 2013. The challenge for the marketing manager is to explain this trend. Perhaps the slight decline between 2003 and 2011 was due to the sluggish economy, which then shot up in 2012 and 2013. Or perhaps in the early years (1995–2003) the upward sales trend was due to increased promotion expenditures, which decreased a little between 2003 and 2011, and increased again in 2012 and 2013. In this case we have two hypotheses: (1) Sales of residential homes for the retired may have been influenced by changes in the gross national product (GDP, a measure of economic well-being), and (2) sales of residential homes may have been influenced by changes in promotion expenditures directed to retired consumers. The marketing manager at this point should test these two hypotheses by gathering GDP data and promotion expenditure data (campaigns directed toward retired consumers) from 1995 to 2013. These two trends (GDP and expenditures) should then be plotted. If the expenditures trend curve looks more similar to the sales trend than the GDP trend, then one can conclude promotion expenditures may have influenced the sales trend. This information is important because such an inference is likely to prompt the marketing manager to allocate a higher level of promotion expenditures to jack up sluggish sales. Of course, this analysis is highly superficial, in the sense that marketing scientists would say that one cannot make a casual inference based on such an "eye-ball inspection" of the trends. A more rigorous analysis has to be conducted using multiple regression, in which sales could be treated as the criterion variable and GDP and promotion expenditures as predictor variables, with a host of other covariates. Even then one has to be very careful in assuming causality because such tests are correlational in nature, not experimental. However, given the nature of the data and in the absence of other information, the marketing manager could rely on such analysis, as long as he is cognizant of the uncertainty of the causal associations.

Customer Analysis

Let us now turn to *customer analysis*. This type of analysis focuses on existing customers— customers that have bought a property from the real estate development firm. Marketing

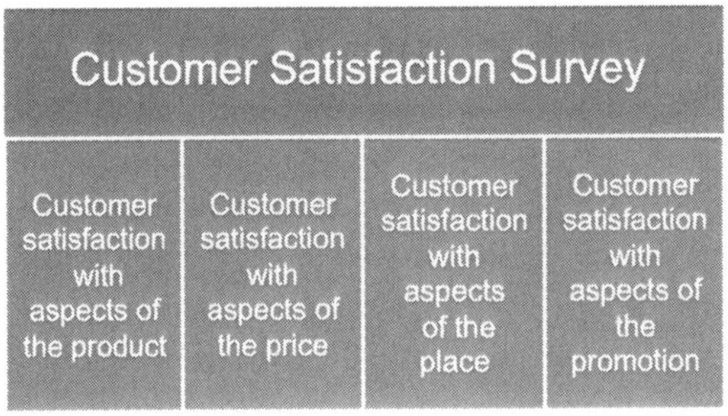

Customer Satisfaction Survey			
Customer satisfaction with aspects of the product	Customer satisfaction with aspects of the price	Customer satisfaction with aspects of the place	Customer satisfaction with aspects of the promotion

Figure 1.6
Elements of a Customer Satisfaction Survey

researchers typically conduct customer satisfaction surveys (see Figure 1.6). The goal is to capture the extent to which customers felt satisfied or dissatisfied about four major areas:

- the product (e.g., housing design, residential site, and energy conservation)
- the place (e.g., location in the neighborhood and the community, neighborhood, and community features)
- the price (e.g., price affordability, value for the money, and price negotiations)
- the promotion (e.g., quality and quantity of information provided to customers to help them with the buying decision)[5]

This type of strategic analysis is vitally important to the marketing manager because it allows the manager to identify problem areas and take corrective action. For example, suppose the real estate developer has built a retirement community. The customer satisfaction survey reveals that there are specific housing design features that customers are not happy with (e.g., many of the housing features are not designed to be sufficiently user friendly for those who have arthritis, such as too many stairs, no railings, toilets that are too low, no handle bars in the bathtubs, kitchen cabinets that are too high, etc.). Knowing that a significant segment of the retired customers are likely to have some form of physical disability, it would be important to make changes in the housing design to accommodate their housing needs.

External Analysis

An assessment of the external environment (i.e., external analysis) involves two types of assessments: market analysis and competitive analysis.

Market Analysis

Market analysis refers to collecting and analyzing data about issues directly related to target marketing, product strategy, place strategy, pricing strategy, and promotion strategy.[6] Let us discuss these different types of analyses (see Figure 1.7).

Figure 1.7
What Is Market
Analysis?

Market Analysis				
Market analysis related to prospect strategy	Market analysis related to product strategy	Market analysis related to place strategy	Market analysis related to price strategy	Market analysis related to promotion strategy

Market analysis related to target marketing focuses on collecting data about specific population segments to assess their market potential and the extent to which the firm may want to target these segments. For example, suppose a real estate developer has the ability to develop five types of residential communities:

1. college student communities (e.g., apartment complexes well-suited for college students)
2. communities for small families (e.g., starter homes)
3. gated communities for affluent families (e.g., luxury homes)
4. communities for the newly retired (residential homes for the retired)
5. communities for the frail elderly (assisted-living homes)

The developer is located in Northern Virginia and would like to focus its real estate development efforts on specific municipalities in the Northern Virginia region. To assess the market potential of college student communities, the developer focuses on those municipalities that contain higher education institutions and identifies several areas. The developer then gathers data from secondary sources related to the growth of the college student population in these colleges and universities. Those municipalities that have colleges and universities with growing student populations are selected for further review. Housing data are then collected from the selected colleges and universities to examine how these institutions are meeting the housing demand of their students—both on-campus and off-campus. Those institutions in which a significant segment of the student population seek off-campus housing are then selected for further review. In those areas that have a significant off-campus student population, additional data are collected regarding the occupancy rate of apartment complexes that are known to house college students. Data showing that the occupancy rate is very high signal market demand for additional off-campus student housing. Thus, those areas are identified for further real estate development. Similar analyses can be conducted in relation to the other potential target market segments (small families, affluent families, the newly retired, and the frail elderly) to assess market demand. Once these analyses are completed, the real estate developer is in a better position to make target marketing decisions—how the developer will allocate resources toward the development of certain types of residential communities. See Cases/Anecdotes 1.1, 1.2, and 1.3 for a glimpse of market analysis related to the assisted living, immigrant segments, and Generation Y home buyers.

CASE/ANECDOTE 1.1. IMMIGRANTS KEY TO HOUSING (AND ECONOMIC) RECOVERY[i]

Research by Housing America and the Mortgage Bankers Association found that immigrants are helping with the housing recovery. The study findings underscored the following: (1) homeownership has been increasing among immigrants and falling among native-born Americans, (2) immigrants are likely to account for 35.7% of the homebuyer market segment from 2010 to 2020, (3) the growth of the immigrant homebuyer market is projected to remain strong across the U.S., and (4) this surge of immigrant buying homes is projected to account for $100 billion in the specified period.

[i] This case/anecdote is based on: Commonwealth. (2013). Immigrants key to housing (and economic) recovery. *Commonwealth*, June/July issue, 8.

CASE/ANECDOTE 1.2. THE ELDERLY AND THE ASSISTED LIVING MARKET[i]

The market segment for assisted living seems to be decreasing in size. This may be due to the fact that adult children are distressed financially and cutting down on spending because of the poor economy. Placing one's parents in an assisted-living facility is considered "discretionary spending." Instead of placing their parents in assisted-living facilities, many adult children are making do using home care services and technology to assist their parents at home. Such technology includes home monitoring devices and long-distance caregiving. There seems to be a rise in demand for compensatory services, including emergency room services and home- and community-based services. In addition, delaying the use of assisted living means that as older people become increasingly frail, they seek more acute services directly related to their frailty, such as memory care services and nursing care.

[i] This case/anecdote is based on: Adami, P. (2011). Marketing assisted living today. *Long-Term Living*, April 14, 2011. Accessed from www.ltlmagazine.com/article/marketing-assisted-living-today, on May 10, 2013; IBIS World. (2009). Retirement communities in the US. May 2009. IBIS World, Santa Monica: CA; University of Texas at San Antonio, John Peace Library. San Antonio, TX. Accessed from www.ibisworld.com on May 10, 2013.

CASE/ANECDOTE 1.3. GENERATION Y HOME BUYERS[i]

Generation Y home buyers are tech savvy, armed with digital spreadsheets, instant communication, and an inclination to buy homes that are technologically friendly. Gen Y home buyers are young and are establishing roots to raise a family. Many are cautious because they've been adversely affected by the economy, have much student debt, and yet to feel a sense of stability in their new jobs. Many have been renting and waiting to seize the right moment to buy a home. How might real estate developers target this market?

What do Gen Y buyers want in their homes? They may share some characteristics with Gen X buyers, such as a preference for walk-in closets. They also differ from Gen X buyers in that they are less interested in state-of-the-art kitchens but more in having fun. Gen Y buyers most likely prefer a high-tech entertainment center, a game room, a home gym, and quite possibly a swimming pool.

Gen Y buyers are accustomed to using tablets, laptops, and mobile phones. Gen X buyers use e-mail, whereas Gen Y buyers text on mobile phones or communicate through Facebook. Gen Y buyers are likely to use the Internet for everything, such as researching their future home. The marketing manager should communicate with these buyers accordingly. To appeal to this group, the marketing manager has to have a complete, user-friendly, mobile-enabled Web site to help these buyers access information on their own time, whether they're lying in bed or brushing their teeth with smartphone in hand.

[i] This case/anecdote is based on: Forrest, J. (2013). Is your brokerage ready for Gen Y buyers? *Realtor Magazine*, September 2013. Accessed from http://realtormag.realtor.org/for-brokers/solutions/article/2013/09/your-brokerage-ready-for-gen-y-buyers?om_rid=AAAuJU&om_mid=_BSXZc5B81yRx$3&om_ntype=BTNMonthly, on October 15, 2013.

Market analysis related to product strategy involves collecting and analyzing data dealing with housing design, residential sites, and energy conservation. For example, if the real estate developer is likely to target the college student market to build off-campus student housing in selected municipalities in Northern Virginia, then identifying the various housing design options and collecting secondary data regarding the popularity of certain options can be very helpful in articulating a product strategy that adopts a popular housing design option.

Market analysis related to place strategy is equally important to the preceding analyses (the analyses related to target marketing and product strategy). The focus here is on collecting data on neighborhoods and communities to aid in the decision of where to build. Location, location, and location! It matters a great deal to find the right location to build. The location has to meet the needs and preferences of the target market segment. To assist with this decision, secondary information can be collected about the various neighborhoods and communities that the developer is homing in on (e.g., selected municipalities in Northern Virginia). Data about the physical, social, and economic features of neighborhoods should be collected. Examples of data on the physical features include the location of the neighborhood vis-à-vis shopping, employment, and highways; the extent of upkeep of property and landscaping in the neighborhood; and the presence of any environmental hazards. Examples of data on the social features include statistics on crime in the selected neighborhoods and the socio-demographic profile of the neighborhood residents. Examples of data on the economic features include rate of occupancy in the neighborhood and the market value of the homes in the neighborhood based on tax assessment records. Such data on the physical, social, and economic features of neighborhoods considered for development can play an important role in site selection.

Market analysis related to pricing strategy involves collecting and analyzing data about housing costs and prices. For examples, if the real estate developer is leaning towards building apartment complexes for off-campus housing for college students, then collecting data about rental fees of various types of apartments (e.g., studio, 1-bedroom apartment, 2-bedroom apartment, 3-bedroom apartment) could be very helpful in developing a rental fee policy once the units are available for rent. If the real estate developer is selling these units, instead of renting, then collecting data concerning recent sales of comparable units should help determine the asking price once the units are made available for sale.

Market analysis related to promotion strategy involves data collection of the media habits of the target market segment. For example, if the target market were to be college students, then information is needed to identify those mass media outlets they use most frequently (e.g., radio stations, newspapers, social media sites, television channels and shows they watch frequently, outdoor bulletin boards they see most frequently, etc.). Information about media habits assists in the development of a media schedule, which is an important element of promotion planning. What is also important in promotion planning is identifying product benefits that the target market perceives as very important in their decision to adopt the product offering (e.g., product, price, and place features that college students believe are most important in their decision to rent an apartment off-campus—features such as number of bedrooms, the sizes of the bedrooms, access to high-speed internet, price, location of the apartment complex to the campus, access to public transportation, etc.). Real estate developers cannot list all benefits in a single promotional message. Developers have to zero in on the most important benefits and use only these in their promotion. Part of this process involves generating a slogan that captures a core benefit.

Competitive Analysis

Similar to market analysis, *competitive analysis* involves data collection that can assist the marketing manager with target marketing, product strategy, pricing strategy, place strategy, and promotion strategy (see Figure 1.8).[7]

Competitive analysis related to target marketing involves gathering information about key competitors in an attempt to identify their target market. Such information is collected from competitors' promotional material (e.g., newspaper advertising, television commercials, radio advertising, promotional brochures, messages embedded in their website). For example, if a key competitor is targeting college students, messages directed to college students are likely to be quite evident in their promotional material. Finding out which market segments key competitors are targeting should be helpful in assisting the real estate developer make a decision whether to occupy the same market space (i.e., engage in fierce competition) or look for a market niche that is absent of key competitors (i.e., avoid competition altogether). Of course, the answer to this important question is whether there is sufficient growth in the competitors' market segment to allow the developer to penetrate the same market and still be profitable. That is, can they survive and prosper in spite of the competition? Thus, additional information

Figure 1.8
What Is Competitive Analysis?

is required about the rate of growth of the target market to fully answer this question. To avoid competition, the rate of growth of the target market that is untouched by the competition also has to be assessed to determine market potential. In other words, the developer may decide to target a market segment that is not targeted by the competition if the data show that it is of sufficient size and profitability and may grow significantly in the foreseeable future.

Competitive analysis related to product, price, place, and promotion strategy involves data collection about four areas:

- the competitors' product line (i.e., information about layout, residential site, and energy conservation related to all the property types of the key competitors targeting a specific market segment, such as college students)
- the competitors' place of their offerings (the location and neighborhood characteristics of the offerings related to the market segment in question)
- the competitors' pricing of their offerings (price of the offerings related to the target market)
- the competitors' promotion material (the message directed to the target market and the media vehicles carrying the message)

Such information should assist the marketing manager to effectively position one's offerings vis-à-vis key competitors' offerings.

PROSPECT STRATEGY

Prospect strategy (or *target marketing*) is a very important strategic decision. It is essentially about market selection. Armed with the information from strategic analysis, the marketing manager is now in a position to make a target market decision—which market segment to target in the firm's marketing program. Target marketing is foundational in the essence that all product, place, pricing, and promotion decisions are made with a target market in mind. That is, one cannot make effective four Ps decisions without knowing exactly who is being targeted (i.e., the market segment that the firm should cater to). See Figure 1.9.

Target marketing involves identifying all the possible market segments for the real estate development firm and then making a deliberate decision to go after certain segments while ignoring others. In doing so, the marketing manager attempts to segment the market using certain segmentation criteria and analyzes the viability of each segment. Market selection is thus a decision based on prioritizing the identified market segments and targeting the most viable ones.

Figure 1.9
Target Marketing as a Foundational Decision

Prospect Strategy (Target Marketing)			
Product decisions with target market in mind	Place decisions with target market in mind	Pricing decisions with target market in mind	Promotion decisions with target market in mind

Market Segmentation

How do marketing managers in real estate development firms go about segmenting the market in prospect segments? A typical market segmentation model involves technologies, customer needs, and customer groups.[8] *Technologies* refer to those particular products that can be defined and distinguished from other products through technological features. For example, suppose we have a real estate developer who would like to develop homes at a nearby lake. A home on the nearby lake can be built with or without various technological features, such as a dock. *Needs* are the functions or purposes a product (e.g., house on the lake) serves for prospects. Homes on a lake can serve three functions: (1) all-purpose living, (2) summer vacation home, and (3) get-away home. *Groups* are "homogeneous" sets of prospective buyers. A customer group can be viewed as an identifiable group of prospects that have the same generic need that is well-suited to a specific technology. For example, a home on a lake with a dock (technology) that serves year-around living (need) may best appeal to people who are retired and who enjoy water sports. A home that has a dock (technology) serving as a summer home (need) may best appeal to high income professionals who enjoy water sports. A home that has no dock (technology) serving as year-around living may be best suited for retired couples who are more sedentary (not actively engaged in water sports; they simply like the serenity of being at the lake). Finally, a house without a dock (technology) serving as a summer home (need) is well-suited for high income professionals who enjoy entertaining guests and holding social events at the lake house. See Figure 1.10.

The product-market matrix shows the possibility of 12 different segments. The question now is which one or more of these segments should the real estate developer target? We will answer this question in the next section.

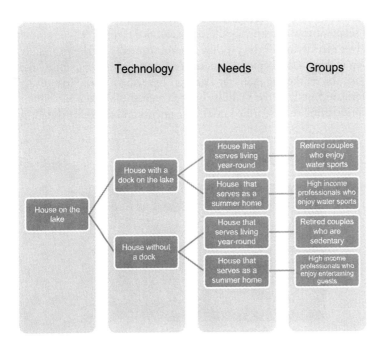

Figure 1.10 An Application of Market Segmentation in Relation to a Real Estate Developer Whose Goal Is to Build Homes at a Nearby Lake

Market Selection

Marketing managers select market segments using a set of criteria:

- size of each segment
- anticipated growth of each segment
- identifiability and reachability of each segment
- responsiveness of each segment
- ethical and societal considerations related to each segment[9]

To better understand these market selection criteria, let us apply them to the example of the real estate developer whose goal is to build homes at a nearby lake. Four concrete market segments were identified: (1) retired couples who enjoy water sports and who prefer a house built for year-around living that has a dock; (2) high income professionals who enjoy water sports and who prefer a house built for summer vacation that has a dock; (3) retired couples who are sedentary who prefer a house built for year-around living that does not have a dock; and (4) high income professionals who enjoy entertaining guests and who prefer a summer home that does not have a dock. See Table 1.1.

Size of Segment

Let's consider the first criterion: *Size of each segment*. The question then becomes: Can we do a market analysis to determine the size of these four segments? Perhaps a regional survey of households around the lake in question may be in order. Adults would be contacted and asked about their income, retirement status, and their preference for water sports

Table 1.1 *Applying the Market Selection Criteria to the Lake House Example*

	Segment 1 (retired couples who enjoy water sports)	Segment 2 (high income professionals who enjoy water sports)	Segment 3 (retired couples who are sedentary)	Segment 4 (high income professionals who enjoy entertaining guests)
Size of the segment	3	5	4	3
Anticipated growth of the segment	3	5	3	5
Identifiability and reachability of the segment	5	5	5	5
Responsiveness of the segment	4	4	4	4
Ethical and societal considerations	1	1	5	5

Notes: Rating scale:
  5 = Excellent
  4 = Very good
  3 = Good
  2 = Poor
  1 = Very poor

(e.g., fishing, boating, water skiing, snorkeling, scuba diving, swimming, etc.), their preference for serenity with nature without active engagement in water sports, and their interest in the purchase of a lake house. Based on the survey results, one can estimate the size of these four segments. The idea here is that the selected segment should be sizable to enhance the marketability of the lake homes.

Applying this market selection criterion to the developer whose plan is to build a community of lake homes, let us assume that the developer has conducted a survey to estimate the size of the four segments and, as such, rated them as shown in Table 1.1. Based on these ratings, the segment that seems to be the most viable for market selection is Segment 2 (high income professionals who enjoy water sports), followed by Segment 3 (retired couples who are sedentary), and Segments 1 and 4 (retired couples who enjoy water sports and the high-income professionals who enjoy entertaining guests).

Anticipated Growth of Segment

Market segments that are anticipated to grow in size are customarily considered to be more attractive than those that are anticipated to remain the same or decline in size. This is because real estate development has a specific time horizon. If the real estate developer is to build a community by the lake, then the time horizon is likely to be at least five years. That is, the marketing manager has to anticipate the size of the market segment five years out. In other words, it is not sufficient to estimate the size of potential market segments as they exist today; the manager must estimate its size at the time the real estate offerings will be available.

How do marketing managers in real estate development estimate anticipated growth of specific market segments? In many cases they use quantitative forecasting techniques such as *trends analysis*.[10] This technique traces the upward or downward movements in a time series as a result of basic developments in the population. For example, to predict changes in household income, the marketing manager plots household income changes in the region (2–3 hours driving distance from the lake region) over the last 20 years or so and examines the trend. Is the trend going up or down? Is the trend stable over time? An improvement on this technique is the use of regression analysis in forecasting. Regression analysis allows the marketing manager to predict the size of a specific market segment at a future point in time (e.g., average household income in 2018) by identifying specific predictors of income in the region (e.g., number of jobs in the area over the past 20 years, the average wages of jobs in the area over the past 20 years, retail sales in the area over the past 20 years). Again, much of the data involving these predictor variables are obtained through the census or other secondary sources.

In the absence of quantitative data, marketing managers rely on qualitative information. A common technique is the *judgment method*.[11] This method relies on experts to render their opinion about the trend of a particular variable (e.g., household income in the area around the lake—geographic distance of 2–3 hours driving time). For example, to predict changes in household income the marketing manager may seek the input of economic development experts in the region or newspaper articles reporting on the current and future economic development of the region. Such information may provide the marketing manager with hints as to whether household income is likely to increase, decrease, or remain the same over the next five years or so.

For the purpose of illustration, let us refer back to the marketing manager's ratings of the four aforementioned segments as shown in Table 1.1. The focus here is on rating the anticipated growth of these four segments. The marketing manger rates Segments 2 and 4 higher than Segments 1 and 3. In other words, the marketing manager anticipates more growth of the high income professional segments than the retired couple segments in the area around the lake. In other words, based on the anticipated growth criterion, the high income professional segments (Segments 2 and 4) seem to be more attractive than the retired couple segments (Segments 1 and 3).

Identifiability and Reachability of Segment

The second criterion is the *identifiability and reachability of the segments.* "Identifiability" refers to the extent that the segment is clearly profiled in demographic, geographic, and/or psychographic terms. *Demographic variables* are population characteristics such as age, sex, education, income, education, marital status, and so on. *Geographic variables* are characteristics defining a population in terms of physical location such as their zip code; neighborhood residence; residence in rural, suburban, or urban community; and so on. *Psychographic variables* are characteristics related to activities, opinions, and lifestyles of the population. Examples of activities related to the lake house case include boating, fishing, water skiing, scuba diving, hiking, lake viewing, entertaining guests, and so on. Examples of opinions related to the lake house case include favorable or unfavorable opinions of water sports, purchasing a house on the lake for year-around living versus seasonal living, opinions about the absence or presence of a dock attached to the lake house, and so on. Lifestyle variables refer to characteristics that reflect a set of activities, values, and beliefs related to a particular way of living. Examples related to the lake house may include a fishing and boating lifestyle; a lifestyle of the rich and famous—those who have second homes and use second homes to entertain guests; a lifestyle that reflects vigorous water sports activities that may include water skiing, swimming, snorkeling, and scuba diving; and a sedentary lifestyle that involves lake viewing, bird watching, and periodic hiking. Now let us revisit the four segments that were identified in relation to the lake house: (1) retired couples who enjoy water sports and who prefer a house built for year-around living that has a dock; (2) high income professionals who enjoy water sports and who prefer a house built for summer vacation that has a dock; (3) retired couples who are sedentary who prefer a house built for year-around living that does not have a dock; and (4) high income professionals who enjoy entertaining guests and who prefer a summer home that does not have a dock. Note that these four segments are clearly identifiable in demographic and psychographic terms. They are not identified in geographic terms. Nevertheless, all four segments are clearly identified, making them all subject to market selection. But then let's look at the "reachability" criterion. This criterion refers to the extent to which the segment can be contacted for survey and promotion purposes. If the decision comes to selecting a particular segment (e.g., retired couples who enjoy water sports and who prefer a house built for year-around living that has a dock), then can a sample of these people be contacted to learn their preferences related to the design of the lake house, the site orientation, energy conservation features, neighborhood characteristics, and pricing and affordability? How do marketing managers *reach* prospects for survey purposes? Customarily market researchers conduct surveys by several methods:

■ phone (researcher contacts a prospect by phone and conducts an interview over the phone)
■ e-mail (researcher contacts a prospect by e-mail, asking the prospect to complete a survey online)

- postal mail (researcher contacts a prospect by postal mail that includes a cover letter asking the prospect to complete an attached survey questionnaire and mail it back in a stamped and addressed envelope)
- face-to-face (research interviews prospect in person, possibly at the prospect's home)
- focus group (researcher recruits a small group of prospects by phone, e-mail, or postal mail; once they are recruited, the recruits assemble in a place chosen by the researcher and easily accessible to the recruits; the researcher then moderates a group discussion in which the recruits express their feelings and preferences about many aspects related to product, place, price, and promotion)

Can the same segment be reached for promotion purposes? That is, when the homes are ready for sale, what media vehicles can be used to reach this segment? If a segment is deemed difficult to reach for survey and promotion purposes, then this segment should not be targeted. It is important that the selected segment be reachable if they are clearly identifiable demographically, geographically, and psychographically. The marketing manager can use these criteria to identify the best mix of media vehicles to reach the target segment.

Going back to the lake house case, the question becomes: To what extent are these four segments reachable for survey and promotion purposes? There are many marketing research companies that specialize in creating lists of people that have a certain demographic profile. In the case of the lake house example, the demographic criteria are essentially age (55+ to identify retired couples) and income (to identify the high income professionals). Knowing the geographic radius of the promotion campaign is important to identify the various media vehicles that can be used to select the best mix of vehicles for promotion. The same can be said in relation to psychographics. Understanding the psychographic profile of a segment allows the marketing manager to select the best media mix.

In the case of our lake community, the marketing manager rates the four segments as equally attractive (see Table 1.1). In other words, the four segments are judged to be equally identifiable and reachable.

Responsiveness of Segment

Can a market segment be influenced by a good marketing campaign? What confidence does the marketing manager have regarding the effectiveness of the marketing campaign? For example, if the target market were to focus on the retired couples who enjoy water sports (Segment 1), can the marketing manager guide the developer to design lake homes suitable for year-around living that have docks sufficiently attractive for these prospects? Can the marketing manager price those properties affordably for these prospects? Can the marketing manager place those homes in an attractive location at the lake with lakefront view and access with a dock? Can the marketing manager effectively promote this offering to the target prospects who reside within 2–3 hours of the lake? If the answer to these questions is YES, then the marketing manager would rate the responsiveness of Segment 1 as high. On the other hand, if the marketing manager does not feel that he may be able to configure an effective marketing program for these prospects, then his rating of the responsiveness criterion for Segment 1 would reflect his lack of confidence and be a low one.

Of course, to rate each segment on the responsiveness criterion requires a great deal of experience. That is, it is not likely that a rookie marketing manager will be able to rate the degree of responsiveness of potential market segments. Only the experienced marketing manager can do so. To complete the ratings in Table 1.1, we will assume that the marketing manager has sufficient confidence that these four segments are likely to be responsive to his good marketing efforts. As such, the manager rates the four segments at 4 to reflect his confidence.

Ethical and Societal Considerations

Similar to the last criterion (responsiveness), the ethics criterion of market selection is equally "soft"; it relies on the subjective judgment of the marketing manager. The focus here, however, is on the extent to which catering to one segment or another may be considered unethical. Consider the following scenario. The real estate developer is focusing on developing a subdivision located at part of the lake that has considerable fluctuations of water levels. The water level of that part of the lake increases considerably when it rains but drops considerably during droughts. During the drought season, the water next to the docks is so shallow that no boats could operate from these docks. The marketing manager thinks that if he were to build homes with docking facilities in that part of the lake, he would promote the lake properties only when the water elevation level is high. That is, he would not disclose the problem about water elevation. This is of course unethical. In this situation, the marketing manager should rate Segments 1 and 2 unfavorably because marketing to these two segments may involve a breach of ethics. These two segments (retired couples who enjoy water sports and the high income professionals who enjoy water sports) would require homes that have docking facilities. Hence, it is not ethically wise to focus on those two segments. The marketing manager rates these two segments (Segments 1 and 2) low while rating Segments 3 and 4 high (because these two segments do not require docking facilities).

In sum, once the marketing manager completes the rating matrix (Table 1.1), he is in a position to make a determination as to which segment he should target. As shown in Table 1.1, the ratings are quite varied. Should the marketing manager simply sum up the ratings of each segment and choose the segment that has the highest summative scores—the higher the summative scores, the more attractive the market segment to the real estate developer? In this case, Segment 1 has a total score of 16, Segment 2 a score of 20, Segment 3 a score of 21, and Segment 4 a score of 22. Based on this analysis, the marketing manager has rated Segment 4 most favorably, followed by Segment 3, Segment 2, and Segment 1, respectively. However, doing so assumes that these market selection criteria are equally important. This may not be the case. If there is variability in importance, then the marketing manager should weigh those criteria that are more important more heavily than those of lesser importance. Perhaps a weighted summative score or a weighted average may work better in this situation.

SUMMARY

This chapter covered many topics related to marketing strategy of real estate development firms. The chapter began with a discussion on the marketing concept and how it is different from the selling concept in the context of real estate development firms. Marketing driven by the selling concept treats the marketing function as a selling function—nothing more, nothing less. A real estate development firm that takes the marketing concept to heart treats selling as only one small element of the marketing function. Marketing involves marketing research that guides the formulation of an integrated marketing plan. The goal is to achieve customer satisfaction, which ultimately leads to profitability.

The chapter then presented a formal definition of real estate marketing guided by the marketing concept. As stated, real estate marketing involves anticipating, managing, and satisfying demand via the exchange process between buyer and seller of a property. As such, marketing encompasses all facets of real estate buyer/seller relationships. Specific marketing activities include strategic analysis, target marketing, property planning, site selection, pricing of the property, promotion planning, and marketing management.

The chapter then discussed strategic analysis, which involves an assessment of the internal and external environment. An assessment of the internal environment involves an analysis of sales and customers, whereas an assessment of the external environment involves an analysis of the market and the competition. A sales analysis focuses on plotting sales trends over the last several years (or as far back as data is available) and making an attempt to explain sales fluctuations. Customer analysis focuses on existing customers—customers that have bought a property from the real estate development firm. Marketing researchers typically conduct customer satisfaction surveys. Market analysis refers to collecting and analyzing data about issues directly related to target marketing, product strategy, place strategy, pricing strategy, and promotion strategy. Specifically, market analysis related to target marketing focuses on collecting data about specific population segments to assess their market potential and the extent to which the firm may want to target these segments. Market analysis related to product strategy involves collecting and analyzing data dealing with housing design, residential site, and energy conservation. Market analysis related to place strategy focuses on collecting data on neighborhoods and communities to aid in the decision of where to build. Market analysis related to pricing strategy involves collecting and analyzing data about housing costs and prices. Market analysis related to promotion strategy involves data collection of mostly the media habits of the target market segment.

Competitive analysis can best be understood in relation to target marketing, product strategy, place strategy, pricing strategy, and promotion strategy. Competitive analysis related to target marketing involves gathering information about key competitors in an attempt to identify their target market. Competitive analysis also involves data collection about the competitors' product line, the competitors' place of their offerings, the competitors' pricing of their offerings, and the competitors' promotion material. Such information should assist the marketing manager to effectively position his or her offerings vis-à-vis key competitors' offerings.

The chapter then discussed prospect strategy (how the marketing manager makes target marketing decisions). Prospect strategy (or *target marketing*) is essentially is about market selection. Target marketing involves identifying all the possible market segments for the real estate development firm and then making a deliberate decision to go after certain segments while ignoring others. In doing so, the marketing manager attempts to segment the market, using certain segmentation criteria, and analyze the viability of each segment. The chapter covered a basic market segmentation model involving three dimensions: technologies, customer needs, and customer groups. Once the market segments are identified, the marketing manager selects segments, using a set of criteria such as size of each segment, anticipated growth of each segment, identifiability and reachability of each segment, responsiveness of each segment, and ethical and societal considerations related to each segment.

DISCUSSION QUESTIONS

1. Describe the marketing concept and distinguish it from the selling concept.

2. Provide a formal definition of real estate marketing guided by the marketing concept.

3. What is strategic analysis in the real estate development industry?

4. Distinguish between internal and external analysis in real estate development.

5. What is sales analysis and how is it conducted in a real estate development firm?

6. What is customer analysis and how is it conducted in a real estate development firm?

7. What is competitive analysis and how is it conducted in a real estate development firm?

8. What does prospect strategy mean in a real estate development firm?

9. Describe a typical market segmentation model used in real estate development.

10. How does the marketing manager go about selecting one or more market segments to target?

NOTES

1. Webster, F.E., Jr. (1994). Executing the new marketing concept. *Marketing Management*, 3, 9–16.
2. This definition of real estate marketing is based on the following sources: Webster, Jr., F.E. (1994). Defining the new marketing concept. *Marketing Management*, 2, 23–31. Webb, D., Webster, C., & Krepapa, A. (2000). An exploration of the meanings and outcomes of a customer-defined market orientation. *Journal of Business Research*, 48, 101–112.
3. Aaker, D.A. (2008). *Strategic marketing management* (8th ed.). New York: Wiley. Collins, D.J., & Rukstad, M.G. (2008). Can you say what your strategy is? *Harvard Business Review*, 86, 82–90.
4. Mentzer, J.T., & Cox, J.E., Jr. (1984). Familiarity, application, and performance techniques. *Journal of Forecasting*, 3, 27–36.
5. Rigby, D.K., Reichheld, F., & Dawson, C. (2003). Winning customer loyalty is the key to a winning CRM strategy. *Ivey Business Journal*, 67, 1–5. Singer, M. (2008). What makes customer satisfaction research useful? Accessed from http://industryweek.com/ReadArticle.aspx?ArticleID=16027, on March 31, 2008.
6. Barrett, C.B. (1996). Market analysis methods: Are our enriched toolkits well suited to enlivened markets? *American Journal of Agricultural Economics*, 78, 825–829. Fleisher, C., & Benoussan, B. (2002). *Strategic and competitive analysis: Methods and techniques for analyzing business competition*. New York: Prentice-Hall.
7. Porter, M.E. (1985). *Competitive advantage: Creating and sustaining superior performance*. New York: Free Press.
8. Wedel, M., & Kamakura, W. (2002). *Market segmentation: Conceptual and methodological foundations* (2nd ed.). Norwell, MA: Kluwer Academic Publishers.
9. Clapp, J.M., & Messner, S.D. (1988). *Real estate market analysis: Methods and applications*. New York: Greenwood Publishing Group.
10. Ferguson, G.A. (1965). *Nonparametric trend analysis*. Oxford, England: McGill Press.
11. Plake, B.S., Hambleton, R.K., & Jaeger, R.M. (1997). A new standard-setting method for performance assessments: The dominant profile judgment method and some field-test results. *Educational and Psychological Measurement*, 57, 400–411.

Chapter 2

Product and Place Strategy

LEARNING OBJECTIVES

This chapter covers topics related to product and place strategy in real estate development. Product strategy is mostly about home design (in residential real estate) and corporate real estate (in commercial real estate). Home design focuses on issues such as layout and floor plan of the residential structure, the residential site, and energy conservation. Corporate real estate is essentially designing the commercial real estate structure to serve the core mission of the business. Place strategy is about site selection and the many factors real estate developers use to guide optimal site selection decisions—both residential and commercial. Although these topics seem to be the domain of the professional builder, marketing managers in real estate development firms should have general knowledge of product and place issues to guide the builders to meet customers' property needs and preferences. Similarly, real estate agents and brokers have to know much about product and place strategy to guide buyers and sellers in making effective transactions. This chapter is designed to help students of real estate marketing learn:

- what housing design is and the elements of housing design
- what a floor plan is and how one judges the quality of a floor plan
- quality standards in the design of the living area of a residential structure
- quality standards in the design of the sleeping area of a residential structure
- quality standards in the design of the service area of a residential structure
- how to judge the type of architecture of a residential structure
- how to judge the style of architecture of a residential structure
- how to judge the quality of topography of the residential site
- how to judge the quality of the site orientation
- how to judge the quality of landscaping surrounding the site
- how to judge the energy efficiency of a house
- how to select a site for development

PRODUCT STRATEGY

"Know thy product" is an important principle of product strategy. That is, the real estate marketer has to know the "ins" and "outs" of the focal product, whether it is a shopping center, office building, multi-unit apartment complex, subdivision of single-family homes, or mountain resort. The real estate marketer has to become intimately familiar with the physical, technological, economic, psychological, social, and cultural features of the property in question. This knowledge should help the marketer guide the developer in designing the property in question. If the marketer is a sales agent or broker, such knowledge goes a long way in serving clients, whether they are buyers or sellers. Hence, intimate familiarity with the product requires knowledge about product quality, *period*.

In discussing product quality in real estate, we first have to make a clear distinction between residential and commercial real estate. Quality issues in commercial real estate are highly complex. In commercial real estate (i.e., corporate real estate), real estate structures are built to support the basic core function of the "business." For example, a shopping mall is designed to enhance the overall shopping experience of shoppers as well as productivity of the retail stores and their employees in serving their patrons. A corporate office is designed to enhance employee productivity. A school is designed to enhance the learning experience of students as well as the productivity of teachers and the administrative staff in teaching students. A college campus is designed to enhance the teaching and research functions of the faculty and the learning function of the students. A restaurant is designed to enhance the eating experience of patrons and to support the staff in food preparation and servicing the patrons, and so on. Quality standards are established for each type of commercial real estate venture. In other words, quality standards are established for the design of college campuses, hospitals, banks, restaurants, medical offices, elementary schools, secondary schools, etc. Professionals in real estate development and building and construction specialize as a function of the type of commercial real estate. Of course it is beyond the scope of this book on real estate marketing to delve into these highly specialized topics. Suffice it to say that real estate marketers in specialized commercial real estate sectors should become intimately familiar with design and quality issues of the focal product. Such knowledge is extremely important in guiding the architects and design staff to create building structures in ways to effectively serve the core mission of the business.

Residential real estate is categorically different from commercial real estate. The quality standards are well-defined but differ significantly across various types of residential real estate (e.g., single-family homes, apartment complexes, beach homes, homes on a lake, mountain resort homes, luxury homes, and mansions). In discussing product quality we will focus on the most common and widely used examples of single-family housing units. The student should be aware that these are only examples to illustrate central concepts. Further reading and research is necessary to master the issues of product quality as a direct function of the type of real estate product. Hence, we will focus on residential single-family housing units in discussing product quality, and we will break down this discussion into five parts: (1) floor plan of the housing unit(s), (2) housing architecture, (3) the site of the housing unit(s), (4) energy conservation and green building issues, and (5) neighborhood features.

The Floor Plan

The most important decision in home design is the floor plan. The floor plan reflects the design of the residential structure to reflect three main functions: (1) living area (living room, dining room, family room, den, recreation room, enclosed porches, etc.), (2) sleeping area (bedrooms, bathrooms, dressing rooms, etc.), and (3) service area (kitchen, utility room, laundry room, other specialized work areas, etc.). See a typical floor plan of a home in Figure 2.1. Also see Zooming-In Box 2.1 for information about deficiencies in a floor plan. See Case/Anecdote 2.1 for an interesting exposition of micro housing as a trend in large cities. See Case/Anecdote Box 2.2 for information about housing design for Generation Y residents.

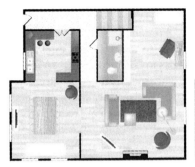 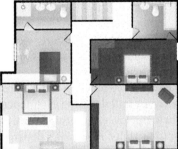

Figure 2.1
A Typical Floor Plan of a Residential Structure

Source: © Thinkstock

ZOOMING-IN BOX 2.1. DETECTING DEFICIENCIES OF A FLOOR PLAN[i]

Common floor plan deficiencies include the following:

- absence of a guest closet
- lack of an entrance from the front door to the living room
- the kitchen does not have enough space for a breakfast nook
- lack of a separate dining room—separate from the kitchen
- the dining room does not have immediate access to the kitchen
- the bedrooms are visible from the foyer
- lack of an outside entrance to the basement
- the family room is not highly accessible
- too many doors and windows limiting the placement of furniture
- no garage, car port, or not enough space to park motor vehicles
- rooms that are not in the right proportion to the total size of the house
- lack of adequate storage space for the size of the house
- the doors and windows do not seem to be properly placed
- the play area is not within view of the kitchen

[i] Based on Skenkel, W. M. (2004). *Marketing real estate*. Mason, OH: South-Western Educational Publishing (p. 160 and p. 163).

CASE/ANECDOTE 2.1. MICRO HOUSING[i]

Micro housing has become very trendy. A micro apartment is a high quality compact that is most suitable for one or two inhabitants. The space includes the same amenities found in larger homes, and the design and layout makes the micro apartment functional and appealing. The micro apartment has open space for sleeping, working, cooking, and eating. One micro apartment, for example, is about 300 sq ft. It includes a bathroom, a kitchenette (stove top, refrigerator, and sink), storage space (for a bike, storage under the seats, and an open closet), seating, wrap-around counters (for workspace and dining), and a bed alcove. There is also a large window next to the bed for natural light and a view of the outside world. The floor is polished concrete that can easily be cleaned and is aesthetically pleasing, too. In some cases, the kitchen area may be shared among several micro apartments. Micro apartments are built in housing projects that are typically six stories high and have forty units each.

Micro housing is becoming increasingly popular in large cities such as New York, Boston, Seattle, Washington, DC, and San Francisco. Drivers of micro housing include limited space and cost of living in large cities. What is its target market? Mostly young professionals and college students are attracted to it.

Of course, there are disadvantages. Residents and neighbors complain that micro housing can lead to cleanliness and hygiene problems. Also, the slightest bit of clutter can overwhelm a tiny space. When people are cramped together, they get on each other nerves quite easily—all it takes is one person speaking loud or turning up the music. There can also be parking problems if the residents have personal transportation vehicles.

[i] Based on the following articles: How small can you go? Big cities embrace micro housing trend. *Realtor.com Blog.* Accessed from www.realtor.com/blogs/2013/04/11/how-small-can-you-do-big-cities-embrace--micro-housing-trend/, on May 5, 2013. Living large in tiny apartments. *Harvard Magazine.* Accessed from http://harvardmagazine.com/2013/05/living-large-in-tiny-apartments, on May 5, 2013. Micro housing boosters show-and-tell at Building Boston 2030 Forum. *Biz Journals.* Accessed from www.bizjournals.com/boston/real-estate/2013/03/micro-housing-Suffolk-show-and-tell.html?page=all, on May 5, 2013. Micro-housing: Good or bad? *Houselogic Blog.* Accessed from www.houslogic.com/blog/issues-affecting-home-owners/micro-apartment-living, on May 5, 2013. Micro-units: A new trend? *Harvard Magazine.* Accessed from http://harvardmagazine.com/2013/04/micro-units-a-new-trend, on May 5, 2013. Nation: Seattle at forefront of micro-housing trend. *Star Tribune.* Accessed from www.startribune.com/nation/205518631.html?refer=y, on May 5, 2013. Micro apartments a more affordable but controversial housing option. *Real Estate Matters.* Accessed from https://blogs.stthomas.edu/realestate/2013/05/03/micro-apartments-a-more-affordable-but-controversial-housing-option, on May 5, 2013. Seattle's 'micro-housing' boom draws criticism, support. *KOMO News.* Accessed from www.komonews.com/news/local/Seattles-micro-housing-boom-draws-criticism-support-197695711.html?tab=video, on May 5, 2013.

CASE/ANECDOTE 2.2. HOUSING DESIGN FOR GEN Y RESIDENTS[i]

Architects and interior design specialists use open floor plans and all-inclusive clubrooms to appeal and accommodate to the housing needs of Gen Y residents (people born between 1980 and 2001). Gen Y residents prefer to spend time socializing in an impressive common area rather than in individual units. They want to be entertained and they want to see who is around. Hence, amenity areas are designed to be large, whereas individual units are designed small. Visual transparency and connectivity are key concepts here (i.e., open design). There are no separate dining rooms or living rooms. These rooms all mesh into one large area.

Common areas are also designed with open space architecture. The clubhouse is essentially a large room that serves as business room, computer center, and game room. Consider the design of Voyager's clubhouse (at the Space Center in Houston, Texas). The clubhouse is a two-story, open-air space that has a scattering of multiple large-screen TVs, a Wii gaming station, pool tables, and a latte lounge with a constellation light fixture that serves to encourage residents to socialize outside of their individual units. The floor plan provides a balance of space for exercise and socializing over coffee.

Large bay windows overlooking beautiful landscape is part of the open space design too. Outdoor areas are important to Gen Y residents and are designed to be an extension of the interior common space, rather than a separate space.

On the interior, a single-color scheme does not suffice. The walls are painted in different colors—one wall in one color and an adjacent wall in another color. Pastels, blues, and greens seem to be the most popular, though saturated colors seem to be on the rise.

How about flooring? Carpets are out; hard surface flooring is in. It is easier to make it shine and to move furniture around on hard-surface flooring. However, built-in shelves are preferred instead of portable book cases. Lighting is also important. The lighting is designed to make the open space look bright and youthful. Technology, of course, is easily accommodated. Cable and electrical outlets are multiple and conveniently located throughout the open space.

[i] Boston, L. (2011). Room with a view. *Units*, June, 40–48.

Quality Standards in the Design of the Living Area

In designing the living area of a residential structure, real estate developers and building and construction experts are guided by quality standards. These standards involve (1) space requirements, (2) layout, and (3) location.

With respect to *space requirements*, a living room for a three-bedroom house should be at least 12' × 18'. Living rooms are typically large because residents spend a great deal of time in the living room watching television and socializing.[1]

The *layout* of the living room is commonly rectangular, not square. Furniture placement in square living rooms seems problematic. One wall is typically devoted to a fireplace, built-in shelves, or cabinets. One long wall is devoted to sofas and couches.[2]

Where are living rooms *located* within a house? They are located in a dead-end area with no through traffic.[3] A living room should not serve as a hallway. It should also supplement the dining room and outdoor recreational space, such as a screened porch, patio, or deck.[4]

Quality Standards in the Design of the Sleeping Area

How to judge the quality of the sleeping area of a house? The common standards include: (1) minimum space requirements, (2) storage space, (3) bathrooms, and (4) noise.

Builders abide by *minimum space requirements* in designing the bedrooms. For example, in the U.S., the minimum space requirements for a single bed sleeping area is 8'10" × 10'0"; 10'0" × 11'6" for a double bed sleeping area; and 11'6" × 12'0" for a sleeping area with twin beds.[5]

The closets (*storage space*) should have a minimum space of 2' × 3'.[6] Modern homes have much larger closet space. This seems to be the trend in modern living.

How about the *bathrooms?* In homes that price mid-range (e.g., a three-bedroom house) two bathrooms seems to be the minimum standard. One bathroom should serve the master bedroom; another should serve the other two bedrooms and be accessible from a hallway and the bedroom area. Upscale homes require one bathroom for each bedroom plus other bathrooms that serve the living area.[7]

With respect to *noise*, the placement of the bathrooms and closets should provide sound-proofing between the bedrooms.[8] The bedrooms should also be separated from the living and service areas to help cut down on the noise level. People sleeping in the bedrooms do not want to be disturbed by the commotion in the living and service areas.[9] Ideally, the bedrooms should be the quietest rooms in the house.

Quality Standards in the Design of the Service Area

The common standards include (1) kitchen space requirements, (2) kitchen location, (3) kitchen layout, and (4) laundry facilities.

With respect to *space requirements*, the space allotted to the kitchen should be proportional to 10% of the entire house. This is because the typical cost of a kitchen of a three-bedroom house is 10%.[10]

Kitchen *location* is important. The kitchen should be located with direct access to the dining room, so that people do not have to walk far from the kitchen to the dining room. The patio or outside eating area should also be within a close range of the kitchen. In addition, the kitchen should be close to the garage or carport so that bringing groceries into the kitchen is not painstaking. Although the kitchen should be placed strategically next to the garage, dining room, and the living area, it should not be a main traffic route for the rest of the house. Such a design would reduce its utility.[11]

How about the *kitchen layout*? There are standard kitchen floor plans. These include the U-type shape, the L-shape, the Corridor-type shape, the Sidewall shape, and kitchen-family-room combination shape. These are traditional kitchen layouts.[12] Kitchen layouts are increasingly creative.

Typically, laundry facilities (or the utility room) are located next to the kitchen, the attached garage, or possibly in the basement. This room should have its bath, counters, storage space, and possibly space for ironing clothes.[13]

Judging Architecture

In judging the architecture of a residential structure, the real estate marketer has to know the advantages and disadvantages of the type of structures (one-story house, one-and-a-half story house, two-story house, and two-story split-level house). The marketer has to also gauge the preference of his customers regarding architectural style (Cape Cod, Colonial, Ranch, Tudor, Victorian, etc.).

Judging Architecture Type

As previously mentioned, the real estate marketer has to know the advantages and disadvantages of the type of structures (one-story house, one-and-a-half story house, two-story house, and two-story split-level house).

Let's begin by focusing on the *one-story house*. What are the advantages and disadvantages of such a residential structure compared to the other three types? The clearest advantage of a one-story house is the fact that the home dweller does not have to climb up and down stairs. One-story homes are favored by older people who are less mobile, families with small children (who fear that children may fall down the stairs), and families that have handicapped members (e.g., quadriplegics). One-story homes are easier to maintain and repair. They are adaptable to adding rooms, patios, porches, fire places, garages, and carports. Disadvantages include the relative lack of privacy between sleeping and living areas and the fact that such architectural types of residence cost more per square foot (relative to the other three types).[14] See Figure 2.2 for an example of a one-story house.

How about the advantages and disadvantages of a *one-and-a-half story house?* Advantages include the fact that added space is gained at relatively low additional cost by raising the pitch of the roof and by adapting construction to the second story. Heating costs are less because of the smaller perimeter of the enclosed living area. Construction costs per square foot are less than the one-story house. Disadvantages include the fact that shoulder height in the second-story rooms may affect furniture placement. Window space tends to be limited. If the house is not well insulated, the rooms on the second story may have greater temperature extremes (hot in the summer and cold in the winter). Also, the second floor rooms may have limited access to electrical and plumbing facilities.[15] See Figure 2.3 for an example of a one-and-a-half-story house.

Figure 2.2
An Example of a One-Story House
Source: © Thinkstock

Figure 2.3
An Example of a One-and-a-Half-Story House
Source: © Thinkstock

There are advantages and disadvantages of a two-story house too. Advantages include the fact that such a structure has more privacy because the sleeping rooms are usually on the second floor. It is less costly per square foot (compared to the one-story house). Also, the structure is more adaptable to a small lot (compared to the one-story house). Disadvantages include that there are a lot of stairs to climb up and down, which presents a problem to those who are less mobile (e.g., the elderly). Also, the sleeping area has poor access to outside space.[16] See Figure 2.4 for an example of a two-story house.

Finally, we have the two-story split level house (see Figure 2.5). Such a structure does better with sloping lots than the other three types. Functional areas are clearly separated, in that the sleeping areas are confined to the upper or lower levels and the living and service areas are confined to the ground level. There is one main disadvantage to the split-level house: The stairway access to the upstairs and downstairs is a handicap, for much space is lost in the stairways.

Figure 2.4
An Example of a
Two-Story House

Source: © Thinkstock

Figure 2.5
An Example of a
Two-Story Split
Level House

Source: © Thinkstock

29

Judging Architectural Style

There are myriad styles of residential structures. Examples include the Art Deco, the Neo-Classical, the Bungalow, the Prairie, Colonial, Contemporary, Craftsman, Creole, Dutch Colonial, Federal, French Provincial, Georgian, Pueblo, Queen Anne, Ranch, Saltbox, Second Empire, Shed, Shingle, Gothic Revival, Greek Revival, International, Italianate, Monterrey, National, Spanish Eclectic, Stick, Tudor, and Victorian. The judgment of style in this instance is a matter of matching the right style with the preference of the client or customer group. This may entail some marketing research if the development is large scale. However, if the real estate developer is a builder for a single or only a handful of clients, the developer would simply expose the clients to the different styles to gauge their preference.

The Residential Site

To build an effective residential structure, the marketer has to guide the builder to situate the structure in ways to match customer needs. Similarly, the buyer agent and broker have to know much about the residential site to select homes that match the housing needs and preferences of their clients. When real estate developers speak of the "residential site," they refer to three things: (1) topography of the residential site, (2) the orientation of the site, and (3) the landscaping surrounding the site. Let's describe these dimensions of the site in some detail.

Judging the Quality of Topography of the Residential Site

Important elements in topography of real estate are temperature, altitude, wind conditions, and the slope of the site toward or away from the sun. This is because altitude of the residential site affects the temperature experienced by the home dwellers. Residential sites located in areas of high altitude (e.g., mountain top) can experience cold weather—a decrease of one degree Fahrenheit for every 300 feet of elevation. However, a residential site in lower elevation (e.g., in a valley) can also experience cold temperature because cold air flows downward, especially at night, compared with higher elevations. Homes built on a slope facing the sun in colder climates take advantage of the sun to warm up the place, compared to homes built on leveled sites.[17]

Judging the Quality of the Site Orientation

The orientation of the site has a direct effect on living comfort. Here are some guidelines. Houses that are built facing south benefit from more warmth in the winter and coolness in summer. Houses facing southeast and southwest are colder in the winter months and warmer in the summer months than houses facing other directions (i.e., east and west exposures tend to be warmer in the summer months and colder in the winter months). A house should be oriented in ways to take advantage of seasonal variations in air temperature and sunlight. For example, in cooler areas, houses whose longitudinal accesses are oriented east of south provide the best heat distribution, especially if they are located halfway up a slope). In temperate areas, orienting the house further east than south provides the best degree of comfort. Upper locations are preferred to protect from winter winds; in contrast, in hotter areas, lower hillside locations that have afternoon shade benefit from cool air flow. In humid areas, houses situated near the crest of a hill tend to benefit from cooling breezes. To take advantage of warmth from sun radiation, houses are built on southern and northern slopes, not on eastern and western slopes.[18]

Judging the Quality of the Landscaping

A house with established landscaping is usually considered much more attractive than one with less established landscaping. It takes years to establish good landscaping around a house—nice trees and bushes. Good landscaping also has great benefits:

- Established trees have beneficial effects on heating and air conditioning.
- In the winter months, evergreen trees serve to break winds, hence they reduce heat loss and snow drift.
- In the summer months, grass and leaves absorb radiation from the sun, hence cooling the air through evaporation.[19]
- Deciduous trees planted close to the house may provide plenty of shade in the summer months.
- Vines placed in arbor areas can provide shade and can be used on sunny walls to reduce heat in hot weather (through evaporation).[20]

Landscaping quality can be judged by many factors, including the following:

- Water drainage must be away from the house.
- The house must be graded to aid runoff from heavy rains.
- In hot climates, trees should have high (not low) branches so that they do not block the wind.
- Also in hot climates, the exterior of the house should be painted with very light colors to reflect light and heat.[21]

Energy Conservation and Green Buildings

In today's real estate market, developers and builders are more cognizant of energy conservation than ever before. Marketers of real estate development firms, real estate agents, and brokers must know something about the science of energy conservation to judge quality buildings. A house that is judged to be good in energy conservation is one that has good insulation—insulation beneath the floors, between the walls, and above the ceilings. To be able to judge the insulation, the real estate marketing professional should be familiar with the measured quality of insulation: the *R-value*. R refers to the degree of resistance to winter heat loss and summer heat gain. R-values are typically marked on the insulation material used throughout the house. These values range from 11 to 38—the higher the number, the better the insulation.[22]

An older home is well insulated when it meets the following criteria:

- Insulation in the attic has a high R-value.
- The exterior wall is protected with blown or foam insulation.
- There is floor insulation beneath the crawl space.
- There is polyethylene ground moisture barrier in the crawl space.
- Heating and cooling ducts are insulated if they pass through unheated space.
- The water heater is insulated in unheated space.
- There is caulking around all windows and doors, foundation sills, and at points where pipes, wires, and electrical outlets break the wall surface.
- There is weather stripping material around all windows and doors.

- There are storm windows.
- The fireplace has a tight-fitting damper and a glass screen.
- Hot water pipes are insulated in unheated areas to prevent them from freezing in the winter months.[23]

The science of insulation in new homes is highly complex and cannot be adequately covered in this chapter. Suffice it to say that builders of modern homes have a highly sophisticated set of guidelines to ensure that a home is well-insulated. See Zooming-In Box 2.2 for additional tips for making homes more green.

ZOOMING-IN BOX 2.2. GREEN HOME TRENDS[i]

Over the past few years, building improvement has taken a dramatic shift toward retrofitting in green and saving money at the same time. Here are some tips:

- Replacing old appliances with new and more energy efficient ones can save money and make the home more green. For example, upgrade old HVAC equipment with Energy Star rated models.
- Instead of constructing a new home, try renovations. Renovating an old building is a green move. Renovations save on building material and waste; renovations also serve to preserve undeveloped landscape.
- Invest in energy efficient upgrades to reduce heating and cooling loads, such as adding more insulation in the floors, walls, and ceilings. Install new windows.
- If you are building a new home, try small. Less square footage means less energy to heat and cool the space.
- Make the home "net-zero." Net-zero homes combine a variety of passive and active design options. Passive design may include strategically placed windows that use or prevent solar heat gains. Passive design may also involve plenty of natural ventilation and well-insulated walls. Active design refers to the installation of renewable energy systems, such as solar panels, geothermal wells, and wind turbines.
- Install new energy monitoring systems that allow you to monitor your energy use comprehensively and in real time. The goal is to cut back on energy consumption whenever you can. To do so, you need to figure out how much you consume and when.
- Incandescent light bulbs are not energy efficient. Compact fluorescents are. LED bulbs use even less energy.

[i] Bernard, M. (2012). 5 green home trends for 2012. *Buildipedia*. Accessed from http://buildipedia.com/go-green/eco-news-trends/5-green-home-trends-for-2012, on May 5, 2013.

Real estate agents and brokers also judge the relative energy efficiency of a house by asking the home owner to provide receipts of utility bills for the last 12 months and comparing the total cost with other houses considered to be energy efficient. Of course, adjustments have to be made in relation to the size of the house.[24] See Case/Anecdote Box 2.3 for an exposition of LEED (Leadership in Energy and Environmental Design).

CASE/ANECDOTE 2.3. LEED AND CERTIFICATION[i]

LEED stands for *Leadership in Energy and Environmental Design*. LEED is an organization that ranks states in terms of green construction. For example, in 2011 LEED ranked the District of Columbia first in green construction in commercial and institutional projects (18,954,022 total sq ft and 31.50 per capita), Colorado second (13,803.313 total sq ft and 2.74 per capita), Illinois third (34,567,585 total sq ft and 2.69 per capita), and Virginia fourth (19,358,193 total sq ft and 2.42 per capita). The data is provided by the U.S. Green Building Council. LEED gives out awards (e.g., platinum) for buildings that meet green building objectives. A major reason for the District of Columbia achieving the highest LEED ranking is due to the fact that LEED ratings are mandated for many of the city's government buildings.

The U.S. Green Building Council and its LEED program has been the command center for energy and environmental leadership. Beyond evaluating and certifying green buildings, LEED is involved in educating and certifying real estate brokers.

[i] O'Brien, L. (2012). Oregon no longer a leader in LEED. *Daily Journal of Commerce*. Accessed from http://djcoregon.com/wp-content/plugins/dmc-sociable-toolbar/wp-print.php?p=79750, on May 5, 2013. Travers, B. (2012). Brokers catch on to the green movement. *Daily Journal of Commerce: Real Estate Marketplace*. Accessed from www.djc.com/news/re/12035660.html?action=get&id=1203566, on May 5, 2013.

Neighborhood Features

Real estate professionals evaluate neighborhoods along three dimensions: (1) physical features, (2) economic features, and (3) social features. Neighborhoods judged to be of high quality tend to have the following *physical features*:

- good property upkeep
- sound structural building condition
- homogeneous housing type
- well-maintained landscaping
- location convenient to shopping, employment, and highways
- no environmental hazards such as cleaning chemicals, solvents and paints, pipes w/lead solder, lawn and garden chemicals, pesticides, asphalt roofing, and asbestos[25]

With respect to the *economic features*, high quality neighborhoods tend to share four major characteristics:

- a high rate of owner occupancy
- rising resale value
- favorable financing terms
- a sense of optimism and confidence about the neighborhood's future[26]

In relation to the *social features*, high quality neighborhoods are likely to be characterized as having residents:

- with moderate-to-upper incomes
- with high school education and above
- who are family-oriented or childless adults
- who are skilled blue-collared workers or above
- who feel a sense of neighborhood cohesion
- who consider it safe[27]

PLACE STRATEGY

Place strategy is mostly about site selection for real estate development. In selecting a site for development, the developer considers the following factors: (1) location needs of the customer (or target market segment), (2) housing needs of the customer (or target market segment), (3) environmental needs of the customer (or target market segment), (4) financing needs of the customer (or target market segment), (5) zoning issues, (6) subdivision regulations, (7) building-code requirements, (8) deed restrictions, and (9) environmental impact statement. See Figure 2.6.

Location Needs of the Customer (or Target Market Segment)

A builder takes into account the location needs of the customer when selecting a parcel of land to build a home. For example, a dual career couple with jobs in different places may need to build a house strategically located between the two places of employment. Similarly, a real estate developer considers the location needs of the target market segment. For example, a real estate developer trying to locate an optimal parcel of land to build a nursing home would consider the distance between the nursing home and the nearest hospital, given the fact that many elderly in nursing homes need emergency medical services that can be provided only in a hospital setting. Similarly, a commercial real estate developer would consider the location needs of the customer population of the core business. For example, in selecting an optimal site for building a mall, the real estate developer would have to consider the distance between the home location of target shoppers and the location of the shopping mall. Shoppers are not likely to travel more than a few miles to get to a shopping mall.

Housing Needs of the Customer (or Target Market Segment)

In selecting a site to build a home for specific customers, a builder considers the housing needs of the customers. For example, consider a family with three children looking to build a house in a subdivision that allows only septic systems that support three-bedroom houses.

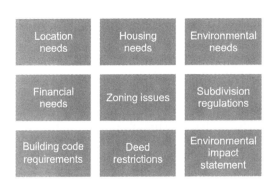

Figure 2.6
Factors
Considered by
the Builder
(or Real Estate
Developer) in
Site Selection

The family wants to build a five-bedroom house. Similarly, a real estate developer planning to develop a shopping mall may have to consider the corporate real estate needs of the retail tenants. How many anchor stores? What are the sizes of these anchor stores? How many other stores? What are the space dimensions of these stores? The space dimensions of the common facilities? The parking area? Computing the space requirements of the shopping mall plays an important role in the decision to select from among optional sites.

Environmental Needs of the Customer (or Target Market Segment)

Real estate developers take into account the environmental needs of their clients. For example, there are many people now who are very conscious of the need to preserve the natural environment. The environmentally conscious home buyers will likely prefer a subdivision that has green space shared by homeowners. In this case, the real estate developer targeting the environmentally conscious home buyers will have to select a site with much green space.

In commercial real estate, many developers take into account the surrounding environment in selecting sites. For example, many corporate headquarters are placed at sites next to a big pond or lake with biking and walking trails around the body of water in a wooded area. Such locations tend to meet the needs of business executives who appreciate the wellness impact of the green areas.

Financial Needs of the Customer (or Target Market Segment)

Builders and real estate developers are guided by financial considerations in site selection, too. In residential real estate, the builder selects a parcel of land that meets the financial requirements of the customer. Consider a home buyer who is not willing to spend more than $500,000 to build his dream home. The builder estimates, based on the architectural design that the home buyer has selected, that the cost of the building is around $400,000. This means that $100,000 is the upper limit for the purchase of the needed parcel of land. Of course, parcels are likely to differ in price as a function of size, location, market demand, etc. Hence, the pool of optional sites to choose from would be limited by the dollar amount that the customer would want to pay for a site.

Another consideration that falls under the umbrella of financial needs is real estate property tax. Some home buyers are keenly aware of differences in real estate taxes in various locations. Hence, if the home buyer does not want to pay high real estate taxes, he may guide the builder to choose a lot in a town, city, or county that meets his financial prerequisites. For example, many home buyers look for homes in the suburbs of large cities (and perhaps in the rural areas) to avoid the high cost of real estate property tax in the city. Similarly, commercial real estate developers are guided by financial considerations of their clients. One real-world example is a car dealership that had long fought county officials over paying taxes. The county assessed the dealership's taxes by the number of vehicles the dealer had on the lot over a specified period. The car dealer did not feel this was fair, so he moved his car dealership to a neighboring town that did not levy such taxes.

Zoning Issues

Zoning issues play an important role in site selection too. The builder or developer has to seek parcels of land available for sale that are located in areas that are properly zoned. In other words, a residential home builder would seek lots in a residential zone; a real estate developer of an industrial plant would seek a parcel of land in areas zoned for industrial development, etc. In many cases, a developer may be interested in a specific parcel of land that is not zoned for the target development (e.g., a developer wants to build an apartment complex in an area zoned for agriculture). Hence, the developer applies to the planning commission of the designated municipality to rezone the site to suit the development purpose. This is of course a long, drawn-out process that requires study by the planning council, a public hearing to allow all those affected by the possible rezoning to voice their concerns, and a recommendation by the planning commission to the town/city council, which in turn votes on this petition. See Zooming-In Box 2.3 for an understanding of zoning ordinances (in the U.S.A.) and how they work. Also, see Case/Anecdote Box 2.4 for an example of a zoning issue in Blacksburg, Virginia (U.S.A.).

ZOOMING-IN BOX 2.3. ZONING ISSUES[i]

Zoning Symbols

- Residential (R)
  - Single-family residences
  - Two-family residences
  - Low-rise apartments
  - High-rise apartments
- Commercial (C)
  - Small stores
  - Shopping center
- Residential-office (RO)
- Industrial or Manufacturing (I or M)
  - Light industrial
  - Smoke-free industrial
  - Heavy industry
- Agriculture (A)
- Overlay zoning categories
  - Planned unit development (PUD)

Zoning: Land Use Restrictions

- Zoning ordinances are essentially *restrictions on land use*.
- For example, in the U.S. land zones for low-density apartments have to have at least 1,500 square feet of land per living unit, a minimum of 600 square feet of living space per unit for 1 bedroom, 800 square feet for 2 bedrooms, and 1,000 square feet for 3 bedrooms. The building should be positioned at least 25 feet from the street, 10 feet from the sides of the lot, and 15 feet from the rear lot line. The building's height should not exceed 2 and 1/2 stories and 2 parking spaces for each dwelling unit.

Zoning Enforcement

- Real estate developers have to obtain building permits from city or county government. Before a permit is issued, the plan must conform with government-imposed structural standards and comply with zoning requirements. A landowner has to obtain a permit to build. The builder has to obtain permission if the original plan (that was approved) is changed. Otherwise, the government can force the landowner to tear down the building.

Zoning: Nonconforming Use
- When a building does not conform with a new zoning law, it is said to be in *nonconforming use* (or "grandfathered-in"). The owner can continue to use the building even though it does not conform to the new zoning ordinance. However, the owner is not permitted to make any significant changes to the building or extend its life.

Zoning Amendment
- A zoning ordinance can be changed by *amendment*. An amendment can be initiated by either the landowner of the area to be rezoned or by local government. In doing so, notice of the proposed change in the zoning ordinance has to be given to all property owners in and around the area, and a public hearing must be held so that property owners and the public at large may voice their concerns.

Zoning: Variance
- A *variance* is a permit that a landowner may seek to deviate from current zoning requirements for his parcel. Variances are granted only when compliance with a zoning ordinance may cause undue hardship to the landowner. Although a variance can be used to change the use of a parcel, it must not change the basic character of the neighborhood.

Zoning: Conditional-Use Permit
- A *conditional-use permit* allows land use that does not conform with an existing zoning ordinance, provided the use is within certain limitations.
- For example, a neighborhood grocery store operating under a conditional-use permit can only be a neighborhood grocery. The building cannot be used in some other capacity (e.g., an auto parts store).

Zoning: Spot Zoning
- *Spot zoning* refers to the rezoning of a small area of land.
- For example, a neighborhood convenience store can be built in a residential neighborhood provided it serves a useful purpose for the neighborhood residents.

Zoning: Downzoning
- *Downzoning* means that land previously zoned for higher-density use is rezoned for lower-density use. In doing so, the property value may fall.

Zoning: Taking
- *Taking* refers to action by the municipality to regulate the property such that it has little or no economic value. Thus, a municipality can "take" a property and condemn it. In doing so, the municipality pays the fair market value of the property to the landowner.

Zoning: Buffer Zone
- A *buffer zone* is a strip of land that separates one land use from another.

- For example, between a large shopping center and a neighborhood of single-family homes, there may be a buffer zone in the form of a strip of land with grass and trees.

Zoning: Legality

- A zoning ordinance can be changed or struck down if it can be proved in court that it is unclear, discriminatory, and unreasonable. Similarly, the ordinance can be challenged in court if it can be demonstrated that the ordinance does not serve public health or is a threat to safety or the general welfare of the residents.

Zoning: Value

- To the extent that zoning can increase market demand for certain parcels of land and decrease demand for others, zoning does have a powerful impact on property *value*.

[i] Based on Bassett, E. M. (1940). *Zoning.* New York: Russell Sage Foundation; Stephani, C. J., & Marilyn, C. (2012). *ZONING 10* (3rd ed.). Washington, DC: National League of Cities.

CASE/ANECDOTE 2.4. APARTMENT PLAN REJECTED BECAUSE OF ZONING ISSUE[i]

A plan to develop an apartment complex in Blacksburg (Virginia, U.S.A.) was tabled by the town council because of traffic concerns. The town council denied the rezoning needed for University City Center, a planned five-story, 495-bedroom structure with an attached 672-space parking deck. The apartment complex targeted Virginia Tech college students. The town council voted down the proposal after a public hearing in which 21 of the 34 speakers complained about the project. The main concern seemed to be that the big apartment complex would have been out of sync with the surrounding homes, which are mostly single-family. Others expressed concerns about the possible traffic problems (both congestion and safety concerns) that the complex is likely to create.

The developers also own a neighboring hotel that they intend to renovate in a major way. Town council members wanted to see the plan for the neighboring hotel and how this will affect the neighborhood. The developers were granted a one-month delay in their request for rezoning to address the concerns raised at the public hearing.

[i] Gangloff, M. (2013). Apartment plan rejected in Blacksburg. *Roanoke Times,* October 9, 1 & 18.

Subdivision Regulations

Before a building lot can be sold, the real estate developer (or subdivider) must comply with government regulations concerning street construction, curbs, sidewalks, street lighting, fire hydrants, storm and sanitary sewers, grading and compacting of soil, water and utility lines, minimum lot size, and so on.

In addition, the developer may be required to either set aside land for schools and parks or provide money so that land for that purpose may be purchased nearby. These are often referred to as *mapping requirements*.

Building-Code Requirements

Before a building permit is granted, the design of a proposed structure must meet *building-code requirements*. During construction, local building department inspectors visit the construction site to make certain that the codes are being observed. When the building is completed, a *certificate of occupancy* is issued to the owner to show that the structure meets the building code. Without this certificate, the building cannot be legally occupied.

Deed Restrictions

Although property owners tend to think of land-use restrictions as a product of government control, it is possible to achieve land-use restrictions through private means. Private land-use controls take the form of *deed* and *lease restrictions* (also known as *restrictive covenants*). For example, a developer can sell lots in a subdivision subject to a restriction written into each deed that the land cannot be used for anything but a single-family residence containing at least 1,200 square feet of living area.

Environmental Impact Statements

The purpose of an *environmental impact statement* (EIS), also called an *environmental impact record* (EIR), is to gather into one document information about the effect of a proposed project on the total environment surrounding the planned structures. This should allow a neutral decision maker to judge the environmental benefits and costs of the project.

For example, a city zoning commission considering a zone change can request an EIS that will show the expected impact of the change on such things as population density, automobile traffic, noise, air quality, water and sewage facilities, drainage, energy consumption, school enrollments, employment, public health and safety, recreation facilities, wildlife, and vegetation.

SUMMARY

This chapter covered topics related to product and place strategy in real estate development. Specifically, product strategy is mostly about product design and how the real estate marketer works closely with the building architects to design structures that meet the expectations of the client or target customers. This chapter covers examples of quality product issues in residential single-family housing. Specifically, we discussed product quality in relation to the floor plan of the housing unit(s), housing architecture, the site of the housing unit(s), energy conservation and green building issues, and neighborhood features.

The floor plan reflects the design of the residential structure and its three main functions in its living area, sleeping area, and service area. In designing the living area of a residential structure, real estate developers and building and construction experts are guided by quality standards such as space requirements, layout, and location. Quality of the sleeping area is judged based on minimum space requirements, storage space, bathrooms, and noise. Quality of the service area is judged based on kitchen space requirements, kitchen location, kitchen layout, and laundry facilities.

In judging architecture of a residential structure, the real estate marketer has to be familiar with the advantages and disadvantages of the type of structures (one-story house, one-and-a-half story house, two-story house, and two-story split-level house). The marketer has to also gauge the preference of his customers regarding architectural style (Cape Cod, Colonial, Ranch, Tudor, Victorian, etc.).

The residential site refers to the topography of the site, the orientation of the site, and the landscaping surrounding the site. Important elements in topography are temperature, altitude, wind conditions, and the slope of the site toward or away from the sun. The orientation of the site also has a direct effect on living comfort. A house with established landscaping is usually considered much more attractive than one with less-established landscaping.

Marketers of real estate development firms are familiar enough with the science of energy conservation and green buildings to be able to judge product quality; many of these guidelines are discussed in the chapter. Real estate marketers also judge the relative energy efficiency of a house by asking the home owner to provide receipts of utility bills from the last 12 months and comparing the total cost with other houses considered to be energy efficient.

Real estate professionals evaluate neighborhoods along three dimensions: physical features, economic features, and social features. Examples of quality features include good property upkeep (physical), rising resale value (economic), and a safe neighborhood (social).

The second part of the chapter covered place strategy, which is mostly about site selection for real estate development. In selecting a site for development, the developer considers factors such as location needs of the customer, housing needs of the customer, environmental needs of the customer, financing needs of the customer, zoning issues, subdivision regulations, building-code requirements, deed restrictions, and the environmental impact statement.

DISCUSSION QUESTIONS

1. What is product strategy in real estate development?

2. How do real estate developers judge quality of an overall floor plan of a single-family house?

3. How do real estate developers judge quality of the living area in a single-family house?

4. How do real estate developers judge quality of the sleeping area in a single-family house?

5. How do real estate developers judge quality of the service area in a single-family house?

6. What are the advantages and disadvantages of a one-story house compared to other architectural types?

7. What are the advantages and disadvantages of a one-and-a-half-story house compared to other architectural types?

8. What are the advantages and disadvantages of a two-story house compared to other architectural types?

9. What are the advantages and disadvantages of a split-level house compared to other architectural types?

10. How do real estate developers judge the quality of the topography of the site of a single-family house?

11. How do real estate developers judge the quality of the orientation of the site of a single-family house?

12. How do real estate developers judge the quality of the landscaping of the site of a single-family house?

13. How do real estate developers judge the quality of the energy conservation features of a single-family house?

14. How do real estate developers judge the quality of the physical features of the neighborhood of a single-family house?

15. How do real estate developers judge quality of the economic features of the neighborhood of a single-family house?

16. How do real estate developers judge the quality of the social features of the neighborhood of a single-family house?

17. Describe the role of the location needs of the customer as a decision criterion in site selection for developers.

18. Describe the role of the housing needs of the customer as a decision criterion in site selection for developers.

19. Describe the role of the environmental needs of the customer as a decision criterion in site selection for developers.

20. Describe the role of the financing needs of the customer as a decision criterion in site selection for developers.

21. Describe the role of zoning issues as a decision criterion in site selection for developers.

22. Describe the role of subdivision regulations as a decision criterion in site selection for developers.

23. Describe the role of building-code requirements as a decision criterion in site selection for developers.

24. Describe the role of deed restrictions as a decision criterion in site selection for developers.

25. Describe the role of the environmental impact statement as a decision criterion in site selection for developers.

NOTES

1. Skenkel, W. M. (2004). *Marketing real estate*. Mason, OH: South-Western Educational Publishing (p. 161).
2. Skenkel, W. M. (2004). *Marketing real estate*. Mason, OH: South-Western Educational Publishing (p. 161).
3. Skenkel, W. M. (2004). *Marketing real estate*. Mason, OH: South-Western Educational Publishing (p. 161).
4. Skenkel, W. M. (2004). *Marketing real estate*. Mason, OH: South-Western Educational Publishing (p. 161).
5. Skenkel, W. M. (2004). *Marketing real estate*. Mason, OH: South-Western Educational Publishing (p. 160).
6. Skenkel, W. M. (2004). *Marketing real estate*. Mason, OH: South-Western Educational Publishing (p. 161).
7. Skenkel, W. M. (2004). *Marketing real estate*. Mason, OH: South-Western Educational Publishing (p. 161).
8. Skenkel, W. M. (2004). *Marketing real estate*. Mason, OH: South-Western Educational Publishing (p. 161).
9. Skenkel, W. M. (2004). *Marketing real estate*. Mason, OH: South-Western Educational Publishing (p. 160).
10. Skenkel, W. M. (2004). *Marketing real estate*. Mason, OH: South-Western Educational Publishing (p. 161).
11. Skenkel, W. M. (2004). *Marketing real estate*. Mason, OH: South-Western Educational Publishing (p. 161).
12. Shenkel, W. M. (1978). *Modern real estate appraisal*. New York: McGraw-Hill Book Company (p. 405).
13. Skenkel, W. M. (2004). *Marketing real estate*. Mason, OH: South-Western Educational Publishing (p. 162).
14. Skenkel, W. M. (2004). *Marketing real estate*. Mason, OH: South-Western Educational Publishing (p. 164).
15. Skenkel, W. M. (2004). *Marketing real estate*. Mason, OH: South-Western Educational Publishing (pp. 164–165).
16. Skenkel, W. M. (2004). *Marketing real estate*. Mason, OH: South-Western Educational Publishing (p. 165).
17. Skenkel, W. M. (2004). *Marketing real estate*. Mason, OH: South-Western Educational Publishing (p. 166).
18. Skenkel, W. M. (2004). *Marketing real estate*. Mason, OH: South-Western Educational Publishing (p. 167).
19. Skenkel, W. M. (2004). *Marketing real estate*. Mason, OH: South-Western Educational Publishing (p. 167).
20. Skenkel, W. M. (2004). *Marketing real estate*. Mason, OH: South-Western Educational Publishing (p. 167).
21. Skenkel, W. M. (2004). *Marketing real estate*. Mason, OH: South-Western Educational Publishing (p. 167).
22. Skenkel, W. M. (2004). *Marketing real estate*. Mason, OH: South-Western Educational Publishing (pp. 168–169).
23. Skenkel, W. M. (2004). *Marketing real estate*. Mason, OH: South-Western Educational Publishing (pp. 168–169).
24. Skenkel, W. M. (2004). *Marketing real estate*. Mason, OH: South-Western Educational Publishing (pp. 169–170).
25. Skenkel, W. M. (2004). *Marketing real estate*. Mason, OH: South-Western Educational Publishing (pp. 171–173).
26. Skenkel, W. M. (2004). *Marketing real estate*. Mason, OH: South-Western Educational Publishing (pp. 171–173).
27. Skenkel, W. M. (2004). *Marketing real estate*. Mason, OH: South-Western Educational Publishing (pp. 171–173).

Chapter 3

Price and Promotion Strategy

LEARNING OBJECTIVES

This chapter covers topics related to price and promotion strategy in real estate development. Price strategy is mostly about methods used for real estate appraisals, both residential and commercial. Promotion strategy is about message and media decisions—what message to disseminate and through what channels of communications. This chapter is designed to help students of real estate marketing learn:

- the factors that influence the price of real estate
- how real estate appraisers go about assessing market value of real estate properties
- key message decisions in real estate promotion
- how real estate marketers use positioning by aspects of the customers
- how real estate marketers use positioning by aspects of the product
- how real estate marketers use positioning by aspects of the firm
- key media decisions in real estate promotion
- how to develop a real estate promotion plan
- how to make an audience selection decision in the context of the promotion plan
- how to identify goals for the promotion plan
- how to identify the core message and slogan of the promotion campaign
- how to select among the multitudes of media possibilities to disseminate the message to the target audience
- how to develop a media schedule in the context of the promotion plan
- how to select measures for evaluating the effectiveness of the promotion campaign and how to use these measures to set concrete objectives for the promotion campaign

PRICE STRATEGY

In this section we will discuss two main topics: (1) factors influencing the price of real estate, and (2) how real estate appraisers go about assessing market value of real estate properties.

Factors That Influence the Price of Real Estate

What are the factors that influence the price of residential real estate? Research has uncovered three sets of factors: (1) proximity factors, (2) neighborhood factors, and (3) structural factors.[1] See Figure 3.1.

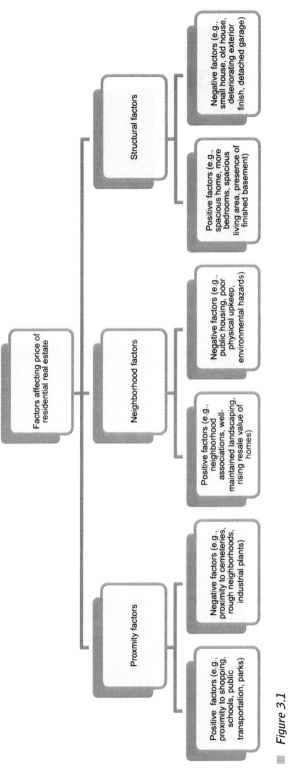

■ *Figure 3.1*
Factors Affecting Prices of Residential Homes

Proximity Factors

Proximity factors are factors related to entities close to the home in question and how these entities enhance or deflate the market value of the target home. Proximity factors can be divided in two major categories: positive and negative factors. Examples of positive factors include proximity to:

■ churches (i.e., homes close to churches are likely to have higher market value)
■ shopping (i.e., homes close to shopping facilities are likely to have higher market value)
■ schools (i.e., homes close to primary and secondary schools are likely to have higher market value)
■ public transportation (i.e., homes close to bus stops, train stations, and within a reasonable drive to an airport are likely to have higher market value)
■ parks (i.e., homes close to neighborhood parks and other family-friendly types of recreational centers are likely to have higher market value)

Examples of negative proximity factors include proximity to:

■ cemeteries (i.e., homes close to cemeteries are likely to have lower market value)
■ railroads (i.e., homes close to railroads, not train stations, are likely to have lower market value)
■ rough neighborhoods (i.e., homes close to neighborhoods with high incidence of crime are likely to have lower market value)
■ industrial plants (i.e., homes close to manufacturing plants and facilities are likely to have lower market value)
■ condemned buildings (i.e., homes close to buildings that are severely distressed, abandoned, or declared condemned are likely to have lower market value)

Neighborhood Factors

Neighborhood factors are factors associated with aspects of homes and the property of other homes in the same neighborhood. Examples of positive neighborhood factors include:

■ neighborhood associations (i.e., neighborhoods that have homeowners' associations are likely to have higher market value)
■ voting percentages (i.e., neighborhoods that have a higher percentage of residents voting in local, state, and national elections are likely to have higher market value)
■ high rate of owner occupancy (i.e., neighborhoods that are mostly occupied by residents and homeowners are likely to have higher market value)
■ historically designated housing (i.e., neighborhoods designated as having homes that have historical significance for the global community are likely to have higher market value)
■ well-maintained landscaping (i.e., neighborhoods that have aesthetically pleasing yards are likely to have higher market value)
■ rising resale value of homes (i.e., neighborhoods in which most of the homes have rising market value are likely to have higher market value)

Examples of negative neighborhood factors include:

■ public housing (i.e., neighborhoods that are essentially public housing neighborhoods are likely to have lower market value)

- poor physical upkeep, trash, and graffiti (i.e., neighborhoods that have poor upkeep are likely to lower market value)
- environmental hazards (i.e., neighborhoods that have environmental hazards are likely to have lower market value)
- old homes (i.e., neighborhoods that have more older than newer homes are likely to have lower market value)
- crime (i.e., neighborhoods that have higher incidences of crime are likely to have lower market value)
- low-SES residents (i.e., neighborhoods in which most of the residents have low socioeconomic status are likely to have lower market value)

Structural Factors

Structural factors are aspects inherent in the physical property. Examples include the overall size of the house and yard, number of bedrooms, size of the living area, age of the property, presence of a finished basement, the exterior finish, number of full baths, and condition of air conditioning and heating systems. Positive structural factors include:

- The house is sizable with a big yard (i.e., spacious home).
- The house has many bedrooms (i.e., home with 4–5 bedrooms or more).
- The house has many bathrooms (i.e., a bathroom for every bedroom, a bathroom next to the living and dining rooms, a bathroom in the basement and next to the den, etc.).
- The house has a sizable living area (i.e., the size of the living room is large).
- The house is built in recent years (i.e., the age of the property is young).
- The house has a finished basement (i.e., the basement can be easily used in many ways without further construction).
- The exterior finish of the house is good (i.e., the exterior of the house is nicely painted and the roof is in good shape).
- The garage is sizable, enclosed, and attached to the house (e.g., a 2–3 car garage with adequate parking space on the driveway, as well).
- The house has modern air conditioning and heating units.

Negative structural factors include:

- The house is small in size with a small yard.
- The house has 1–2 bedrooms.
- The house has 1–2 bathrooms.
- The house has a small living area.
- The house is old.
- The house does not have a finished basement.
- The exterior finish of the house is in poor condition.
- Parking is difficult (e.g., no garage, carport, or a detached garage; inadequate parking space in the driveway).
- The house has old air conditioning and heating units.

Approaches to Pricing Real Estate

There are three common approaches to pricing real estate properties: (1) the market approach, (2) the costing approach, and (3) the income approach.[2] See Figure 3.2.

▦ *Figure 3.2*
Three Approaches
to Pricing
Real Estate

The Market Approach

The market approach to pricing real estate is commonly used in residential real estate, especially in relation to the buying and selling of single-family homes. Here real estate appraisers and real estate agents use Comparative Market Analysis (commonly referred to as CMA). This pricing method guides the real estate marketing professional to determine the price of a residential property as a direct function of the price of comparable homes ("comps") sold in the last six months or so. In other words, the appraiser identifies properties that are close to identical to the house being appraised in the same neighborhood (or comparable neighborhoods in the same town or city) that were sold recently. For example, if the house that is appraised is a three-bedroom house with a 1/2 acre yard, then the appraiser looks for homes sold in the last sixth months with three bedrooms and a 1/2 acre yard in the same neighborhood as the house being appraised. Once these "comps" are identified, the appraiser obtains information about these comparable properties through the sales record (i.e., date of sale, sales price, financing terms, location of the property, and a description of its physical characteristics and amenities). Usually these "comps" are limited to three properties. Once the information about these comps is obtained, the appraiser proceeds to make adjustments; he or she adjusts the price of each comp to further match the house being appraised. For example, consider the following three comps (Comp A, Comp B, and Comp C). Comp A was sold for $141,500 six months ago, Comp B was sold for $136,000 three months ago, and Comp C was just sold for $140,000. A very rough estimate of the appraised house in relation to these three properties would be the average of the sale price of these properties, or $139,167. This is, of course, a crude estimate. The appraiser can do better by adjusting the price of the comps by time—the price of Comp C is probably most accurate as a referent price because it was just sold; Comp B has to be slightly adjusted because it was sold three months ago. In other words, if Comp B had been sold today, it would likely have sold for more than $136,000 because of appreciation. How much more? Perhaps 1% more of $136,000, which amounts to $137,360. In other words, the appraiser adjusts the price of Comp B from $136,000 to $137,360 to make it more "comparable" to the house being appraised. The same type of adjustment has to be made to Comp A. Remember Comp A was sold six months ago for $141,500. Considering a 2% appreciation in value, that house would

47 ▦

have sold today at $144,330. Therefore, a more accurate estimate of the appraised house is the average of the adjusted prices of these comps ($144,330, $137,360, and $140,000), or $140,563. The appraiser continues to make adjustments to each comp by examining differences in house size, garage, age, upkeep and overall quality, landscaping, lot size, and other features. See Table 3.1 to further examine the example described above. As shown in the table, the adjusted market price for Comp A is $133,940; $138,960 for Comp B; and $137,300 for Comp C. Taking the average of the three comps, we estimate the market value of the house in question to be $136,733.

See Case/Anecdote 3.1 for information about home valuation websites and their credibility.

Table 3.1 An Example of a Comparative Market Analysis

Item	Comp A	Comp B	Comp C
Sales price	$141,500	$136,000	$140,000
Time adjustment	Sold 6 months ago, add 2% = +2,830	Sold 3 months ago, add 1% = +1,360	Just sold = 0
House size	160 sq ft larger at $60/sq ft = -9,600	20 sq ft smaller at $55 per sq ft = +1,200	Same size = 0
Garage/carport	Carport = +4,000	3-car garage = -2,000	2-car garage = 0
Other	Larger patio = -900	No patio = +1,800	Built-in bookcases = -2,000
Age, upkeep, & overall quality	Superior = -2,000	Inferior = +1,200	Equal = 0
Landscaping	Inferior = +2,000	Equal = 0	Superior = -700
Lot size, features, & location	Superior = -3,890	Inferior = +900	Equal = 0
Terms & conditions of sale	Equal = 0	Special financing = -1,500	Equal = 0
Total adjustments	-5,360	+960	-2,700
Adjusted market price	$133,940	$138,960	$137,300

CASE/ANECDOTE 3.1. THE CREDIBILITY OF HOME VALUATION SITES[i]

There are many home valuation sites in cyberspace, but how credible are they? Here's a list of some of the most popular services and methods these sites use to price a home.

- CyberHomes (www.cyberhomes.com): CyberHomes uses data mostly on sales, mortgage and new ownership records, demographic information, and property appraisals in calculating home price.

- **HomeGain (www.homegain.com):** This program calculates the home price based mostly on recent sales in the immediate area and comparable sales.
- **HouseValues (www.housevalues.com):** Home owners interested in assessing the value of their home complete a form that describes the home. The completed form is then sent to a local appraiser who responds with a CMA report.
- **HouseFront (www.housefront.com):** This is a free service using MLS sales data and public records in computing home valuations. Valuations are rated A, B, or C, to indicate the company's confidence in the accuracy of the price.
- **ValueMyHouse (www.ValueMyHouse.com):** This service is a free home value report to home owners; the homeowner completes a form on the Web site. The information is then sent to local appraisers who conduct a CMA.
- **Zillow (www.zillow.com):** Home values are calculated using data from public records, tax assessment, past sales history, and recent sales in the area.

As useful as these services may seem to buyers and sellers of real estate, the home valuations are usually far from accurate. The credibility of each site and accuracy of the price estimates should be questioned. Home valuation websites are motivated by two main business goals: lead capture and traffic driver. What does *lead capture* mean? The site visitor, assumed to be a prospective buyer or seller, submits a form with contact information. The request is passed along to a real estate agent who pays a membership fee for affiliation with the site. The real estate agent responds with a CMA. The CMA provides an introduction that could turn into professional representation. In other words, real estate marketers use the valuation site as a method for prospecting new clients. The second goal, *traffic driver,* refers to the use of the home valuation service to draw traffic to the site in order to advertise other services to buyers and sellers. Home valuation sites in this category allow buyers and sellers to obtain a price estimate anonymously—just enter an address and click. The resulting report is based on a formula.

[i] Antoniak, M. (2008). Story behind home valuation sites. *Realtor Magazine,* September. Accessed from http://realtormag.realtor.org/technology/tech-watch/article/2008/10/story-behind-home-valuation-sites, on May 5, 2013.

The Costing Approach

The costing method to pricing real estate focuses on the actual costs expended to build the appraised property. This method is commonly used by builders in both residential and commercial real estate. Let's use an example to illustrate this pricing method. Suppose we would like to appraise a home that will sit on a parcel of land worth $130,000. The buyer asks the builder to construct a 5,000 square foot home. The home builder normally charges $100/square ft. Hence, the price of the home structure is structure would be $500,000. Adding the price of the structure ($500,000) to the price of the lot ($130,000) brings the total price of the home to $630,000.

Another costing method can be used based on past costs of building similar property. For example, suppose that we are trying to estimate the price of a three-bedroom house that has

2,500 square feet. We have identified a similar house that was built two years ago and ascertained that the customer paid $320,000 to the builder. Now we need to adjust that figure by depreciation. In other words, we need to subtract a certain amount of money because the price of the house depreciates by the age of the house. In this case, the depreciation is for two years (1% depreciation per year, which is $3,200/year, or $6,400 for two years). Therefore, the adjusted cost becomes $313,600 (or $320,000 minus $6,400). We then adjust this cost figure by inflation. In other words, the $313,800 two years ago is actually $344,960 in today's dollars (5% inflation annually, or $313,600 + $15,680 + $15,680).

The Income Approach

The income method to pricing real estate focuses on the revenue generation potential of rental properties. This method is commonly used by property management companies (real estate firms that lease or rent apartments, office space, retail space, etc.). Let us consider the example of a small apartment complex (of 10 units). What might be the appraised value of that property? The appraiser estimates the net income generated by the apartment complex per year and divides this figure by the cap rate—you can determine the cap rate by dividing the net operating income of a comparable apartment project sold recently and dividing it by the sold price. Let us plug in some numbers. Suppose an apartment complex generates a net income of $48,730 per year. Divide this figure by the cap rate—let's say 9%—and we end up with $541,000. In other words, the appraised market value of this apartment complex is $541,000. See Figure 3.3.

How does the property manager calculate the net income figure? Suppose the apartment complex is scheduled to generate $84,000 in annual revenue (10 units × $8,400 per unit annually, or $700/month per unit per month). However, the property manager has to allow for vacancy and collection losses. This is estimated to be $4,200 per year. Therefore, the gross income from the apartment complex amounts to $79,800 (i.e., $84,000 – $4,200). This is gross income, not net income. To get to the net income figure, the property manager takes the gross income figure ($79,800) and subtracts from that figure operating costs (property taxes, hazard and liability insurance, property management, janitorial services, gardener, utilities, trash pickup, repairs and maintenance, reserves for replacement of furniture and appliances, etc.). Operating costs per year are estimated, in this case, to be $31,070 (see Table 3.2). Hence, the net income is derived by subtracting the gross income from the operating costs, which amounts to $48,730 (or $79,800 – $31,070). This net income figure allows us to appraise the apartment complex when we divide it by the rate of return.

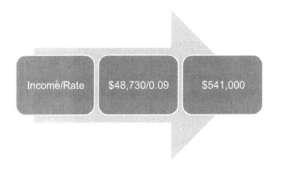

Figure 3.3
The Income Approach to Pricing Real Estate Properties

Income/Rate $48,730/0.09 $541,000

Table 3.2 *Deriving Net Income from a Rental Property*

Gross annual income	$84,000	
Vacancy allowance and collection losses	4,200	
Effective gross income		$79,800
Operating expenses		
-Property taxes	7,000	
-Hazard and liability insurance	2,100	
-Property management	4,200	
-Janitorial services	1,500	
-Gardener	1,200	
-Utilities	3,940	
-Trash pickup	850	
-Repairs and maintenance	4,000	
-Other	1,330	
Reserves for replacement		
-Furniture & furnishings	1,200	
-Stoves & refrigerators	600	
-Furnace &/or air conditioning	700	
-Plumbing & electrical	800	
-Roof	750	
-Exterior painting	900	
Total operating expenses		$31,070
Net annual income		$48,730

PROMOTION STRATEGY

In this section we will discuss three main topics: (1) message decisions in real estate promotion (what to say in a promotional message and why), (2) media decisions in real estate promotion (what form of media should carry the message), and (3) developing a promotion plan (putting things together).

Message Decisions

Message decisions in real estate marketing are all about what to say in a promotional message and why. When marketers speak of message strategy they use the term *positioning*. Positioning can best be viewed as key mental associations with the focal product.[3] In other words, what is the key concept with which the real estate marketer would like to associate? In many cases, this positioning is reflected in the *slogan* of all promotional messages.

There are three types of positioning common in real estate promotion: (1) positioning by an aspect of prospective customers, (2) positioning by an aspect of the product, and (3) positioning by an aspect of the firm. Let's discuss these in some detail.

Positioning by an Aspect of Prospective Customers

Consider a real estate developer of an assisted living facility. The target market involves senior citizens who are physically frail. A promotion campaign is launched to appeal to these prospective customers. Central to the promotion campaign is a message addressed to the senior citizens who have physical difficulties. The real estate marketer for the assisted living facility develops a slogan: "We are here to help you with your daily living." This slogan captures the major benefit of the assisted living facility in addressing the needs of the housing needs of senior citizens. The slogan appears clearly on the company website, all promotion brochures and mail communications, television commercials, radio commercials, newspaper ads, outdoor advertising, etc. In essence, the real estate marketer is positioning the assisted living facility in the minds of prospective customers by associating the facility with the key benefit: helping senior citizens who are physically frail with daily living. Below are other examples of positioning by customer type:

- appealing to prospective home buyers with a certain demographic or psychographic profile (e.g., college students, young/single professionals, the affluent, golfers)
- appealing to opportunistic home buyers to think about taking advantage of certain opportunities to buy a new home (e.g., decline in mortgage interest rate; availability of many new homes, creating a buyers' market)
- appealing to home owners who may be thinking about buying a new home as a viable solution to a deteriorating home, instead of opting for home renovation
- appealing to people going through a specific life stage to induce them to think about buying a new home as a viable solution to changing circumstances (e.g., graduating from college, getting married, family size is getting bigger with more kids, kids leaving home to go to college, retirement)

Positioning by an Aspect of the Product

The message here is the type of real estate or something inherent in the physical or tangible aspect of the property. Consider the following examples:

- appealing to prospective customers by emphasizing the price of the house as affordable—a bargain
- appealing to prospective customers by emphasizing certain house design features (e.g., architectural style, floor plan, quality construction, space)
- appealing to prospective customers by emphasizing energy conservation features (e.g., well insulated, consumes little electricity, water conservation devices installed, gas heat)
- appealing to prospective customers by emphasizing neighborhood features (e.g., proximity to shopping facilities, friendliness of the neighbors, outdoor playground facilities for the children, neighborhood safety, value of homes in the neighborhood)
- appealing to prospective customers by emphasizing community features (e.g., quality of public schools, crime rate in the community, opportunity for employment, recreational programs and facilities)

Positioning by an Aspect of the Firm

In this case, positioning by an aspect of the firm involves an appeal to prospective customers based on a positive aspect of the firm, such as reputation longevity or long-time service to the community. Consider the following examples:

- appealing to prospective customers by emphasizing the longevity of the organization in service of the community (e.g., "we've helped people in this community buy and build their dream homes for the last 30 years")
- appealing to prospective customers by emphasizing results (e.g., "for the past five years, we have been able to help 1,000 families find their dream homes," "the average turnover of homes for sale was 30 days over the past five years," and so on)
- appealing to prospective customers by recognizing certain real estate brokers who have achieved high levels of sales (e.g., names and pictures of those who have won the platinum award in the last quarter, as well as gold, silver, bronze, and so on)
- appealing to prospective customers by emphasizing certain services (e.g., "we help with mortgage financing," "we buy the house from you if we can't sell it in 30 days," "we provide relocation services," and so on)
- appealing to prospective customers by highlighting the firm's mission to serve certain populations (e.g., the elderly, the poor, college students, upscale clientele, farm families, or another demographic)

Media Decisions

Once the real estate message is formulated, it has to be communicated to a target audience through media. There are two major forms of media: interpersonal and mass media. Interpersonal forms of communications involve personal selling. Mass media forms of communications in real estate marketing involve advertising and public relations. We will describe these various forms of media communications below.

Interpersonal Forms of Marketing Communications

As previously mentioned, interpersonal forms of communication are essentially personal selling. Personal selling is the most potent form of marketing communications in the marketing of real estate development. Personal selling refers to promotion conducted by a marketing agent, such as the real estate agent, using face-to-face encounters with prospective customers in the form of "open house" (i.e., presenting a model house, apartment, office, etc.), responding to questions from prospective clients, and providing them with information about the product in appealing and persuasive means.

Interpersonal communication in real estate development can be most effective when the selling agent has the following characteristics: (1) referent power, (2) expert power, (3) legitimate power, and (4) reward power.[4] *Referent* power refers to the power related to identification. If the selling agent belongs to a referent group that the prospective customer may identify with, the selling agent is said to have referent power. For example, a prospective customer is a Korean American woman. The selling agent happens to also be Korean American woman. The prospective customer feels a special bond with the selling agent—they are both Korean Americans, and both are women, too. The selling agent is likely to be highly persuasive because of this social bond. The selling agent can also be highly persuasive if he comes across as having special knowledge and expertise in the type of real estate being sold. For example, the selling agent is showing a model of a medical clinic for a prospective physician. He has much experience in the building and construction of medical facilities. Such expertise can easily be leveraged for effective interpersonal communication. Selling agents who impress their prospective clients with their expertise are in a better position to effectively respond to the prospect's detailed questions and make them feel at ease. This is expert power, and such power adds an important dimension to effective marketing communications. *Legitimate power* is also an important dimension of effective interpersonal communications, especially in commercial real estate development. Suppose the selling agent is showing corporate real estate to business executives of a prospective firm. In this situation it may help much if the selling agent is a high-level business executive of the real estate development firm, not merely a low-level sales agent. The social power in this instance comes from holding a position of authority—a high-level position in the real estate development firm. Such power exudes influence and persuasion. Finally *reward power* comes from those instances in which the selling agent can offer inducements and incentives to the prospective client to nudge him towards purchase. For example, a selling agent may offer to present detailed information about the "product" over dinner in a nice restaurant. This is essentially rewarding the prospective customer for being receptive to additional information about the product. Reward power, combined with other forms of social power, can enhance the persuasiveness of the selling agent and the information provided.

Mass Media Forms of Marketing Communications

Mass media forms of marketing communications include advertising and public relations. *Advertising* in real estate development tends to be mostly in *broadcast advertising* (e.g., television advertising, radio advertising), *print advertising* (e.g., newspaper advertising, local real estate publications, direct-mail advertising, yellow pages, and local and regional consumer or trade magazines), *out-of-home advertising* (e.g., bulletin board advertising, transit advertising, airport advertising, banners, building illumination), *specialty advertising* (e.g., mugs, calendars, t-shirts, pens, letter openers, key chains), *digital advertising* (e.g., website, banner ads on related sites, e-mail messages, and video marketing), *online classified advertising* (e.g., Zillow, Trulia, Realtor. com, Yahoo!, Yahoo!-Zillow Real Estate Network, Craigslist, online classified ads in major national newspapers such as the *New York Times*[5]), *mobile advertising* (banner ads displayed on mobile phone apps), and *social media* (the use of Facebook, Twitter, LinkedIn, Pinterest, etc.).

Recent research has found that e-mail campaigns, social media (Facebook, Twitter), promotional mailers, and featured listings to be highly popular among real estate marketers. Real estate professionals reported listed networking, e-mail campaigns, and online featured listings as the most effective tools for promotion, while print listings and print advertising were least effective.[6] See information about mobile advertising in Zooming-In Box 3.1, social media in Zooming-In Box 3.2, and digital advertising in the form of video marketing in Zooming-In Box 3.3. Zooming 3.4 focuses on the use of product placement in the context of online social games. See Case/Anecdote 3.2 to read about a social game for kids designed to enhance brand awareness in the very young generation. Also see Zooming-In Box 3.5 on how the online real estate market is segmented and how these segments can be targeted.

ZOOMING-IN BOX 3.1. MOBILE ADVERTISING[i]

Market research firm comScore has conducted a study that has found that mobile searches will surpass desktop searches in 2014. In 2012, research has shown that mobile is how 55% of adult cell phone users access the Internet. More than 80% of younger consumers (25–34 years old) use their cell phone to go online, compared to 68% of the slightly older crowd (35–44 years old).

Mobile traffic has been increasing, with almost 48% of Homes.com's traffic coming from mobile devices. Consumers who visit Homes.com on their mobile phones to search for available properties in target areas can see banner ads that help connect them with sellers. Mobile local ads include a picture or logo, a company name, a call to action, and a phone number. The mobile ads appear at the top of listing search results pages.

To take the mobile search results further, Corcoran Group Real Estate in New York has designed an iPhone app that offers interactive floor plans for all listings. The company has around 60,000 iPhone users who interact with the app. Foursquare offers an app that integrates open-house check-ins on a real estate firm's website as well as on Facebook.

Some real estate firms are putting Quick Response (QR) on their signage. Prospective homebuyers scan the QR code at the location and obtain detailed information on the listing. In addition, the real estate firm can track who is scanning the QR codes.

[i]Homes.com launches mobile local advertising program, *Watchlistnews.com*. Accessed from www.watchlistnews.com/2013/05/10/homes-com-launches-mobile-local-advertising-program, May 10, 2013. O'Brien, S. (2012). More real estate marketing trends for 2013. *Realestate.com*. Accessed from www.realestate.com/advice/more-real-estate-marketing-trends-for-2013/, May 5, 2013.

ZOOMING-IN BOX 3.2. SOCIAL MEDIA[i]

Many real estate development firms are shifting their promotion dollars to social media, such as Facebook, Twitter, and LinkedIn. A study conducted by Postling in 2012, a social media research firm, found that 79% of real estate professionals are using Facebook to promote themselves and their listings, 48% are using Twitter, and 29% are using LinkedIn.

One company, N-play, creates real estate applications for Facebook. The application allows sellers to list their properties and prospective buyers to view and find listings in their geographic area. The Multiple Listing System's (MLS) listings can easily be imported directly into the app too. Real estate marketers can obtain data analytics: When a consumer clicks on your ad to view it, you are notified immediately—who the consumer is, when he visited your website, and how long he stayed at the site.

[i]Huffman, M. (2013). Advertising is going social in a big way. *Consumer Affairs*, May 6.

ZOOMING-IN BOX 3.3. VIDEO MARKETING[i]

Research has shown that mobile ad engagement can be significantly improved with the addition of video. Mobile-rich ads containing video score higher in audience engagement rates. It is estimated that on Facebook alone there are 20 million videos uploaded and 2 billion videos viewed every month. Consumers are spending a lot of time watching video content online; research indicates that Americans watched approximately 21.8 hours of online video in April 2012 alone. Consumers typically watch video before they read text or view photos. Google Analytics enable marketers to monitor and measure traffic and user engagement on the marketer's website.

Many apartment real estate marketers are increasingly using video marketing as an effective promotion tool. The videos are placed on the property management's website or on general websites such as YouTube. The goal of video marketing is to bring the selling message to life in the minds (and eyes) of prospective residents. Apartment real estate marketers are also using search engine optimization to direct consumers looking for apartments in a specific geographic locale to a site where the apartment complex can be viewed online. Some marketers stream videos continuously in their leasing offices and common areas, or make them viewable through a window, especially when the office is closed. They show them in exhibit booths at career fairs; they use them as trailers at movie theatres. Business executives use them in their recruitment and management meetings.

In producing a video promotion for an apartment complex, some marketers try to convey a community's culture (i.e., what kind of people live in the apartment complex). Typical residents are used in the video to convey the culture. Some marketers use a personality to host the video tour. The host speaks and describes various parts of the apartment complex, its amenities, and its residents. Some include testimonials from residents.

[i] Bergeron, III, P. R. (2012). Apartment marketers get 'real' with video. *Units*, July, 48–52.

ZOOMING-IN BOX 3.4. ONLINE SOCIAL GAMES[i]

Century 21 launched a marketing campaign within the online games SimCity Social and The Sims Social. SimCity Social and The Sims Social are online games designed to interface with Facebook. Facebook users access these games and develop characters and cities through avatars. Sims Social prompts the user to build and expand property and earn money through their avatars. SimCity Social focuses on building and maintaining cities.

In Sims Social, there is a fireplace add-on branded with the Century 21 logo that players may purchase for their avatar's home. This is akin to product placement in movies and shows. But who plays social games online? There are 10 million social gamers in the U.S. alone; a third are in the 24–25 age-bracket, and these people are first-time home buyers. Research estimates that social gamers are 27% more likely to buy a home within a year than those who do not play social games online.

[i] REMI applauds Century 21's the Sims social ad campaign. *Prweb.com*. Accessed from www.prweb.com/releases/century-21-sims-social/real-estate-marketing/prweb10333384.htm, on May 5, 2013.

ZOOMING-IN BOX 3.5. SEGMENTING THE ONLINE REAL ESTATE MARKET[i]

Research shows that 86% of home buyers started the buying process by using the Internet as part of their search tool before contacting a real estate agent. In other words, the online real estate market is huge. Online media is the Number One ad spending medium in the real estate industry, capturing more than 30% of total real estate ad spending in 2012.

One 2012 study segmented the online real estate market into four major segments: (1) passionate, (2) conventional, (3) actives, and (4) future prospects. The *passionate segment* constitutes 6% of the online real estate market. They are highly engaged and interested. They are Internet savvy and are heavy users of the Internet for real estate information. Passionates do their own research before contacting a real estate agent. They use multiple sites (i.e., they are not "one stop shop" consumers). How to reach them? One strategy may be to author your own blog, establishing a presence in mainstream social media and real estate communities (e.g., participating in Q&A sites). It goes without saying that having your own website is essential to communicate with the passionate. Make sure that your website is search engine optimized.

The *conventionals* are traditional and are looking for guidance. They constitute 14% of the online real estate market. The demographic profile of these consumers is as follows: 43-year-old married males with an average income of $69,000. They own property with an average value of $349,000 and may own more than one property. Their properties are financed (not owned). They use offline sources, such as newspapers, to look for properties and agents. They turn to the Internet for additional information. Most of them use MLS sites as the first place to do research. They heavily rely on real estate agents for information and guidance. How to reach the conventional? Use offline sources (e.g., newspaper advertising) to make contact. Use online sources to drive home the message.

The *actives* make up 19% of the market. They like to be engaged online. Actives tend to be highly educated. The average demographic profile is a 44-year-old married male, income of $87,000, owning property with an average value of $410,000; a large proportion finance. They seek out information mostly through online sources to go beyond what is handed to them by real estate agents. They use multiple sites and multiple online tools for researching real estate. How to target the actives? Establish your presence on multiple sites. Partner with mortgage lenders to get on their websites. Try to get listed in the emerging category of online real estate directories. Establish presence in local MLS sites, newspaper websites, and large real estate search and portal sites.

The *future prospects segment* makes up 61% of the online real estate market. They are "just looking right now." They are certainly not heavily engaged in the buying or selling of real estate. They are likely not to own property. However, they are slightly more engaged in real estate than the general population. They mostly go online to check out certain listings. Demographics: 45-year-old married female with an average income of $69,000, owning property with an average value of $324,000. More than one-third are renters. They enlist the assistance of real estate agents. They use search engines for information on mortgage rates, school information, and maps.

How to target the future prospects? Building awareness is most important here. Put your brand front-and-center. Use good graphics and visuals to attract attention. Make sure that your listings and profile are search engine optimized.

[i] Embracing the online real estate market. *Yahoo! Real Estate*. Accessed from http://l.yimg.com/d/lib/yre/embrac-the-online-re-mkt.pdf, on May 5, 2013.

CASE/ANECDOTE 3.2. PROMOTING BRAND AWARENESS TO KIDS[i]

A large real estate agency in Australia, L.J. Hooker, is trying to enhance awareness of its brand to the next generation of future home buyers—kids as young as four years old. The agency launched a free educational app, Mr. Hooker Bear's Letter Pop, and a special Web site *(kids.ljhooker.com.au)*, directed to children 4–9 years old. Mr. Hooker Bear, the mascot for the agency, is featured on both the app and the website.

The app takes the children on a journey (through city, beach, and country) to help them with word play, spelling, and hand/eye coordination. The player attempts to find letters, create words, and jump over animated obstacles to progress through three levels to win.

The CEO of the agency, Georg Chmiel, asserts that it is a way to raise brand awareness of the firm in the minds of children and build a positive relationship. He claims that 2,500 have signed up for the app so far, and he says that the app serves real estate marketers in open houses and at inspections; it is given to children who accompany their parents to open houses and house showings so that they have fun and do not get bored and cranky. Thus, the app serves two goals: to entertain the children and to build brand awareness.

[i] Agency targets kids as future home buyers. *Daily Real Estate News*, May 28, 2013. Accessed from http://realtormag.realtor.org/daily-news/2013/05/29/agency-targets-kids-future-home-buyers, on November 11, 2013.

Public relations is corporate-form communications that come from the CEO's office, not necessarily from the marketing department (but in many cases both are conjoined). Public relations typically include *press releases* (e.g., the CEO of the real estate development issues a statement to the press announcing the completion of a major real estate development), *corporate sponsorships of events* (e.g., the real estate firm sponsoring local sports events in which their corporate slogan appears repeatedly and in various forms throughout the event), and *publicity* (e.g., the real estate marketer has his own blog, or the corporate communications officer writes an article in a regional magazine about a nursing facility and how it is designed with up-to-date technologies that meet the housing needs of the elderly). See Zooming-In Box 3.6 for an exposition of *advertorial columns*.

ZOOMING-IN BOX 3.6. ADVERTORIAL COLUMNS[i]

Advertorial columns in real estate are essentially editorials of listings and agents. These publicity articles serve to inform and engage consumers by profiling selected real estate agents and displaying their credibility and expertise. Unlike the classified ads that are paid advertising (paid by real estate agents), advertorials are written by third parties who are supposed to be unbiased in their reporting. These authors are professional journalists who work with real estate firms to gain accurate information about the firm's listings.

These articles become columns in prominent consumer magazines, newspapers, and real estate-related news websites. Many real estate agents, brokers, and marketers of real estate development firms communicate with one organization, advertorialagency. com, to provide them with their listings, and the journalists at advertorialagency.com do the publicity. This form of marketing communications is the wave of the future.

[i]Real estate advertising annuals. *Aimgroup.com.* Accessed from http://aimgroup.com/combo-2013-2012-real-estate-advertising-annuals/, on May 2, 2013. Arshad, S. (2013). Real estate companies turn to advertorial marketing. *The Canadian,* 10 (May). Accessed from www.agoracosmopolitan.com/news/headline-news/2013/05/01/5825.html, on May 5, 2013.

Promotion Planning

The real estate developer assembles a promotion plan by putting all the pieces together (i.e., message and media decisions). The plan involves audience selection, goal selection, message selection, media selection, media schedule, and performance evaluation. See Figure 3.4.

Audience Selection

Audience selection is the first step in promotion planning. That is, the marketer of a real estate development attempts to promote the development first and foremost by identifying

Figure 3.4
Promotion Planning
in Real Estate
Development

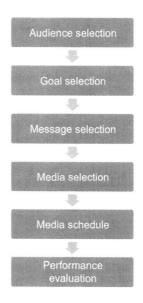

the audience of communications. Who should be subject of the promotion campaign? If the real estate development is college student housing located in a university town, then obviously the audience should be mostly sophomore-level and junior-level college students. Freshmen-level students tend to be housed in college dorms. Marketing to seniors (college students) is not practical because most seniors already have signed leases during their junior years. Graduate students tend to have families or special housing for graduate students only.

Goal Selection

Goal selection refers to the desired outcomes of a media campaign. Some campaigns are designed to simply create an awareness of a particular development (i.e., name recognition is important in this case); some campaigns are designed to generate good will or positive image of the development; other campaigns focus on inducing the audience to action (e.g., advertising an open house, or offering a special discount to those who sign a lease before a specified deadline). Campaign goals vary as a function of what management is trying to accomplish in what time frame. For example, a campaign directed toward college students in the latter part of the summer (in the U.S.) to induce freshmen and sophomore students to sign a one-year lease before the beginning of the Fall semester is likely to focus on action (i.e., providing the students with an inducement to sign up as soon as possible). A campaign launched throughout the year may focus on name recognition and good will, not action.

Message Selection

Message selection is the next step in promotion planning. The marketer has to select the core message to position the product in the audience's minds. Recall the discussion we had on positioning by customer, product, or firm. The marketer has to first decide on what type of positioning. Ultimately, what will be the slogan that carries the core message or positioning? The slogan will have to be featured in all forms of promotions. This decision is made by considering the goals selected. For example, if the campaign is to induce action, then the focus of the message should not be the firm or the product. It should be the customer. Messages that focus on aspects of the customer are likely to be more effective in inducing action than messages focusing on the product or firm. This may be due to the level of ego involvement that customer-type messages have over the audience compared to product- and firm-type messages. Consider the following promotional message placed in the local newspaper two weeks before the beginning of the Fall semester: "Apartments for lease designed specifically for sophomore- and junior-level college students. Those who sign up before August 15th will receive a 30% discount on the lease price." Such a message is likely to induce much personal involvement and motivate prospective customers toward action. Compare this message to a firm-type message: "Lease your apartment with Edgewood Real Estate, the firm with 20 years experience in college housing." This message can induce name recognition and a positive attitude toward the firm, but it is not likely to induce prospects to action. Or how about this product-level message: "Edgewood Estate is a beautiful apartment complex located within walking distance from campus. It has a swimming pool and a club that allows students to hold parties." Again, this message is designed to create a positive attitude toward the real estate development, but not action.

Media Selection

How will the message be communicated? That is, will the marketing communications campaign involve interpersonal communications, mass media communications, or both? Typically, mass-media communications are used in the early stages of the marketing communication campaign to generate awareness and create a positive attitude of the product. Interpersonal communications follow once a prospective client expresses an interest and makes contact with the real estate development firm. In deciding among the media categories and vehicles of advertising, the marketer considers a variety of factors such as marketing communication goals, market concentration, type and amount of information in the ad, advertising frequency, and cost. If the marketing communications goal is to create an awareness and positive image, then television advertising, radio advertising, and out-of-home advertising are effective for this purpose. Out-of-home advertising and newspapers are most effective if the real estate development firm is trying to reach a local audience (i.e., a geographically-concentrated audience). Television, radio, and regional magazines are effective if the audience is geographically dispersed. If the ad requires detailed information, then magazine and newspaper media are most effective in this case. How frequent the advertising? Media that are adept at frequency include television, radio, and outdoor advertising. Advertising in print (newspapers and regional magazines) is not well-suited for frequency. Cost is also a consideration. Television advertising is most expensive, followed by outdoor advertising (e.g., bulletin board advertising), and radio advertising.

Much research in marketing communications and real estate marketing has shown that promoting real estate listings through a mix of various media types is much more effective than doing so using one medium alone. In other words, relying on personal selling exclusively is not necessarily a good thing. The effectiveness of personal selling is enhanced when it is accompanied by all forms of advertising using a variety of media. Prospective buyers exposed to a promotional message about a listing, for example, are likely to notice it and process it fully when the same message is contextualized in a variety of media. For example, data show that homes for sale that are advertised both in print and online are 20% more effective (i.e., they sold faster and had lower discounts) than homes advertised in one medium. This research was based on 503,000 properties that were studied in 2012.[7]

Media Schedule

Once the media categories (as well as the media vehicles) are selected, the real estate developer assembles all these elements in a media schedule. A media schedule is essentially a document showing what ad is placed in what media, when will the ad be released, and at what cost. That is, the advertising is "scheduled" to be shown, displayed, or aired at certain days and certain times, and the media schedule spells out this information very concretely. In developing a media schedule, the marketer takes into account the extent of audience reach of each media vehicle, the optimal level of frequency of advertising within the selected vehicles, and the cost of this advertising in each vehicle.

The marketer develops alternative media schedules and evaluates the effectiveness of these options using at least three criteria: reach, frequency (or exposure opportunities), and cost.[8] *Reach* is a performance measure that captures the extent to which the media vehicles selected in the media schedule are likely to "reach" the target audience (i.e., present an opportunity for the target audience to be exposed to the ad embedded in the media vehicle). For example, a real estate developer of upscale homes located in Roanoke, Virginia (U.S.A.) wants to promote a gated community of affluent homes for the year 2014. He identifies several media vehicles that can reach this target audience: *Roanoker Magazine* (regional magazine that many high income households in and around Roanoke subscribe to—let's say, hypothetically, 40% of them), Roanoke *Times* (major newspaper in the area that has a high subscription rate among high income households in the area—let's say, hypothetically, 30% of them), the local news on WFXR (a FOX television channel affiliate that many high income households watch—let's say 25% of them), the local news on WDBJ (a CBS television channel affiliate that many high income households watch—let's say 21% of them), the local news on WSLS (NBC local affiliate that has 18% of its viewers from high income households), and the local news on WSET (ABC local affiliate that has 22% of its viewers from high income households).

Will the marketer choose to promote the gated community through all these media vehicles? Most likely not! The marketer selects among these media vehicles in a way to maximize reach/frequency and minimize cost. So he assembles alternative configurations of these media vehicles and compares and contrast them in terms of these three criteria (reach, frequency, and cost). He selects the "best" configuration.

With respect to *frequency*, the marketer believes that an ad placed in the *Roanoker Magazine* (a monthly publication) should be placed every month (for a total of 12 ad insertions in 2014). With respect to the Roanoke *Times*, the ad should be placed in every Sunday edition (for a total of 48 ad insertions in 2014). Turning to ad insertions in local news channels, he believes that the television commercial should be run once a week throughout the year (for a total of 48 spots in 2014). He then tabulates the *cost* of these ad insertions in all the prospective media vehicles.

Media planners use a specific measure to help them compare the reach/frequency effectiveness of alternative configurations: *total gross rating points*. This is a measure that multiplies reach by the frequency of each vehicle and then sums up these figures. Thus, the media planner chooses the best configuration that has the highest total gross rating points at the lowest cost.

Performance Evaluation

Performance evaluation is the last step in promotion planning. The marketer plans to survey the customer population, asking customers whether they recognize the product (brand awareness), whether they have a favorable opinion of the product (brand attitude), and the extent to which they have formed intentions to buy the product (brand conviction). The results are compared with pre-test data and the stated objectives of the promotion campaign. If the results fall short of the real estate developer's communications' objectives, then corrective action should follow suit. If the results meet the objectives, then the marketer feels rewarded that his many decisions (audience selection, goal selection, message selection, media selection, and media schedule) were correct and optimal.

For example, consider the marketer of a newly built affluent and gated community who decides to focus on building a positive image of the community. What is the exact survey measure he may use to gauge the level of performance of the campaign after having run the campaign for the entire year of 2014? The goals for the promotion campaign that were selected early on should be translated into measurable objectives. This can only be done when the marketer is in a position to know the exact measures he will be using to capture the results of the promotion campaign. Once the exact survey measures are selected, a random sample of the target audience is contacted and asked to respond to questions such as:

"Have you heard of gated community XYZ? [Yes or No]"
"If yes, then what is your overall impression of the community?
[1 = Very poor; 2 = Poor; 3 = So/so; 4 = Good; 5 = Very good]"

The two questions above are designed to measure awareness (awareness of the real estate development or "product") and attitude (favorable opinion of the product). Such a procedure is referred to as a pre-test. In other words, to evaluate the performance effectiveness of a promotion campaign, the marketer has to establish a referent (or a base line) through the pre-test. Suppose the results of the pre-test show that the sample of high income households in the area in question are 40% aware of the development, and from the 40% (those who responded "Yes" to the survey question, "Have you heard of gated community XYZ?"), their attitude registered is 3.5 ("What is your overall impression of the community? [1 = Very poor; 2 = Poor; 3 = So/so; 4 = Good; and 5 = Very good]). In other words, we now have a baseline of 40% awareness and 3.5 attitude.

Objectives for the promotion campaign can now be set. The marketer may set an objective of 90% awareness and 4.5 attitude. That is, the goal is to increase awareness from 40% to 90%, and attitude from 3.5 to 4.5, after one full year of promotion in 2014. Specifically, the marketer has to use the same survey instruments (i.e., survey measures) and conduct another survey of the target audience (high income households in the designated area). The results of the survey should then be compared against both the objectives and the baseline data. This is the essence of what marketing researchers call "post-test." Suppose the post-test results reveal 80% awareness and 4.1 attitude. This means that the promotion campaign succeeded in increasing the level of awareness of the gated community from 40% to 80%, but it fell 10% short of the stated objective of 90%. Similarly, the promotion campaign succeeded in enhance the favorability (i.e., attitude) of the gated community from a baseline of 3.5 to 4.1, but it fell short 0.4 points of the stated objective of 4.5. In sum, based on this performance evaluation, was the promotion campaign successful? It was mostly successful in increasing brand awareness and attitude, yet the campaign may be judged as not being entirely successful because the post-test results indicate a slight underperformance compared to the stated objectives.

SUMMARY

This chapter covered topics related to price and promotion strategy in real estate development. Price strategy focused mostly on methods used for real estate appraisals, whereas promotion strategy focused mostly on message and media decisions.

With respect to pricing of real estate, we started out by identifying and describing the factors that influence market value of residential real estate. These include proximity, neighborhood, and structural factors. We also described three common approaches to pricing real estate properties: the market approach, the costing approach, and the income approach.

We then turned to promotion strategy. In this context, we discussed three main topics: message decisions in real estate promotion, media decisions in real estate promotion, and developing a promotion plan. Message decisions in real estate marketing are all about what to say in a promotional message and why. We described three types of positioning common in real estate promotion: positioning by an aspect of prospective customers, positioning by an aspect of the product, and positioning by an aspect of the firm.

We then turned to media decisions. That is, once the real estate message is formulated, it has to be communicated to a target audience through media—interpersonal and mass media. Interpersonal forms of communications involve personal selling. Mass media forms of communications in real estate marketing involve advertising and public relations. We argued that personal selling is the most potent form of communication in the marketing of real estate development. Personal selling refers to promotion conducted by a marketing agent, such as the real estate agent, using face-to-face encounters with prospective customers in the form of hosting an "open house," responding to questions by prospective clients, and providing them with information about the product in appealing and persuasive means. Interpersonal communication in real estate development is most effective when the selling agent has referent, expert, legitimate, and reward power.

Mass media forms of marketing communications include advertising and public relations. Advertising in real estate development tends to be mostly in broadcast advertising (e.g., television advertising, radio advertising), print advertising (e.g., newspaper advertising, local real estate publications, direct-mail advertising, yellow pages, and local and regional magazines), out-of-home advertising (e.g., bulletin board advertising, transit advertising, airport advertising, banners, building illumination), specialty advertising (e.g., mugs, calendars, t-shirts, pens, letter openers, key chains), digital advertising (e.g., website, banner ads on related sites, e-mail messages, and video marketing), online classified advertising (e.g., Zillow, Trulia, Realtor.com, Yahoo!, Yahoo!-Zillow Real Estate Network, Craigslist, online classified ads in major national newspapers such as the *New York Times*), mobile advertising (banner ads displayed on mobile apps), and social media (the use of Facebook, Twitter, LinkedIn, Pinterest, etc.). Public relations involve corporate-form communications that come from the CEO's office, not necessarily from the marketing department. Public relations typically include press releases, corporate sponsorship of events, and publicity.

The real estate developer then assembles a promotion plan by putting all the pieces together (i.e., message and media decisions). The plan involves audience selection, goal selection, message selection, media selection, a media schedule, and a performance evaluation.

DISCUSSION QUESTIONS

1. Describe some of the important proximity factors that influence market value of residential real estate.

2. Describe some of the important neighborhood factors that influence market value of residential real estate.

3. Describe some of the important structural factors that influence market value of residential real estate.

4. Describe the market approach to real estate appraisals.

5. Describe the costing approach to real estate appraisals.

6. Describe the income approach to real estate appraisals.

7. Describe the three positioning strategies commonly used in marketing communication campaigns of real estate developers.

8. Distinguish between the use of interpersonal and mass media approaches in marketing communication campaigns of real estate developers.

9. Social power can enhance the persuasive effectiveness of interpersonal communication by sales agents of real estate development firms. Explain.

10. Real estate developers commonly use advertising and public relations. What is the difference between these two forms of marketing communication approaches?

11. Advertising in real estate development tends to involve broadcast advertising. Explain and provide examples.

12. Advertising in real estate development tends to involve print advertising. Explain and provide examples.

13. Advertising in real estate development tends to involve out-of-home advertising. Explain and provide examples.

14. Advertising in real estate development tends to involve specialty advertising. Explain and provide examples.

15. Advertising in real estate development tends to involve digital advertising. Explain and provide examples.

16. Advertising in real estate development tends to involve online classified advertising. Explain and provide examples.

17. Advertising in real estate development tends to involve mobile advertising. Explain and provide examples.

18. Do real estate development firms use public relations as a tool of marketing communication? If yes, then how?

19. How do marketers in real estate development firms use press releases, corporate sponsorship of events, and publicity in marketing communications?

20. Describe the process of audience selection in promotion planning in a real estate development context.

21. Describe the process of goal selection in promotion planning in a real estate development context.

22. Describe the process of message selection in promotion planning in a real estate development context.

23. Describe the process of media selection in promotion planning in a real estate development context.

24. Describe the process of media scheduling in promotion planning in a real estate development context.

25. Describe the process of performance evaluation in promotion planning in a real estate development context.

NOTES

1. Des Rosiers, F., Theriault, M., Kestens, Y., & Villeneuve, P. (2002). Landscaping and house values: An empirical investigation. *Journal of Real Estate Research*, 23, 139–161. DiPasquale, D., & Wheaton, W.C. (1996). *Urban economics and real estate markets*. Englewood Cliffs, NJ: Prentice Hall. Jackson, T. (2001). The effects of environmental contamination on real estate: A literature review. *Journal of Real Estate Literature*, 9, 93–116. Leichenko, R.M., Coulson, N.E., & Listokin, D. (2001). Historic preservation and residential property values: An analysis of Texas cities. *Urban Studies*, 38, 1973–1987. Simons, R.A., Quercia, R., & Maric, I. (1997). The value

impact of neighborhood transition on residential sales price. *Journal of Real Estate Research*, 15, 147–161. Simons, R. A., & Saginor, J. D. (2006). A meta-analysis of the effect of environmental contamination and positive amenities on residential real estate values. *Journal of Real Estate Research*, 28, 71–104. Strand, J., & Vagnes, M. (2001). The relationship between property values and railroad proximity: A study based on hedonic prices and real estate brokers' appraisals. *Transportation*, 28, 137–156.

2. Messick, L. P., & Huber, W. P. W. (2012). *Real estate appraisal: Principles and procedures* (4th ed.). New York: Educational Textbook Company. Pagourtzi, E., Assimakopoulos, V., Hatzichristos, T., & French, N. (2003). Real estate appraisal: A review of valuation methods. *Journal of Property Investment & Finance*, 21, 383–401.

3. Gwin, C. F., & Gwin, C. R. (2003). Product attributes model: A tool for evaluating brand positioning. *Journal of Marketing Theory and Practice*, 11, 30–42.

4. French, J. R. P., & Raven, B. (1959). The bases of social power. In D. Cartwright (Ed.), *Studies in social power* (pp. 45–65). Ann Arbor, MI: University of Michigan Press.

5. Mira, P. (2011). Mobile real estate classified: Winning model brings revenues back. *INMA*. Accessed from www.inma.org/blogs/blogPrint.cfm?blog=mobile&id=14DE6BEC-1E4F-37DA-FF9CB76D21E0D711E, on May 5, 2013.

6. eCampaignPro releases marketing trends within the real estate industry, survey thousands. (2013). *Online.WJS.com*, April 18, 2013. Accessed from http://online.wjs.com/article/PR-CO-20130418–909547.html?mod=googlenews-wjs, on May 5, 2013.

7. Davis, K. (2012). Research backs up print + digital real estate strategy. *INMA*. Accessed from www.inma.org/blogs/blogPrint.cfm?blog=ideas&id=AAA4A3A2–1EAF-37DA-FF1D30874557BF753, on May 10, 2013.

8. Grace, I. (2012). Media match: Make all your ad media work together. *Realtor Magazine*, October 2012. Accessed from http://realtormag.realtor.org/sales-and-marketing/feature/article/2012/10/media-match-make-all-your-ad-media-work-together, on May 5, 2012.

Part 2

Personal Selling—the Real Estate Agent

Chapter 4

Personal Selling in Real Estate
The Seller Representative

LEARNING OBJECTIVES

This chapter is designed to help students of real estate marketing learn:
- the personal selling cycle for the real estate seller representative
- how seller representatives prospect for clients
- how seller representatives determine the market valuation of the property of a prospective seller
- how seller representatives analyze the needs of prospective sellers
- how seller representatives make a sales pitch to prospective sellers
- how to negotiate specific services that seller representatives can provide to prospective sellers and the signing of a contract
- how seller representatives go about servicing their clients
- how seller representatives follow up with their clients

THE PERSONAL SELLING CYCLE FOR THE SELLER REPRESENTATIVE

Figure 4.1 shows the personal selling cycle for the seller representative (i.e., the listing agent). The first stage of the cycle is prospecting. The seller representative starts out the cycle by identifying and contacting prospective clients. The second stage involves gathering information and assessing the market valuation of the target property (the property an individual seller is trying to sell). The third stage is need assessment. That is, the seller representative sits down with a prospective client, asks questions about the prospective seller's needs—asking price, timeliness, commission, etc. This is followed by the seller representative making a sales pitch to the prospective seller (stage four). At this stage, the seller representative tries to impress the prospective seller of the agent's ability to sell the property and meet specified goals. Once the prospective seller expresses a certain degree of comfort with the seller representative, the agent negotiates the type of service that can be provided (stage five). From this point on, the prospective seller becomes a formal client by signing a contract with the agent. Stage six involves implementing the types of services that the listing agent was contracted for (i.e., type of advertising, inclusion in the Multiple Listing Service, open house, showing, etc.). The last stage of the seller representative personal selling cycle is follow-up. Here the listing agent follows up with his client to ensure that all went according to plan and to develop the relationship further.

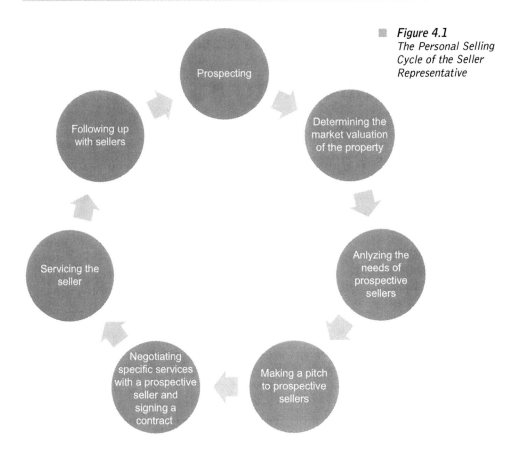

PROSPECTING FOR CLIENTS

The first stage of the seller representative cycle is prospecting. Prospecting refers to the methods used to identify and contact prospective clients. These methods are categorized in two groups: active and passive. See Figure 4.2. *Active prospecting* involves methods and ways to identify and contact prospective clients that require the seller representative to actively seek out prospects, rather than sitting back and having prospects contact him. The latter is, of course, referred to as *passive prospecting*.[1]

Let's review the active prospecting techniques first, before we turn our attention to the passive techniques. These techniques include (1) direct mail campaign, (2) canvassing, (3) For Sale by Owner (FSBO), (4) expired listings, (5) obituaries, (6) non-occupant owners, and (7) builders. *Direct mail campaign* involves mailing prospective sellers a brochure with a cover letter. The cover letter is typically personalized, in the sense that the prospective seller is addressed by name. The letter is designed to introduce prospective sellers to the listing agent by including information about the brokerage firm and its selling record and the agent's background and his selling record. The brochure typically describes the agent's full set of services that are typically provided to sellers (e.g., listing in the Multiple Listing Service, advertising in the local newspapers, constructing an exclusive webpage of the property shown on the agent's website, development of a virtual tour of the property, preparing

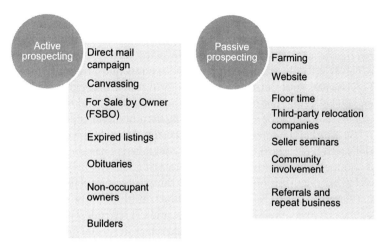

■ *Figure 4.2*
Active versus
Passive Methods
of Prospecting
Commonly
Used by Seller
Representatives

the property for showing, conducting open house). A list of prospective sellers (names and addresses) is assembled from directories such as telephone directories or staff directories of local companies.

Canvassing refers to door-to-door solicitations. This is typically done when the listing agent has served one or more sellers in the neighborhood. The agent goes door to door introducing himself to the neighborhood, saying something like, "I have helped your neighbor so and so sell his house, and I would like to introduce myself to you and offer my services."

For Sale by Owner (FSBO) are prospective sellers trying to sell their property on their own (without the assistance of a professional real estate agent). FSBOs can be persuaded to use a listing agent if they are convinced that hiring a listing agent may save them time and money. A listing agent may be able to expedite the selling process because he has the means to do so. A listing agent can save the seller money by using professional negotiation skills to extract from the buyer the highest price possible.

Expired listings are listings in the Multiple Listing Service that have expired after a specified time (usually six months). With a little homework, the agent makes an attempt to find out why specific expired listings were not sold—perhaps one property was overpriced, another property was not professionally prepared for showing, and so on. Thus, the listing agent can approach the sellers of those expired listings and offer remediation services. For example, if the listing agent believes that the property was not sold because it was overpriced, then he would try to convince the seller that the asking price is out of line compared to other properties recently sold in the neighborhood and that if the seller hires him, he can sell the property with a certain price reduction. If the listing agent believes that the property was not sold because it was not well prepared for a decent showing, then he may be able to recommend a staging company that can do a good preparation to make the home more appealing to prospective buyers.

One can also identify prospective sellers from the *obituaries* (newspaper articles publicizing the death of people in the area). Often a family's housing needs change after the death of a family member (e.g., when an elderly man dies, his widow may feel that she no longer wants to live in the same house alone, and she would rather move in with one of her children). The listing agent may be able to approach the surviving relatives and offer his services to help the surviving relative sell the home of the deceased.

There are also many homeowners that rent out their property, perhaps because these are second homes or because of life changes; for example, a widower might decide to move in with his children and rent out his home instead of selling it outright. These people are referred to as *non-occupant owners*. In many instances, renting homes can be tiring and expensive; the homeowner has to find tenants and physically maintain the house. Many homeowners start out thinking that renting their non-occupied homes is a good idea, but end up believing that it is not worth the time, effort, and money. Hence, the non-occupant owners may be highly disposed to selling their homes if and when prompted by a good agent.

Builders build homes and other properties. Many builders do not have the organizational skills to sell their own buildings once completed. These builders are likely to be inclined to do business with a listing agent to more easily sell their newly built homes.

Let us now turn our attention to the passive methods of prospecting. These include: (1) farming, (2) website, (3) floor time, (4) third-party relocation companies, (5) seller seminars, (6) community involvement, and (7) referrals. *Farming* is a method in which a real estate agent establishes expertise in real estate in specific neighborhoods. The agent is most likely to be known by most long-time residents of those "farmed" communities. The agent's real estate signs have been plastered all over the neighborhood, creating a high degree of awareness among neighborhood residents. When a real estate agent has done much business in specific neighborhoods, her reputation alone can drum up new business. Prospective sellers in the neighborhood are likely to be aware of the real estate agent who has long farmed the community and established a reputation as an agent devoted to the housing needs of that community. Farming is also used in commercial real estate: certain real estate agents specialize in niche markets (e.g., land for agriculture, land for development of retirement communities, real estate space for clinics and medical establishments). By farming, a real estate agent's reputation alone can provoke owners to call on him when they are ready to sell.

The *website* method of prospecting is probably the one method that is considered indispensable in the era of the Internet and digital marketing. Ninety-one percent of people look at houses on the Internet first before contacting a real estate agent or seller.[2] Every seller representative must have his own website, which is part of the brokerage firm. The website should promote seller representative services, claim certain specialized services, and provide much evidence to substantiate these claims. Both content and style are important in promoting the seller representative. If the content and style are impressive, prospective sellers are likely to call on him.

Floor time refers to the availability of a real estate agent when he is manning the office. That is, many real estate brokerage firms assign various real estate agents "floor time." When a call comes in to the office from prospective sellers, the agent who is "manning the floor" attempts to assist the person who calls in. These callers can be prospective sellers.

Third-party relocation companies are firms that specialize in helping company executives and their families move to another location—the location of job transfer. These third-party relocation companies typically try to connect with certain agents in the communities in which their client firms have a presence. Seller representatives can identify these relocation companies and attempt to establish a relationship with them. Once a relationship is established, these relocation companies are likely to call on the seller agent to help certain executives and their families by selling their homes.

Seller representatives sometime hold *seller seminars* for community residents to help them become better consumers of real estate. The selling of a home is indeed highly complex. A seller representative may organize and offer a seminar to prospective sellers to help them understand the selling process. Prospective clients attending these seminars may be impressed with the expertise and capabilities of the seller representative (who is the instructor of the seminar). If so, some of these seminar attendees may call on the seller representative to represent them in their efforts to sell their homes.

Many experts of real estate marketing say that the first step in successfully managing a real estate business is to realize the local nature of this business. Real estate business is a local business. It other words, its success is very much dependent on the extent to which its serves the local area. If a real estate brokerage firm establishes a bad reputation in the community, this can cause its demise. A community involves a social network of residents and businesses that interact and serve one another in various ways; real estate agents can become successful by building a reputation of service to the community. Part of that effort is *community involvement*. Real estate agents become involved in serving their communities in various volunteer roles through their churches, synagogues, mosques, and temples; through schools and other educational institutions, through law enforcement and the judicial system, through the community chamber of commerce and other public charity organizations, etc.

Last but not least are *referrals* and *repeat business*. According to National Association of Realtors (U.S.), 43% of the buyers typically come from referrals and 11% tend to be repeat clients. In contrast, only 38% of the sellers come from referrals and 26% come from repeat business.[3] This is a often testimony of the impact of the agent's longstanding involvement in the community and his reputation. The maxim is: Agents who are firmly entrenched in their community and have acted with honor and integrity over many years are likely to be highly successful in drumming up new business, and much of this new business is likely to come from referrals and repeat business.[4]

DETERMINING THE MARKET VALUATION OF THE PROPERTY OF THE PROSPECTIVE SELLER

Before approaching a prospective seller, it is wise to know about the property he may be trying to sell. The possible selling (or asking) price of the property is probably the most important piece of information the seller representative needs. To establish an asking price for the property, the listing agent should use one of the established pricing methods discussed in Chapter 3 on pricing. Recall the three methods: (1) comparative market analysis, (2) costing, and (3) income. See Figure 4.3.

As a refresher, the *comparative market analysis* (CMA) is a pricing method common in residential sales. This is a process that collects recent sales data of comparable properties ("comps") and compares them to the property being appraised. Specifically, the appraiser

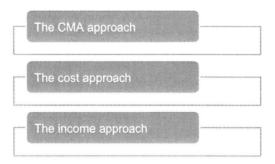

Figure 4.3
Three Different Approaches to Determine the Market Value of a Property

The CMA approach

The cost approach

The income approach

gathers information about recent sales of comparable properties. He makes price adjustments based on the differences between the focal property and those comps and derives a market valuation of the property based on a comparison with each comp. Then he takes the average (or weighted average) of these market valuations to derive the selling price (possibly the asking price).

The *cost* method of market valuation is mostly used by builders and real estate developers. In this case the market valuation of the property (i.e., price) is determined by summing up the cost of all the parts. For example, the cost of building a residential house may involve the cost of the land plus the cost of building a certain number of bedrooms, a certain number of bathrooms, a living room, a dining room, a kitchen, a breakfast nook, a patio, a garage, etc. The builder knows that the cost of building a bedroom is A per square/foot, a bathroom is B per square/foot, a living room is C per square/foot, etc. If we know the dimensions of each room, then we may be able to compute the total cost of the property. This would be the selling price (or asking price).

The *income* method is typically used in rental property. The net operating income generated by the property is scheduled gross annual income generated from the property minus vacancy allowance/collection losses and operating expenses. Once this figure is determined, the appraiser divides the net operating income generated by the rental property by the interest rate to derive the market valuation (i.e., selling price) of the property.

ANALYZING THE NEEDS OF PROSPECTIVE SELLERS

To win an agency contract from a seller, the seller representative has to make an effective presentation to the prospect. This presentation focuses on personal results, company results, specific marketing related to the property in question, and the selling or asking price of the property. Providing information about personal and company results to prospective sellers is fixed, while information regarding marketing the property and the price is variable. In other words, information about the seller representative's personal record of achievement (i.e., number of sales in the last x number of years, number of sellers served in the last x number of years) and the company's record of achievement (i.e., number of years the brokerage firm has served the community, the number of listings of the firm over the past x number of years, the dollar amount of sales that was brought into the agency over the last x number of years) is fixed, in the sense that once this information is compiled it is presented to all prospective sellers. That is, this information does not vary from seller to seller. However, the information concerning the marketing plan and the price of the target property varies from one seller to another; hence, information about marketing and price is considered "variable."

The seller representative has already collected data to determine the market value of the property (i.e., asking price). Now, we need to focus on the marketing plan of the target property. For the seller representative to present a viable marketing plan about the property in question, the agent has to sit down with the seller and perform a needs assessment. In this vein, the agent would ask the seller questions such as:

- Do you have a price? If so, how did you arrive at your price?
- Which of the following is most important to you: price, timing, or convenience?
- What would it do to your plans if you just couldn't sell?
- When do you want to or have to move?
- What major improvements have you made to the house recently?

- Can you move without selling the house first?
- Would you consider owner financing?
- Do you have any concerns about selling your house?
- Who else are you talking with about the sale of your house?

MAKING A PITCH TO PROSPECTIVE SELLERS

Armed with information about the market valuation of the property and information about marketing that property, the seller representative is now ready to make a pitch to the seller. This pitch is usually done in the form of a "presentation." The seller representative meets with the prospective seller at the seller's home (or some other convenient meeting place) and makes a presentation. The presentation typically focuses on five aspects: (1) personal results, (2) company results, (3) specific marketing, (4) price, and (5) closing costs.[5] See Figure 4.4.

With respect to *personal results*, the seller representative presents to the seller information about his personal selling record. The agent has to convince the prospect that he has the experience, the expertise, and character to do a good job selling the property in question. As such, the agent presents information about his personal record of achievement, such as the number of sales he has made during the last five years, the number of clients he served, any achievement awards he has received recently, and any media publicity he has received about his work in the community. In many cases, the listing agent provides references from past satisfied clients to demonstrate credibility.[6]

Similarly, the agent presents information about *company results*. That is, he wants the prospect to be impressed with not only his personal record of achievement but also with the record of achievement of his employer (i.e., his brokerage firm). In that regard, he presents information about how long the brokerage firm has served the local community, the number

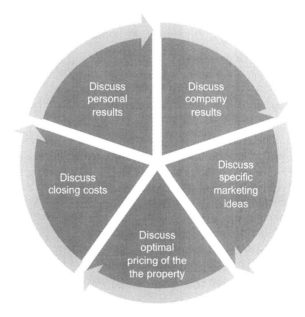

Figure 4.4
Making the Sales Pitch to Prospective Sellers

of sales made every year for the past five years or so, the number of customers served, recognition awards given to the firm for various activities and community service, etc.

Then the agent turns to the property in question. The seller has to be convinced that the agent can sell his property at the right price and in a timely fashion. As such, the agent presents his idea of how the property should be *marketed*. He shares his ideas with the seller about how the property should be advertised, how it should be presented on the brokerage firm's website, how it should be staged for showing, how many open house events should be held, how it will be presented to other buyer agents, and how he can negotiate effectively with prospective buyers and buyer representatives. In some countries, such as Australia and New Zealand, marketing a home may involve an auction,[7] which is used as a marketing tool of last resort in the U.S. In the U.S., auctions are used in selling distressed residential properties and, in some cases, commercial real estate involving plant closings and company bankruptcies. See Zooming-In Box 4.1.

ZOOMING-IN BOX 4.1. REAL ESTATE AUCTIONS: THREE TYPES[i]

Real estate auctions are becoming increasingly popular. They are most popular in commercial real estate and foreclosures of residential real estate. Three common types of real estate auctions include the minimum bid, the published reserve, and the absolute auction.

A *minimum bid auction* is an auction in which the auctioneer accepts bids for the auctioned property. The auctioneer strives to attain the minimum bid—the lowest price that the seller will accept for the auctioned property. Once the minimum bid is reached, the property sells to the highest bidder.

In a *published reserve auction*, the seller sets the lowest price for which he will sell his property at auction. The bidding can start wherever the bidders choose. The bidders do not know the lowest price that the seller will accept. If the "published reserve bid" is not reached, then the seller can accept or reject the highest bid.

In an *absolute auction,* the highest bidder buys the property, regardless of the amount of the bid. There is no reservation price below which the seller will not accept a bid. This type of auction is the most popular among the three types.

[i] Defining the terms: The 3 main types of real estate auctions. *PropertyAuction.com*, January 11, 2011, accessed from www.propertyauction.com/blog/index.php/2011/01/defining-the-terms-the-3-main-types-of-real-estate-auctions/, on October 11, 2013.

The seller representative then discusses the *price* of the property. He presents his ideas of an optimal selling price based on an established method of market valuation (i.e., CMA, cost, or income approach). He makes a convincing case that a certain asking price is likely to help expedite the selling of the property.

Finally, the listing agent discusses *closing costs*. Closing costs in the U.S. typically include the following expenses:

- *broker's fee* (the fee of commission that the seller pays the brokerage firm of the listing agent for services rendered)
- *title insurance* (the fee for insurance against unknown liens, encumbrances, or defects to the property)

- *warranty deed* (the fee related to a warranty that offers protection of the deed of the property—a document that demonstrates a person's legal right or title to the property)
- *filing fees per release* (expenses related to filing and recording releases that clear the title of the property)
- *escrow fees* (fees paid to an escrow agent—typically an attorney or a title company—for making sure that all the terms of the earnest money contract are implemented; earnest money is a deposit made by the buyer to the seller demonstrating the buyer's good faith in a transaction)
- *termite inspection* (fees related to the service provided by a licensed exterminator as to the condition of the property regarding termites and other wood-destroying insects)
- *appraisal* (fees paid to a real estate appraiser for determining the market value of the property)
- *loan discount fee* (a fee charged by the mortgage lender to make yield on the loan competitive with other investment options; usually called "points," where 1 point equals 1% of the mortgage loan amount)
- *survey* (a fee charged by a licensed surveyor to determine the exact boundaries of the property)[8]

NEGOTIATING SPECIFIC SERVICES WITH PROSPECTIVE SELLERS AND SIGNING A CONTRACT

Negotiating specific services with prospective sellers can be grouped into five major categories: (1) type of listing, (2) product decisions (fix-ups, staging the property, etc.), (3) pricing decisions (determination of the asking price, authority to negotiate price reductions, etc.), (4) promotion decisions (newspaper advertising, virtual tour and web presence, profile sheets and special brochures, advertising in real estate publications, etc.), and (5) place decisions (open house, caravan tours, etc.). See Figure 4.5.

Figure 4.5
Negotiating Specific Services with Prospective Sellers

Negotiating the Type of Listing

The services that a listing agent can provide to sellers are quite varied as a function of the type of listing. There are several types of listing that are commonly used in the U.S. These are: (1) open listing, (2) exclusive-right-to-sell listing, (3) exclusive agency listing, (4) net listing, and (5) a one-time listing.[9] An *open listing* allows anyone to sell the property, and only the broker of the agent that ends up selling the property gets the sales commission (which in turn is split with the brokerage firm of the buyer agent, assuming that the buyer has an agent). If the seller ends up selling the property himself, the seller does not have to pay a commission to any broker or agent. This type of listing is totally nonexclusive in the sense that it is not tied to any one specific seller agent. This type of listing is most advantageous to the seller and least advantageous to the seller representative. The seller can list the property with multiple brokers at one time. The seller agent who initially services the seller may end up with no compensation at all if he cannot find a buyer.

The *exclusive-right-to-sell listing* is most advantageous to the seller representative. Here the brokerage firm of the seller agent who listed the property is the firm that gets paid the commission (which in many cases is shared with the brokerage firm of the buyer representative, assuming that there was a buyer agent). This type of listing is totally exclusive to the seller agent. That is, only the brokerage firm belonging to the seller agent has the right to sell the property, and all other brokers have to work with the brokerage firm that has this exclusive listing. An exclusive listing prohibits the seller from negotiating with prospective buyers directly; the seller agent has to negotiate on behalf of the seller.

Exclusive agency listing is similar to the exclusive-right-to-sell listing, with the exception that the seller has the right to sell the property alone; if he does, the brokerage firm of the seller agent does not receive the sales commission.

Net listing refers to an arrangement in which the broker of the seller representative agrees to pay the seller a set amount from any sale (plus closing costs). The remaining money goes to the brokerage firm of the seller agent. For example, the seller says to the agent: "All I want is $300,000 plus expenses related to closing; the rest is yours." The agent finds a buyer who forks out $330,000 for the property. The seller is reimbursed $300,000 plus closing costs; the brokerage firm of the seller agent keeps the remainder (i.e., a little less than $30,000).

A *one-time listing* refers to a listing that is limited to one buyer for one time, for one property. This is commonly used when the listing agent has a buyer in mind and does not want to avail the listing to multiple offers. See Zooming-In Box 4.2 for additional information about agency relationships.

ZOOMING-IN BOX 4.2. AGENCY RELATIONSHIPS[i]

The chapter discussion about types of listings is based on the assumption that the listing agent is representing the seller only. In other words, the listing agent has a fiduciary responsibility to look after the interest of the seller and the seller only (i.e., not prospective buyers). However, there are exceptions to this rule. The listing agent can play the role of the *dual agent*. The same agent (who ultimately and legally reports to the broker) can represent both seller and buyer. In this situation, the dual agent must disclose this relationship and get approval from both seller and buyer to move forward in the transaction. Dual agency is very controversial and can lead to ethical and legal trouble. Albeit it is legal, it is wise to avoid dual agency.

Listing agents can still do business in a *nonagency* capacity. The broker of the listing agent may offer to list the property and charge the seller a flat fee for services rendered. This role is well suited for "For Sale By Owners" (FSBO) and new home construction (builders). The FSBO or builder may use a listing agent for some services and not others; perhaps they feel that they can perform certain services on their own. Hence, they contract a brokerage house to perform a specific set of services (e.g., placing the property in the MLS database) and pay a flat fee (not based on the commission of the sold property).

[i] Hamilton, D. (2006). *Real estate: Marketing and essentials.* Mason, OH: Thomson Higher Education (pp. 94–98).

Making Product Decisions

The seller and the listing agent have to further negotiate the type of service that the listing agent will provide for the length of time of the contract. This may involve making product, price, promotion, and place decisions. The first and most important product decision is what fix-ups are needed before the property is ready for showing. The listing agent tours the property and makes suggestions for needed *fix-ups*. The seller is likely to resist many of the fix-up suggestions because fix-ups require that the seller invest time, effort, and money. However, the listing agent wants to make sure that the property can be sold so that he can make a commission. Hence, there may be a "tug of war" in this discussion; the listing agent may "push" the seller to perform certain fix-ups, while the seller is "pushes back" because of the time, effort, and money required in getting the fix-ups done.

The second "product" decision that has to be made (and therefore negotiated with the seller) is *staging*. Of course, the listing agent wants the property to be staged by a professional staging company. This can go a long way to make the property (e.g., home) attractive to prospective buyers. Consider the following survey results signifying the power of staging:

- Over 91% of people surveyed say they believe staging to be a powerful tool when selling a property.[10]
- More than 75% of those 91% will drive by the property before contacting an agent for showing, and half of them will do this after dark.[11]
- Seventy-eight percent of sellers today say they would invest up to $5,000 and more to make their property move-in ready.[12]
- More than 63% of buyers today say they are willing to pay more money for a move-in ready house.[13]
- Ninety-three percent of buyers today look at property on the Internet first to create a list of showings, and the photos posted of a property on the Internet must be attractive (i.e., the photos should be taken by a staging professional).[14]
- Real estate professionals estimate the projected return on investment (ROI) for staging could be as high as 589%.[15]

But again, staging the property is a significant expense. Who will absorb this expense? Will it be the seller or the listing agent? This is a negotiable issue.

Making Price Decisions

Both seller and listing agent have to agree on the asking price and authority to negotiate price reductions. With respect to the asking price for the property, it is very likely that the seller would push for a higher *asking price*, while the listing agent would push for a price that the market can bear.[16] A price that the market can bear is likely to expedite the sale of the property. If the asking price is too high, then the listing agent may not be able to sell the property within the specified time of the contract; therefore, the agent does not get a sale commission. In contrast, many sellers insist on a high asking price and many are willing to wait until the right buyer comes along. This situation again involves a "tug of war" because the interest of the listing agent does not coincide with the interest of the seller. Hence, both parties attempt to negotiate the asking price.

Another important decision that has to be made in relation to pricing is who will negotiate with prospective buyers. Negotiating with prospective buyers is likely to involve making *price reduction* decisions. The listing agent has the expertise to make price reduction decisions. The seller not is likely to have negotiation skills. Therefore, the listing agent asks the seller to give him the authority to negotiate price reductions on his behalf. Of course, the seller will have to agree on the final price. However, as will be discussed in Part 3 of the book (on real estate negotiation), the negotiator has to have the authority to make concessions to be able to extract concession from the other party. Having the full authority to make concessions in the form of price reductions is a requirement in effective negotiation. The listing agent would not want the seller to second guess every concession he makes. Therefore, the exact nature of the authority to make concessions and price reductions has to be agreed upon.

Making Promotion Decisions

Both agent and seller have to agree on various ways to promote the property. Of course, what is given is promotion through the *Multiple Listing Service* ("MLS" as it is called in the U.S.). This is a digital database controlled by the National Association of Realtors (NAR) that allows all "realtors" (real estate agents and their brokers, as members of the NAR) to list properties in the system, thus promoting the property to other realtors who have relationships with prospective buyers.[17]

Besides the inclusion of the property in the MLS database, both listing agent and seller have to agree on other forms of promotion. For example, it is now common for every listing agent to have his own website that is linked to the brokerage firm's *website*. The seller's property can be featured in the brokerage firms' website but also linked to the listing agent's website. Much can be done to the property featured on the brokerage firm's website. For example, *special photography* and a *virtual tour* can be used to enhance the overall appeal of the property. Of course, the seller would want the special photography and the virtual tour without an added expense (i.e., assuming the expense is essentially part of the listing agent's commission). However, not all listing agents provide this service because of the high expense in hiring a video production company or a photographer to do the shooting, edit the footage, and upload it on the website. Therefore, this aspect of promotion is subject to negotiation between the listing agent and the seller.

Customarily, the listing agent develops *profile sheets* for the property to be disseminated widely to other real estate agents and prospective buyers. These are typically left at the property (inside and outside the property) to be picked up by other agents and prospective buyers. A profile sheet is an information sheet that provides details regarding the "specs" of the property (information about the property, seller, size, amenities, etc.). However, many listing agents also develop special brochures and pamphlets showing the property using special photography and artistic displays. Of course, the development of these special brochures and pamphlets requires the use of graphic design specialists, and this comes at a significant expense. Again, sellers expect that this service to be provided to them as part of the commission that the listing agent receives, while the listing agent may argue that the additional expense is unaffordable. Hence, this issue also becomes negotiable.

See the case of listing agents in Australia who are successful in getting sellers to share the cost of advertising (Case/Anecdote 4.1).

CASE/ANECDOTE 4.1. GETTING SELLERS TO SHARE ADVERTISING COSTS[i]

Listing agents in Australia and New Zealand have been experimenting with trying to get sellers to pay for advertising (which typically amounts to 1–2% of the selling price). Only the top-performing agents are those who are successful in obtaining substantial advertising dollars from sellers. Other agents pay all the advertising costs, although they don't spend much because it is coming out of their pocket.

Top-performing agents have been quite successful in conducting very effective advertising campaigns, generating great results. They also become more visible in the community. Their reputation goes through the roof.

Newspapers in Australia and New Zealand are commonly used in advertising, carrying full-color real estate advertising worth hundreds of millions of dollars. For example, the daily newspaper in the Gold Coast of Queensland, Australia, has a Saturday real estate supplemental section that carries 70–200 pages of ads. Some say that newspaper advertising has lost its effectiveness. Not true in Australia and New Zealand! Newspaper advertising becomes most effective when newspaper ads are reinforced by other ads that use other elements of the media mix: websites, social media, and other print advertising.

Research in Australia and New Zealand indicates that the more sellers pay for advertising, the better the results—more sales at higher prices in a shorter amount of time. Sellers do not mind sharing the cost of advertising if they believe they will achieve better results.

Furthermore, the Real Estate Institute of Queensland endorsed the idea of seller-paid advertising several years ago. The public in Australia and New Zealand regard this institute as having clout and credibility, and many Australians and New Zealanders trust their endorsements.

[i] Grace.J.(2013).Getting your sellers to pay for advertising. *Realtor Magazine,* January 2013, accessed from http://realtormag.realtor.org/sales-and-marketing/feature/article/2013/01/getting-your-sellers-pay-for-advertising, on May 12, 2013.

Making Place Decisions

Although open house and office caravan can be considered forms of promotion, we treat them as place decisions because they require "promotion in place." That is, the listing agent promotes the property to prospective buyers at the property site (*open house*) and to other real estate agents through a *caravan tour*—real estate agents inspecting listed properties to become familiar with them and to help find suitable buyers for them. Listing agents customarily provide caravan tours as a basic service to all sellers. This does not require negotiation. However, open house is another matter. Listing agents would rather forgo doing open house and concentrate on other forms of promotion; they tend to limit the number of open house showings to a minimum (i.e., once every two months or so). In contrast, sellers tend to push for as many open house events as possible. Hence, the frequency of open house events becomes a negotiable issue.

SERVICING THE SELLER

Having negotiated with the prospective seller the type of listing and the negotiable issues regarding product, price, promotion, and place, both agent and seller are now ready to sign a contract. The contract reflects all the negotiable issues and binds both seller and agent to the terms of the contract. At that point, the prospective seller becomes a client by signing the contract. The contract typically ties the seller to the listing agent for about six months, after which the contract has to be renewed.

One can group all the selling services provided by a listing agent in terms of the traditional four Ps: (1) product-related services, (2) price-related services, (3) promotion-related services, and (4) place-related services. See Figure 4.6.

Product-Related Services

Product-related services include fix-ups and staging the property. Although the *fix-ups* may have been discussed before the contract was signed, the listing agent helps the seller carry out the agreed upon fix-ups by connecting the seller with contractors, home repair services, plumbers, electricians, painters, etc. Of course, the seller may not need the assistance of the

Figure 4.6
Listing Agents' Services
Provided to Sellers

listing agent to get the fix-ups done, perhaps because he has relied on certain service outfits in the past and decides to use their services. Or perhaps the seller has special expertise that would allow him to carry out those fix-ups himself. In any case, the minimum that the listing agent can do in this situation is to provide advice and suggestions to ensure that the job is well done and the property is ready for staging.

If the contract calls for *staging the property*, then the listing agent (in consultation with the seller) hires a staging company. He also oversees the staging process. What does "staging a home" mean? Staging involves changing the look of the house to neutralize it—to change the personal identity of the place from its past owner to a look more suited to the average prospective buyer. The goal is to prompt potential buyers to envision themselves living there. Staging a vacant house involves the rental of stylish furniture and accessories to meet the housing lifestyle of prospective buyers. Staging an occupied house involves rearranging the homeowner's furniture and accessories, and perhaps bringing in rental furniture, or a combination of both. See Zooming-In Box 4.3 for information on how sellers can be advised to enhance the curb appeal of the property.

ZOOMING-IN BOX 4.3. CURB APPEAL[i]

First impressions about a home are formed based on curb appeal. That is, the prospective buyer drives by the house and looks at it from the outside. Based on this viewing, the prospect forms a first impression. This first impression is very important. Here are 10 things sellers can do to enhance the curb appeal of the property:

1. Make sure that the siding, decks, and walkways are visibly clean. Consider pressure washing.
2. Make sure the windows and gutters are visibly clean.
3. Make sure that trees and bushes are well-trimmed.
4. Make sure there is no mold or mildew around the property.
5. Make sure that the lawn is mowed, get rid of the weeds, and rake and dispose of leaves.
6. Make sure that the front door looks nice. Consider a paint job and choose a nice color that fits (or possibly contrasts aesthetically) with the color of the siding.
7. Make sure the porch has a "welcome" sign (e.g., welcome rug). Place flowerpots on the porch.
8. Make sure the mailbox and the house number look aesthetically pleasing.
9. Make sure that the back yard is also in good shape. Prospects may attempt to see the back yard during a drive by.
10. Outdoor lighting can make a big difference, especially at night. Consider placing lighting along the sidewalks, driveway, or in other places that will make the landscape attractive. In many cases, prospects driving by at night may be able to see aspects of the interior from the windows. Consider placing articles next to the window (e.g., Christmas tree during the Christmas season) and using aesthetically pleasing lighting.

[i] Rae, C. (2009). First impressions: Get instant curb appeal with these 10 must-dos. *Realtor Magazine*, November 2, 2009, accessed from http://styledstagedsold.blogs.realtor.org/2009/11/02/first-impressions-get-instant-curb-appeal-with-these-10-musts-dos/, on September 20, 2013.

Price-Related Services

Price-related services include *negotiating with prospective buyers* (and their representative agents). This means receiving offers from prospective buyers (or agents), negotiating with them, making concessions, extracting concessions from them, and notifying the seller of certain offers and outcomes of negotiations. This has to be done in a timely fashion. Of course, the listing agent has to adhere closely to the terms of negotiation on behalf of the seller as spelled out in the contract (e.g., make concessions and extract concessions from a prospective buyer, but do not make a final offer without the permission of the seller). See ethics scenarios (Ethics Box 4.1 and 4.2) related to the agent's ethical obligation to pass all offers to the seller and serve the client's interest fully.

ETHICS BOX 4.1. PASS ALL OFFERS TO SELLER[i]

A buyer agent contacts a listing agent and makes an offer of $200,000 for the purchase of a specific property. The listing agent becomes upset, saying something to the effect of, "Don't waste my time; put this offer in writing." The listing agent may have acted unethically in this situation. Why? The code of ethics requires real estate agents to submit offers and counteroffers as quickly as possible (Article 1, Standard of Practice 1–6). Article 1, Standard of Practice 1–7 requires the listing agent to submit *all* offers, not only written offers. The listing agent is ethically obligated to pass all offers to the seller, both oral and written offers.

[i] Legal Lines (2013). Do this, not that. *Commonwealth: A Journal for Real Estate Professionals Published by the Virginia Association of Realtors*, August/September, 13.

ETHICS BOX 4.2. SERVE YOUR CLIENT'S INTEREST FULLY[i]

A listing agent has a client who lost his job. He has found another job in another state. He needs to move quickly, which means that he has to sell his house quickly. The listing agent has lunch with his colleagues at the brokerage house. He explains his client's situation. One of his colleagues says that she knows of a potential buyer who is interested in real estate investment, and that he may be interested if the price is right. The seller manages to sell his house, but at a significantly lower price than he anticipated. In this case, the listing agent may have not served his client's best interest by disclosing his sense of urgency in selling the house. This is a violation of the confidence that the seller places in the listing agent—not to divulge personal or financial information to prospective buyers.

[i] Kantor A., & Hegeman, B. (2013). Do unto others. *Commonwealth: A Journal for Real Estate Professionals Published by the Virginia Association of Realtors*, April/May, 21.

The second price-related service is *price reductions*. If the property is not sold within a specified period, then the listing agent makes an attempt to reassess the market viability of the asking price. He does so by conducting a comparative market analysis (CMA) to examine the feasibility of a price reduction and, if one is needed, how much of one to make.

Promotion-Related Services

The first order of business for a listing agent is to place the seller property in the Multiple Listing Service (MLS). This is part of the basic service that all listing agents in the U.S. provide to sellers. He develops a profile sheet for the property and places these sheets inside the property in a strategic location (e.g., kitchen island in a residential home) and out-side the property (e.g., next to the mailbox). He puts up a "For Sale" sign in the yard outside the property. He installs a lock box to allow other real estate agents (representing buyers) to show the place to their clients. Other services include building a webpage of the property on the website of the listing agent's brokerage firm. In doing so, he may have to contract the services of a special photographer or videographer to enhance the attractiveness of the vir-tual look of the property. He advertises the property in the local newspaper, and he may use a sizable display ad in advertising. Display advertising in the newspaper may require the use of graphic design professionals. He may use the same services of a graphic design professional to develop a special brochure or pamphlet. He may advertise the property in real estate pub-lications that are widely distributed in local stores and in news stands.[18] Advertising can also be done on other online platforms such as Craigslist, Trulia, and eBay.[19] Also, don't forget the social media platforms, such as Facebook, Twitter, LinkedIn, and Google+: Promoting a listing on social media can go a long way. See Zooming-In Box 4.4 for more information regarding the use of effective advertising copy by the listing agent.

ZOOMING-IN BOX 4.4. LISTING AGENTS WRITING EFFECTIVE ADVERTISING[i]

Goals

It is important for the listing agent to realize that an effective ad in print is one that:

- attracts attention (use headlines that attract attention, such as "30% price reduction from the original asking price")
- arouses interest (focus on a story that builds on the headline to further arouse interest, such as "seller has passed away and relatives are trying to sell the prop-erty to pay debts")
- creates desire (show features of the property that are most attractive, such as a beautiful backyard)
- calls for action (instill a sense of urgency by saying something like, "call now; the property will be sold by . . . .")

Copy

Advertising copy is customarily broken down into seven basic elements. These elements have to be in sync:

1. *Photo:* The photo has to match the headline.
2. *Headline:* Make sure that the benefits are explicit or clearly implied—the posi-tive features of "living there" in that property and area.
3. *Qualifying copy:* Use the headline to highlight major benefit(s).
4. *Body copy:* Explain the major benefits in some detail.
5. *Reinforcing copy:* At the end of the ad copy, remind the readers of the major benefits.

---

6. *Action copy:* Provide contact details.
7. *White space:* Avoid too much copy; this can easily clutter up the ad, making it less effective. Inject enough white space to help the reader focus on the major benefits.

Regulations

Make sure that the ad abides by Fair Housing laws. To do this, make sure that the ad does not indicate or imply preference to a group of people based on their race (e.g., "whites only"), color (e.g., "this home is located in an African American neighborhood"), national origin (e.g., "this is located in a Korean American community"), religion (e.g., "Catholics preferred"), sex or sexual orientation (e.g., "gays and lesbians are welcome"), handicap (e.g., "this home is not well-suited for the elderly because of lack of amenities"), or familial status (e.g., "preference for singles").

Make sure that the copy abides by Truth-in-Lending laws. For example, advertisers must inform the audience about all details related to financing. For fixed-rate loans, you need to include information about annual percentage rate (APR), simple interest rate, down payment, monthly payment, and loan term (i.e., length).

[i] Hamilton, D. (2006). *Real estate: Marketing and essentials.* Mason, OH: Thomson Higher Education (pp. 80–88). Grace, I. (2013). Research your advertising to guarantee its effectiveness. *Realtor Magazine*, February 2013, accessed from http://realtormag.realtor.org/sales-and-marketing/feature/article/2013/02/research-your-advertising-guarantee-its-effectiveness, on September 20, 2013.

Consider a few ethics scenarios (Ethics Boxes 4.3 and 4.4) related to the agent's ethical obligation to be honest and disclose information that may harm the buyer, to pass all offers to the seller and serve the client's interest fully, and to make sure that the pictures of the property do indeed reflect the actual state of the property. In addition, Ethics Box 4.5 discusses a controversial case in which an agent used sex in advertising.

ETHICS BOX 4.3. BE HONEST AND DISCLOSE INFORMATION THAT MAY HARM THE BUYER[i]

A listing agent is walking around the house for sale with his client. His client says something to the effect that the creek in the backyard floods every year, around April or May. The listing agent says, "I wish you hadn't told me this, because now I am obligated to disclose this information to prospective buyers." The seller becomes furious with the listing agent, saying that he does not want to share this information with any prospective buyer.

The seller may think that whatever information he confides to his agent should remain confidential, as an attorney would keep it. The listing agent explains by saying, "This is the law. It is called Material Adverse Fact. If I know something that can adversely affect the well-being of the buyer, I am obligated by law to disclose it." The listing agent then adds, "If you have any other information that is not so positive about the property, please don't tell me; otherwise, I will have to disclose it."

In this situation, the listing agent has done one thing right and another thing wrong. He has done the right thing by telling the seller that he has to disclose the information about flooding. However, telling the seller not to inform him about other possible negative things about the property was not ethical.

[i] Kantor A., & Hegeman, B. (2013). Do unto others. *Commonwealth: A Journal for Real Estate Professionals Published by the Virginia Association of Realtors,* April/May, 22.

ETHICS BOX 4.4. THE PICTURES SHOULD REFLECT THE REALITY OF THE PROPERTY[i]

A listing agent has hired a photographer to take pictures of a house. The property is out on the rural side of the county next to a power substation. The photographer takes shots from such an angle that the substation is not visible in the pictures. The listing agent is happy with the pictures. He says, "Good idea, keeping the substation not visible in these pictures. Worse comes to worst, prospective buyers may grumble a bit. But at least I get them to see the place."

The question is, did the listing agent present a true picture of the property in his advertising and other promotion materials? The answer is no. The listing agent, in collusion with the photographer, committed an unethical violation. The right thing to do is to make sure that information and images of the property reflect the reality of the property. The information and images should not mislead nor misrepresent the facts.

[i] Kantor A., & Hegeman, B. (2013). Do unto others. *Commonwealth: A Journal for Real Estate Professionals Published by the Virginia Association of Realtors,* April/May, 23.

ETHICS BOX 4.5. USING SEX IN ADVERTISEMENT[i]

A real estate agent in Calgary (Canada) working for Re/Max stirred up quite a bit of controversy with an ad that has her posing provocatively. The caption reads, "Let me take you home; it's gorgeous inside." The ad has irritated a lot of people. One man, whose daughter was a real estate agent who was murdered at an open house, was particularly upset. He thought that such advertising could heighten the risk of rape and battery of women in real estate; many others agreed. Such advertising is considered unethical.

[i] Let me take you home. It's gorgeous inside: Calgary real estate agent's sexy advertisement stirs controversy. *National Post,* Feb 2, 2012, accessed from http://news.nationalpost.com/2013/02/07/let-me-take-you-home-its-gorgeous-inside-calgary-real-estate-agents-sexy-advertisement-stirs-controversy/, on May 4. 2013.

Promotion-related services may come under different names, depending upon the country; see, for example, Case/Anecdote 4.2, involving an enterprising individual in Cuba listing properties on the Internet and getting paid by the buyers, not the sellers.

Place-Related Services

As previously discussed, the listing agent can further promote the property "in place" by arranging open house events and showing the property on office caravan tours.

FOLLOWING UP WITH SELLERS

The listing agent customarily follows up with his clients by doing several things: (1) closing gift, (2) internet (blog), (3) events, and (4) periodic contact.[20] On the day of closing (i.e., completing a transaction with a buyer), the listing agent gives a nice gift to the seller as a sign of appreciation for doing business with the seller. *Closing gifts* can be a flower arrangement, a pen set, a gift certificate to the seller's favorite restaurant, tickets to a football game, etc. Listing agents can also keep sellers up to date on community events by including them in their newsletters and blogs. Many listing agents have blogs to announce community events and articles of local interest related to real estate. This *newsletter* or *blog* is e-mailed to the seller to help maintain the relationship. See Case/Anecdote 4.3 for how to set up a professional blog.

THE SELLER REPRESENTATIVE

blog is SEO optimized to increase the chances that an agent's blog will be picked up by most search engines. This tool could be appealing to real estate agents who don't have time to spend setting up a blog on a generic platform like *Wordpress*. Many agents don't have the time to figure out what plugins and widgets are; they simply want a clean tool that looks professional, making blogging about real estate very simple.

[i] Attention real estate agents: If you aren't blogging, you don't exist. *www.SFGate.com*, April 24, 2012, accessed from www.sfgate.com/cgi-bin/article.cgi?f=/g/a/2012/04/24/preweb9422454.DTL, on May 11, 2013.

Listing agents also host *social events* (i.e., tailgating parties at football events) and invite their past clients. These social events help foster the relationship between agent and client, which may lead to future referrals. Finally, *periodic contact* is also a good way to follow up. Every few months the listing agent picks up the phone and calls his past clients to see how they are doing, and to ask for referrals.

SUMMARY

This chapter covered the personal selling cycle of the real estate seller representative. The first stage of the cycle is prospecting. Here the seller representative starts out the cycle by identifying and contacting prospective clients. Prospecting methods are customarily categorized in two groups: active and passive. We describe several active prospecting techniques: direct mail campaign, canvassing, FSBO, expired listings, obituaries, non-occupant owners, and builders. Passive methods of prospecting include farming, website, floor time, third-party relocation companies, seller seminars, community involvement, and referrals.

The second stage involves gathering information and assessing the market valuation of the target property (the property an individual seller is trying to sell). To establish an asking price for the property, the listing agent uses established pricing methods such as CMA, costing, and income.

The third stage is need assessment. That is, the seller representative sits down with a prospective client, asks questions about the prospective seller's needs—asking price, timeliness, commission, etc. Ultimately, the information gathered has to help the buyer agent make an effective presentation to the prospect. This presentation focuses on personal results, company results, specific marketing related to the property in question, and the selling or asking price of the property.

Discussion is followed by making a sales pitch to the prospective seller. At this stage the seller representative tries to impress the prospective seller of the agent's ability to sell the property and meet specified goals. The sales presentation focuses on personal results, company results, specific marketing related to the property in question, and the selling or asking price of the property. The listing agent also discusses closing costs.

Once the prospective seller expresses a certain degree of comfort with the seller representative, the agent negotiates the type of services that can be provided. Specific services negotiated with prospective sellers tend to be grouped into five major categories: type of listing (open, exclusive-right-to-sell, exclusive agency, net, one time), product decisions (what fix-ups are needed before the property is ready for showing, staging the property for showing), pricing decisions (both seller and listing agent have to agree on the asking price and authority to negotiate price reductions), promotion decisions (both agent and seller have to agree on various ways to promote the property), and place decisions (open house and office caravan).

The next stage involves implementing the types of services that the listing agent was contracted for: product-related services (assisting with making strategic decisions concerning fix-ups and staging), pricing-related services (actual negotiations with prospective buyers and strategizing in helping the seller make price reduction decisions, if necessary), promotion-related services (placing the seller property in the MLS, developing and distributing profile sheets, putting up a "For Sale" sign, installing lock box to allow other real estate agents to show the place to their clients, building a webpage of the property, and advertising the property in various media outlets), and place-related services (promoting the property "in place" by arranging open house events and showing the property on office caravan tours).

The last stage of the seller representative personal selling cycle is follow-up. Here the listing agent follows up with his client to ensure that all went well and tries to develop the relationship further. The listing agent customarily follows up with his clients by doing several things, such as giving the client a closing gift, including the client on his internet blog or newsletter, inviting the client to social events, and maintaining periodic contact.

DISCUSSION QUESTIONS

1. Describe the personal selling cycle of the seller representative.

2. Distinguish between active and passive prospecting methods commonly used by the seller representative.

3. Describe how the seller representative goes about using a direct mail campaign to prospect for clients.

4. Describe the canvassing method of prospecting.

5. Describe the FSBO method.

6. How about the expired listing method?

7. The obituaries prospecting method?

8. The non-occupant owners prospecting method?

9. The builders prospecting method?

10. The farming prospecting method?

11. The website prospecting method?

12. The floor time prospecting method?

13. The prospecting method related to third-party relocation companies?

14. How about the use of seller seminars?

15. Community involvement?

16. Referrals?

17. How does the seller agent go about gathering information pertinent to a prospective client?

18. The seller agent then makes a sales pitch to the prospective seller. How does he do this?

19. What happens next?

20. The seller agent negotiates the type of listing with the prospect. What does this mean?

21. The seller agent negotiates product-related services. What does this mean?

22. The seller agent negotiates pricing-related services. Elaborate.

23. The seller agent negotiates promotion-related services. Explain.

24. The seller agent also negotiates place-related services. Explain.

25. Describe how the seller agent goes about carrying out the contracted product-related services.

26. Describe how the seller agent carries out pricing-related services.

27. Describe how the seller agent carries out promotion-related services.

28. How does the seller agent carry out place-related services?

29. How does the seller agent follow up with his client after closing?

NOTES

1. Hamilton, D. (2006). *Real estate: Marketing and essentials*. Mason, OH: Thomson Higher Education (Chapter 7).
2. Rae, C. (2009). What does "Britain Got Talent" singer Susan Boyle have to do with staging? *Realtor Magazine*, April 19, 2009. Accessed from http://styledstagedsold.blogs.realtor.org/2009/04/19/what-does-britain-got-talent-singer-susan-boyle-have-to-do-with-staging/, on May 13, 2013.
3. Zeller, D. (2009). Capture the leads. *Realtor Magazine*, July 2009. Accessed from http://realtormag.realtor.org/for-brokers/build-your-business/article/2009/07/capture-leads, on September 20, 2013.
4. As an agent, it is important to build and nurture a referral network. (2011). *Real Town*, May 25, 2011. Accessed from www.realtown.com/articles/view/as-a-realtor-it-is-important-to-build-and-systematically-nuture-a-referral-net, on May 1, 2012.
5. Hamilton, D. (2006). *Real estate: Marketing and essentials*. Mason, OH: Thomson Higher Education (pp. 177–182).
6. Rosenauer, J., & Mayfield, J. (2007). *Effective real estate sales and marketing*. Mason, OH: Thomson Higher Education (p. 91).
7. Grace, I. (2012). Pricing: Take your sellers for a ride. *Realtor Magazine*, December 2012. Accessed from http://realtormag.realtor.org/sales-and-marketing/feature/article/2012/12/pricing-take-your-sellers-for-ride, on September 20, 2013. Evans, M. (2009). Let the bidding begin. *Realtor Magazine*, October 2009. Accessed from http://realtormag.realtor.org/commercial/feature/article/2009/10/let-bidding-begin, on September 20, 2013.
8. Based on Rosenauer, J., & Mayfield, J. (2007). *Effective real estate sales and marketing*. Mason, OH: Thomson Higher Education (p. 94).
9. Hamilton, D. (2006). *Real estate: Marketing and sales essentials*. Mason, OH: Thomson Higher Education (pp. 182–183).

10. Rae, C. (2009). What does "Britain Got Talent" singer Susan Boyle have to do with staging? *Realtor Magazine*, April 19, 2009. Accessed from http://styledstagedsold.blogs.realtor.org/2009/04/19/what-does-britain-got-talent-singer-susan-boyle-have-to-do-with-staging/, on September 21, 2013.

11. Rae, C. (2009). What does "Britain Got Talent" singer Susan Boyle have to do with staging? *Realtor Magazine*, April 19, 2009. Accessed from http://styledstagedsold.blogs.realtor.org/2009/04/19/what-does-britain-got-talent-singer-susan-boyle-have-to-do-with-staging/, on September 21, 2013.

12. Rae, C. (2011). Sex sells! . . . Did I catch your attention? *Realtor Magazine*, June 2, 2011. Accessed from http://styledstagedsold.blogs.realtor.org/2011/06/02/sex-sells-did-i-catch-your-attention/, on September 21, 2013.

13. Rae, C. (2011). Sex sells! . . . Did I catch your attention? *Realtor Magazine*, June 2, 2011. Accessed from http://styledstagedsold.blogs.realtor.org/2011/06/02/sex-sells-did-i-catch-your-attention/, on September 21, 2013.

14. Rae, C. (2011). Sex sells! . . . Did I catch your attention? *Realtor Magazine*, June 2, 2011. Accessed from http://styledstagedsold.blogs.realtor.org/2011/06/02/sex-sells-did-i-catch-your-attention/, on September 21, 2013.

15. Rae, C. (2011). Sex sells! . . . Did I catch your attention? *Realtor Magazine*, June 2, 2011. Accessed from http://styledstagedsold.blogs.realtor.org/2011/06/02/sex-sells-did-i-catch-your-attention/, on September 21, 2013.

16. Yavas, A., & Yang, S. (1995). The strategic role of listing price in marketing real estate: Theory and evidence. *Real Estate Economics*, 23, 347–368.

17. Hamilton, D. (2006). *Real estate: Marketing and sales essentials*. Mason, OH: Thomson Higher Education (pp. 188–202).

18. Hamilton, D. (2006). *Real estate: Marketing and sales essentials*. Mason, OH: Thomson Higher Education (pp. 188–202).

19. Bankrate.com. (2012). Sell your home online, where buyers search. *Newsday*, May 3, 2012. Accessed from www.newsday.com/classified/real-estate/sell-your-home-where-buyers-search-1.3692653, on September 21, 2013.

20. Hamilton, D. (2006). *Real estate: Marketing and sales essentials*. Mason, OH: Thomson Higher Education (Chapter 12).

Chapter 5

Personal Selling in Real Estate
The Buyer Representative

THE PERSONAL SELLING CYCLE FOR THE BUYER REPRESENTATIVE

A survey conducted on homebuyers who search for real estate online show that homebuyers regard the role of the buyer representative as very important in assisting them in the purchase transaction. Specifically, 63% of homebuyers expect agents to educate them on the buying process, 47% expect the agent to help inform them on the market value of a property that the homebuyer might be interested in, and 47% plan to hire an agent to help them with the buying process but they haven't hired one yet.[1]

Figure 5.1 shows the personal selling cycle for the buyer representative. The first stage of the cycle is prospecting. The buyer representative starts out the cycle by identifying and contacting prospective clients. The second stage involves needs assessment. That is, the buyer representative sits down with a prospective client, asks questions about the prospective client's real estate needs and their financial status. Qualifying the prospect is the third stage. At this stage the buyer representative delves into the financial status of the prospective client to establish the maximum amount of money the prospect can afford to pay on a monthly basis to cover the loan mortgage. Once the buyer representative establishes what the prospect can afford, the rep goes about showing the prospect properties that are on sale. When the buyer representative and the prospect home in on one or two properties that best match the prospect's needs, the buyer rep makes an attempt to negotiate with sellers (on behalf of the client). Once the concerns and objections of the client are addressed, the buyer representative proceeds to close the deal. Once the deal is closed, the buyer representative follows up with the client to make sure that the client is happy with the property and

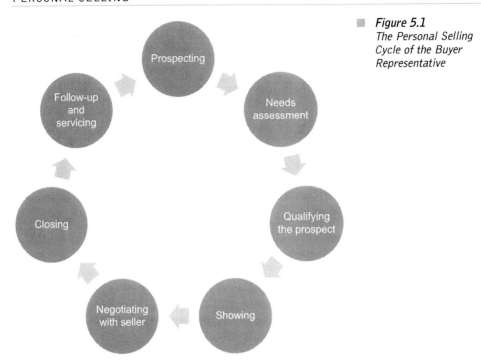

Figure 5.1
The Personal Selling
Cycle of the Buyer
Representative

provides assistance to help the client deal with any further problems or concerns with the property. See Zooming-In Box 5.1 for an in-depth exploration of the benefits that a buyer agent provides his clients.

ZOOMING-IN BOX 5.1. ADVANTAGES AND DISADVANTAGES OF BUYER REPRESENTATION[i]

Buyers of real estate have the option to work with a real estate agent who helps them buy a home that fits their housing needs—a buyer agent who shows them properties and negotiates the price. A buyer agent is essentially a counterbalance to the agent who represents the seller. Based on statistics provided by the National Association of Realtors, 57% of home buyers have buyer representation in 2009, down from 64% in 2006, before the housing bust.

A home buyer going at it alone—trying to buy a house without professional representation—is in a bad position. The buyer should remember that customarily buyers do not pay for representation; the buyer's agent splits the commission with the seller's agent, so the service to the buyer is essentially free. Buyer agents help their clients in many ways. For example, they often access historical price data for home sales in the area. Based on the data, the agent can recommend a bidding strategy in negotiation. Buyers' agents can also suggest home inspectors and financing companies to their clients, and they're not supposed to make money from the referrals. Also, a buyer agent looks for red flags to make sure all the "i's" are dotted and the "t's" are crossed in the contract.

However, many buyers choose to represent themselves. Why? The reasons include:

- Much of the housing data that used to be exclusive to real estate agents are now available online (e.g., real estate listings, median sales prices in a neighborhood, the number of days a home has been on the market, and the number of times a house was reduced in price).
- Buyer representatives have a strong incentive to push for a sale when a sale is not wholly justified—simply because buyer reps want to move on to other clients.
- Some buyer agents may not have their clients' interests at heart when the seller agent of a prospective home is from the same brokerage firm.

[i] Andriotis, A. M. (2001). Home buyers go hunting alone. *Market Watch*, May 10, 2011. Accessed from www.marketwatch.com/story/do-you-need-a-buyers-agent-1304523622290, on July 27, 2013.

PROSPECTING FOR CLIENTS

The first stage of the buyer representative cycle is prospecting. Prospecting refers to the methods used to identify and contact prospective clients. These methods are categorized into two groups: active and passive. See Figure 5.2. *Active prospecting* methods are ways to identify and contact prospective clients that require the buyer representative to actively seek out prospects rather than sit back and wait for prospects to contact him. The latter is, of course, referred to as *passive prospecting*.[2]

Let's review the active prospecting techniques first before we turn our attention to the passive techniques. These techniques include (1) direct inquiry, (2) directories, (3) referrals, (4) canvassing, and (5) renters. *Direct inquiry* refers to a method in which the buyer representative reaches prospective buyers through one of the following methods:

- *direct mail* (e.g., launching a direct mail campaign to people in the community who are close to retirement age; in the direct mail letter and brochure, the buyer representative touts his expertise in serving the retired and asks the direct mail recipients to contact

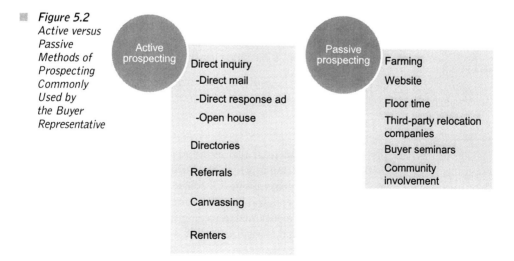

Figure 5.2
Active versus Passive Methods of Prospecting Commonly Used by the Buyer Representative

Active prospecting

Direct inquiry
-Direct mail
-Direct response ad
-Open house

Directories

Referrals

Canvassing

Renters

Passive prospecting

Farming

Website

Floor time

Third-party relocation companies

Buyer seminars

Community involvement

him—the buyer representative—if they will be considering buying a house in the next few months)

- *direct response advertising* (e.g., placing an ad in the local newspaper by advertising the special services the buyer representative can offer to the retired and asking them to call if they are on the market to buy a new home)
- *open house* (e.g., assembling a list of prospective buyers from open house registries and contacting these people by phone, direct mail, or e-mail)

The *directories* method of prospecting involves the use of directories to identify prospective buyers. For example, a buyer representative specializing in serving retiring personnel in a college town may target retiring college professors by going through the faculty and staff directory and identifying those people of the senior ranks (full professors) and perhaps then going through the departments' websites to trim that list to the full professors who have had more than 10 years of service. Full professors who exceed 10 years of service to the university are likely to be in the retiring mode.

Referrals are probably one of the most effective methods of prospecting. Buyer representatives identify prospective buyers by asking former clients to identify people in the community they know who may be thinking of buying a real estate property. According to the U.S. National Association of Realtors, 43% of buyers come from referrals and 11% tend to be repeat clients.[3] Of course, referrals are only forthcoming if the clients are happy with the service provided by the buyer representative. See Case/Anecdote 5.1 regarding the use of a smartphone application as a referral service.

CASE/ANECDOTE 5.1. A TECHNOLOGY-BASED REFERRAL SERVICE[i]

Siri, an iPhone application, is mobile technology that uses voice recognition to provide information that users request. That means it is also a referral service: It serves to identify and locate real estate agents in the area. For example, consider a prospective buyer of real estate who drives through a neighborhood and locates several properties of interest. To pursue these possibilities, the prospect uses the app to identify agents that he may contract with to help him further explore these options. The key feature within this app is that it helps the user identify "credible" agents in the area by using reviews and feedback from social media. In this way, agents in the area are rated based on the feedback that their past clients have posted on social media.

[i] How Siri finds real estate agents. *RealtyBizNews: Real Estate News & More.* April 2012. Accessed from http://realtybiznews.com/how-siri-finds-real-estate-agents/98711829/, on September 15, 2013.

Canvassing typically applies to residential real estate marketing, not commercial real estate. It refers to a method of prospecting in which the buyer representative goes door-to-door introducing himself as a buyer representative. Of course, this method is only effective when the buyer representative can use a referral (e.g., "I just helped your neighbors, the Jones family, sell their house, and Mr. Jones mentioned in passing that you may be on the market for a new home"). See the international Case/Anecdote 5.2 for an interesting example of how canvassing is done in Australia.

CASE/ANECDOTE 5.2. CREATIVE CANVASSING IN AUSTRALIA[i]

Many real estate agents don't like cold canvassing. The public does not appreciate real estate agents knocking on their doors without notice and sometimes at inappropriate times. Some agents invent creative ways to canvass a neighborhood without feeling the trepidation of cold canvassing.

Julie, one Australian real estate agent, purchased a quantity of cheap, small Australian flags. On Australia Day, she placed them in the front yard of every house in her assigned territory with her business card stapled to each flag. Part of her business card says "Happy Australia Day!" Some recipients even called her to ask for more flags to give to their kids.

Julie did other creative things too. During Christmastime, she distributed thank you notes to all the homes in the area with holiday displays, saying, "Thanks for brightening our holiday with your Christmas lights"; her note had her business card with her contact information.

[i] Grace, J. (2013). Never cold canvass again. *Realtor Magazine*, April 2013. Accessed from http://realtormag. realtor.org/sales-and-marketing/feature/article/2013/never-cold-canvas-again, on September 15, 2013.

The *renters* method of prospecting works exclusively in residential, not commercial, real estate. The method targets people who are currently renting, with the assumption that renters will eventually choose to buy their own place. Of course, this crude method of prospecting can be further refined by starting out with a renters' list and within that list identifying those who are older in age, those with families, and those with an adequate household income to afford a house.

Let us now turn our attention to the passive methods of prospecting. These include: (1) farming, (2) website, (3) floor time, (4) third-party relocation companies, (5) buyer seminars, and (6) community involvement. As previously described in Chapter 4, *farming* is a method in which a real estate agent establishes expertise in real estate related to specific neighborhoods. The agent is mostly likely to be known by most long-time residents of those "farmed" communities. The agent's real estate signs have been plastered all over the neighborhood, creating a high degree of awareness among neighborhood residents. When a real estate agent has done much business in specific neighborhoods, her reputation alone can drum up new business. Prospective buyers looking for a home in a specific neighborhood are likely to communicate with residents in that neighborhood and perhaps ask, "Do you know a good real estate agent who may be able to find me a home here in this neighborhood?" The resident, in turn, may respond by referring the real estate agent who has farmed the community and established a reputation as someone devoted to the housing needs of that community. A similar technique can be used in commercial real estate, in which certain real estate agents specialize in niche markets (e.g., land for agriculture, land for development of retirement communities, real estate space for clinics and medical establishments). In other words, once a real estate agent establishes a reputation of having "farmed" a certain community, his reputation alone can prompt prospective buyers to call on him because of his reputation. See Zooming-In Box 5.2 concerning niche marketing.

> **ZOOMING-IN BOX 5.2. NICHE MARKETING AND PASSIVE PROSPECTING[i]**
>
> There is a difference between "niche marketing" and "marketing to a niche." Niche marketing is establishing your brand identity (i.e., your special reputation), and it allows prospects to select you to represent them when your brand identity aligns with what they are looking for in a real estate agent. For example, you may establish a brand identity as highly ethical with high moral standards. Prospects who look for an agent with high moral standards will seek you out.
>
> Marketing to a niche, on the other hand, is developing a marketing program that targets a specific segment, such as retired couples who need assistance because they have limited physical mobility. If you have developed special expertise in understanding their housing needs, you may decide to target them to offer your services.
>
> [i] Cooke, J. (2012). Niche marketing for real estate agents: Finding your target audience. *RISMedia*, June 4, 2012. Accessed from http://rismedia.com/2012-06-04/niche-marketing-for-real-estate-agents-finding-your-target-audience/print, on May 2, 2013.

The *website* method of prospecting was described in Chapter 4 as it is applied to the seller representative. The focus here is on using the same technique to prospect buyers. Every buyer representative has to have his own website, which is part of the brokerage firm. The website should promote the buyer representative's services, claim certain specialized services, and provide evidence to substantiate these claims. Both content and style are important in promoting the buyer representative. If the content and style are impressive, prospective buyers are likely to call on him.

As described in Chapter 4, *floor time* refers to the availability of a real estate agent when he is manning the office. That is, many real estate brokerage firms assign various real estate agents "floor time." When a call comes in to the office from prospective buyers, the agent who is "manning the floor" attempts to assist the person who calls in. These callers can be prospective buyers.

Also discussed in Chapter 4 are third-party relocation companies. *Third-party relocation companies* are firms that specialize in helping company executives and their families move to another location—the location of job transfer. These third-party relocation companies typically try to connect with agents in the communities in which their client firms have a presence. Buyer representatives can identify these relocation companies and attempt to establish a relationship with them. Once a relationship is established, these relocation companies are likely to call on the buyer agent to help executives and their families find homes in suitable neighborhoods. The same method equally applies in commercial real estate. Third-party relocation companies help client companies locate and establish presence in certain communities (e.g., establishing a plant, a subsidiary operation, a branch office).

In Chapter 4, we discussed seller seminars as a method to prospect new clients. Buyer representatives also hold *buyer seminars* for community residents to help them be better consumers of real estate. The purchase of a new house can be mindboggling for many. A buyer representative may organize and offer a seminar to prospective buyers to help them choose wisely. Prospective clients attending these seminars may be impressed with the expertise and capabilities of the buyer representative who leads the seminar. Some of them may call on the buyer representative to represent them in their search for the "dream home."

We also discussed community involvement as a method of prospecting for seller representatives in Chapter 4. The same can be said for buyer representatives. In other words, real estate agents can become successful by building a reputation of service through *community involvement*. Real estate agents become involved in serving their communities in various volunteer roles through their churches, synagogues, mosques, and temples; schools and other educational institutions; law enforcement and the judicial system; the community chamber of commerce; and other public and charity organizations.

NEEDS ASSESSMENT

Real estate marketing professionals representing the interests of their buyer clients play a very important role in every real estate transaction. They match the needs of their clients with available properties. This process is key to good buyer representation. To do this well, the buyer representative has to understand and profile his client's real estate needs. He needs to be equally knowledgeable about all the properties out there for sale and the extent to which each property may fit his client's needs.[4] See Figure 5.3.

Hence, the second stage of the personal selling cycle for a buyer representative is the needs assessment. The buyer representative makes a sincere attempt to document his client's personal and financial characteristics (those that have a direct bearing on some aspect of the real estate transaction) and property needs. In the context of residential real estate marketing, the buyer representative would attempt to develop a personal and financial profile by asking basic questions:

- contact information?
- number in family?
- children's genders and ages?
- possession needed when?
- will you consider renting before purchasing?
- how long have you been looking?
- price range?
- down payment?
- monthly payment?

The buyer representative may ask questions regarding the buyer's preferences and needs regarding property and place:

- architectural style and type?
- new or old house (how old)?

Figure 5.3
Matching the Client's
Needs with Available
Property

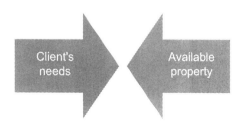

- areas or neighborhoods?
- how many bedrooms?
- sizes of the bedrooms?
- how many baths?
- size and style of living room?
- dining room size?
- kitchen size and layout?
- recreation room?
- den?
- basement (half or full)?
- workshop (or other space)?
- fireplace?
- storage area?
- heat and fuel?
- air conditioning?
- garage, carport, street parking?
- near school? If so, what type of school?
- near shopping?
- near public transportation?
- other requirements?

QUALIFYING THE PROSPECT

The next stage of the personal selling cycle for a buyer representative is qualifying the prospect. That is, the buyer agent has to assess the client's (1) ability to pay the loan, (2) motivation to repay the loan, (3) net worth, and (4) security for the loan.[5] See Figure 5.4.

It should be noted that there are many professionals in real estate marketing who argue that real estate agents should not get in the business of qualifying the prospect. They claim that it is the business of the mortgage company or the institutional lender to do so, and the agent has to work closely with the lender to qualify a prospective client. This may be so. The

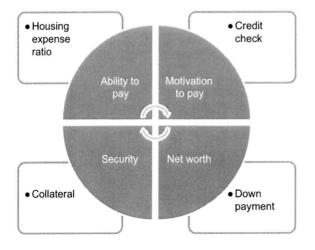

Figure 5.4
Qualifying the Prospect

problem is that in most situations, working with a lender occurs after a suitable property is found. Many lenders feel reluctant to spend time, effort, and money on a prospective client who may not turn out to be a client at all. In such situations, the real estate agent has to do the "qualifying" himself. We should emphasize the fact that when the agent does the qualifying, he does not do this formally. This is done informally to be able to identify the maximum amount the client can afford on a house, which would in turn allow the process to continue to the next stage (showing). Otherwise, both the buyer representative and the client have to wait for the lender to come through with a formal qualification, which may take some time, and this assumes that the client is even able to find a lender who will do the qualifying before the client commits to doing business with that lender. There are many buyer representatives who refuse to take on clients if they are not already qualified. That is, clients have to approach a lender and receive a formal qualification from the lender before they can seek buyer representation. In any case, real estate agents representing buyer clients should be familiar with methods used by lenders to qualify prospective clients. Such knowledge should help the buyer representative better assist his client through the transaction process.

Ability to Pay

The first criterion for qualifying the prospective client is *ability to pay the loan*. Does the prospective client have enough income (on a monthly basis) to be able to make the mortgage payments? Typically, the lender uses a formula referred to as "housing expense ratio" to determine eligibility. Specifically,

Housing expense ratio = housing expense / monthly gross income

Let's use an example to illustrate. Let's say the monthly gross income of the prospective client is $10,000. The mortgage payment for a specific house is $4,000 per month. This means,

$4,000 / $10,000 = 0.4 or 40%

Does the prospective client qualify to buy a house that requires a monthly mortgage payment of $4,000 with a monthly gross income of $10,000? Usually not. Most lenders use the 28% rule, which says that the mortgage payment cannot exceed 28% of monthly gross income. So what is the maximum monthly mortgage payment for which this prospective client qualifies?

0.28 = housing expense / $10,000
$10,000 × 0.28 = $2,800

The housing expense cannot exceed $2,800, considering the gross monthly is only $10,000. If the prospective client would like to buy a house that has a housing expense (a monthly mortgage payment) of $4,000, then he needs to make $14,285 minimum in gross income per month.

0.28 = $4,000/ monthly gross income
$4,000 / 0.28 = $14,285

Motivation to Pay

An important cue that mortgage lenders use to predict the extent to which a client is motivated to pay the loan is the client's history of paying past loans (i.e., *credit check*). That is, the mortgage lender obtains a credit check on the prospective buyer and checks his credit score. If the prospect has a good score, then he meets the motivation-to-pay criterion.

Net Worth

Mortgage lenders inspect the balance sheet of a prospective buyer. A balance sheet is essentially a financial statement that reflects the person's financial assets, liabilities, and net worth. *Net worth* is the difference between assets and liabilities. In other words, if the prospect has high financial assets (savings, investments, income, material possessions, etc.) and low liabilities (i.e., debt in the form of mortgage loans, car loans, credit card debt, etc.), then the prospect is likely to have good net worth. In contrast, if the prospect has low assets and high liabilities, then his net worth is not good. Mortgage lenders like to lend to people with high net worth and are likely to disqualify those with low net worth.

Security

Mortgage lenders like to guarantee the loan repayment in case the client defaults on the loan. A loan guarantee usually comes in the form of *collateral* (i.e., specific assets, such as the property in question, other properties, savings and investments, valuable material possessions). Such guarantees are referred to as *security*.

SHOWING

The next stage of the buyer representative cycle is *showing*. Showing refers to the process in which the buyer representative takes his client on the road to show him selected properties that the buyer representative believes may fit the client's needs. That is, the buyer rep invests time and energy inspecting available properties for sale and selects a handful that may fit the bill.

Once a handful of properties are selected, the buyer representative makes decisions concerning the number of showings, the times of showings, and the method of presentation. Typically, a buyer representative would present three or four properties to the client in a full day. More than four may be overwhelming for the client, both mentally and physically; less than three may not be efficient use of time and energy.

The time of the showing is also important. Typically the buyer rep presents properties during the day to allow the client to see the properties in full daylight. However, there are many clients who have a full work schedule that may allow them to see those properties only after work. If the buyer representative has a choice in the matter, he will usually schedule the showing during daylight hours and possibly on weekends. This, of course, is very common in residential real estate. Commercial real estate showing is different, in the sense that showing is invariably done during the day and during the week. Buyers of commercial real estate usually consider seeing and inspecting a property as part of their daily job and, as such, would not hesitate to schedule to do this during the day and on weekdays.

In presenting properties to a client, buyer representatives consider the presentation method. For example, in residential showing, the buyer representative may select three homes varying in degree of appeal. He may start out with a home that will probably be the least desirable and leave the home that is predicted to be most desirable for last. By showing the least desirable home first, the client uses the first house as an anchor or benchmark. When the buyer then sees the more desirable house last, the contrast may lead the buyer to form a significantly more positive attitude toward it. Though this is a common presentation technique, one should wonder about the ethics of it. Are we psychologically manipulating the client by using this framing technique? You decide. See Zooming-In Box 5.3 for additional tidbits about showing a residential property to prospective clients.

ZOOMING-IN BOX 5.3. GUIDELINES ON SHOWING[i]

Preview the House

The buyer agent should preview the house before actually showing it to a prospective buyer. This allows the buyer agent to plan the showing effectively. In other words, the buyer agent is in a better position to point to certain features of the property that may be well suited to the housing needs of the prospect.

Make an Appointment

For homes that are still occupied by homeowners or tenants, the buyer agent has to make an appointment to show the house. Barging in unexpectedly can be a disaster. The property may be a total mess. This, of course, becomes a moot point if the house is vacant.

Choosing a Route

To show a home in the best possible light, you may want to choose your route to getting to the home carefully. Choose a route that takes you and the prospective buyer through nice neighborhoods with good landscaping. However, don't overdo it. You don't want to drive through an upscale neighborhood to show a home located in a low income neighborhood. While you go through the neighborhood, point out the amenities, such as the restaurants, shopping centers, schools, and public transportation.

Keep the Seller Away and Maintain Control

When the home is still occupied by the seller, make arrangements to have the seller leave the house during the showing. Having the seller at the premises may be a disadvantage because you can't control the situation; you don't know what the seller will say or do, and what is important to the seller may not be important to the prospective buyer. You, the buyer representative, know what is important to the prospective buyer. In addition, make sure any pets are either out of the house or on a leash.

Allow the Prospect Enough Time and Limit the Choices

Give the prospect time to inspect the property. Don't rush him. If you have several homes to visit, then you may have to change your plan. Too many choices can be detrimental. Many people can't process, let alone retain, information about three properties in one day. They get confused and may experience information overload. So limit the number of showings and allow the prospective buyer to linger around the house to get a feel for the place.

Help the Prospect Assess the Property by Making Comparisons and Don't Become Defensive

To help the prospective buyer get a good feel for the place, point out the positive and negative features of the place and make comparisons with other properties the prospect has seen. When the prospect points out negative features or rules out a house, don't take it personally. Don't get defensive. Instead, use this information to get a better understanding of what the prospect is looking for in a home.

Sometimes the only way to find out what a prospect really likes or doesn't like in a home is to observe him as he actually sees, hears, touches, and feels the home. Remember, your job is not to sell a particular property; it is to find the best match for the prospect—to find a home that best matches his housing needs.

[i] Rosenauer, J., & Mayfield, J. (2007). *Effective real estate sales and marketing*. Mason, OH: Thomson Higher Education (Chapter 10).

See Case/Anecdote 5.3 for how buyer agents can use IT software to better assist foreign clients who may feel the need to process the listing information in another language (other than English).

CASE/ANECDOTE 5.3. TRANSLATING LISTINGS INTO OTHER LANGUAGES[i]

Based on statistics provided by the U.S. Census Bureau, today's homebuyers are more diverse than ever before. There are around 55 million Americans, 1 in 5, who speak a language other than English at home. The National Association of Realtors reports that foreign buyers acquired $82 billion of residential real estate in the U.S. in 2011. Proxio is a marketing/networking software that translates real estate listings into 19 different languages. The goal of the software is to bridge geographic and cultural borders, helping agents to better assist clients who feel more comfortable reading real estate listings in their native language (or some other language besides English). Buyer agents using this software may find a competitive advantage in contracting with specific foreign buyers.

[i] Growing number of U.S. real estate firms offer multilingual services through Proxio. *Market Watch, Wall Street Journal*, April 24, 2012. Accessed from www.marketwatch.com/story/growing-number-of-us-real-estate-firms-offer-multilingual-services-through-proxio-2012-04-24, on September 15, 2013.

NEGOTIATING WITH THE SELLER

The buyer representative negotiates with sellers on behalf of his client. In doing so, an effective buyer negotiator has to understand and implement the basic concepts of real estate negotiation. These concepts include target price, reservation price, ZOPA, tradeoffs, and BATNA.[6] Consider the following example. A home seller is negotiating with a buyer agent trying to sell a home with an appraised value of $300,000. The home seller's asking price is $320,000. In other words, the *target price* (i.e., desired price) of the seller is $320,000—this is the price that the seller wishes to make on the transaction. The seller has a *reservation price* of $300,000—the bottom line or the price threshold that he is not willing to go below. See Figure 5.5.

The buyer agent knows that the buyer does not want to pay more than $280,000 (target price) but is willing to go all the way up to $310,000 (reservation price); the buyer actually qualifies for a loan mortgage of $310,000. In this case ZOPA (Zone of Potential Agreement)

Seller:
target price = $320,000

reservation price = $300,000

Buyer:
target price = $280,000

reservation price = $310,000

Figure 5.5
Buyer Representative Negotiating on Behalf of the Buyer

is the price range in which an agreement can be reached, which is the range between the two parties' reservation prices. In this case, ZOPA is $10,000 (the range between the buyer's reservation price, which is $310,000, and the seller's reservation price, which is $300,000).

The first rule of negotiation from the vantage point of the buyer representative is to identify the target price positions of both buyer and seller, the buyer's reservation price (highest price willing to pay), and the seller's reservation price (lowest price willing to accept). The seller's reservation price is most likely concealed and not easily revealed. A good buyer rep will make attempts to "discover" the seller's reservation price by asking for concessions. He then examines the pattern of concessions to determine the seller's bottom line (the lowest price that the seller will accept).

In our example, by pushing the seller for concessions, the buyer agent could note that the seller made the following pattern of concessions: His asking price was $320,000. His first concession was $310,000; his second concession was $305,000; his third concession was $302,000. The pattern ($320,000, $310,000, $305,000, and $302,000) shows decreasing concessions, which may signal the seller's reservation price. In other words, this pattern indicates a reservation price of $300,000; it is not likely that the seller, having made the aforementioned concessions, will go below $300,000.

The second rule of effective negotiation is to determine ZOPA (Zone of Potential Agreement). Once the seller's reservation price is estimated ($300,000, in our example), then the buyer representative should realize that bargaining has to take place between the buyer's reservation price ($310,000, or the price that the buyer will not exceed) and the seller's reservation price ($300,000, or the price that the seller will not go under). This means that bargaining can take place between $310,000 and $300,000 (ZOPA is $10,000, or $310,000 minus $300,000). The buyer representative should make the seller come down in price to $300,000.

To achieve this goal (inducing the seller to come down in price as close as possible to $300,000), the buyer agent attempts to make tradeoff decisions and use BATNA for leverage. Let's explain these concepts. *Tradeoffs* are decisions made by the negotiating parties that help achieve a final settlement. The buyer agent attempts to find out something about the seller's interests in this deal, besides price. Perhaps the buyer agent finds out that the seller has to get out of the house in three weeks because he has to relocate in another city and already has a contract in the works to buy a house in this other city. Therefore, the seller's primary concern is timeliness of the deal and price is secondary. Using the preceding ZOPA example of a $10,000 range, the buyer agent may make tradeoff decisions to conclude the deal. The buyer agent may offer the seller $300,000, with the stipulation that the agent will work very hard to expedite the closing so that the seller may be able to move on time to start his new job in the new city. Therefore, in making this offer to the seller, the agent is not simply making a price offer; he is also making a concession on the timeliness of the deal in expediting the closing in three weeks or less. Therefore, the third rule of effective negotiation is to make tradeoff decisions by addressing the seller's other concerns (besides price) and offering a deal that addresses these concerns.

BATNA stands for Best Alternative to the Negotiated Agreement. It is commonly used as leverage in real estate negotiations. BATNA is used as leverage by exploring other deals to ensure that if negotiations with the current seller do not come through, that the buyer has alternative deals to fall back on. Ensuring that the seller knows that the buyer has other possible deals provides leverage to the buyer. A buyer agent can strike a good deal (for instance, press the seller to accept the $300,000 offer) if the seller knows he can walk away from failed negotiations. To be able to walk away, the buyer agent has to have at least one BATNA to fall back on. Thus, the fourth rule of effective negotiation is to enter into negotiation with a seller holding a BATNA. In Part 3 of the book, we will discuss negotiation concepts in more depth. Stay tuned.

CLOSING

When the buyer agent is satisfied that a particular property is well suited to the client, he moves to close the deal on that particular property. Given that the vast majority of real estate transactions involve a great deal of money (and are sometimes the one investment that is most meaningful to the buyer), most buyers feel somewhat anxious at this stage. They are likely to raise objections and procrastinate. This, of course, is to be expected.

Buyer agents use a variety of techniques to ease the tension and help the buyer resolve any conflicts he may be experiencing at this stage. These techniques include (1) assumptive close, (2) summarizing advantages, (3) the balance sheet approach, and (4) closing on a single objection or minor details.[7] The *assumptive close* refers to when the buyer agent assumes that the client is comfortable with the target property and would like to move to closing. That is, the buyer agent assumes that the buyer is willing to move to the next step, which is to close this deal and process the transaction. Of course, this technique is used when the buyer agent picks up on signals that point to the fact that the client is happy with the focal property and would like to close the deal. Conversely, this technique should not be used when the client seems too anxious and raises objections about certain aspects of the property he finds objectionable.

The *summarizing advantages* technique involves having the buyer agent develop a synopsis of all the advantages related to the purchase of the property. For example, if the buyer has expressed positive aspects of the home (e.g., the home is strategically located next to public transportation, easy access to shopping, good schools in the community, nice neighborhood, enticing architecture, low-maintenance backyard), then the agent reminds the client of these advantages. Doing so should motivate the client to consolidate his positive feelings about the property and move to closing. Of course, one may criticize this technique for focusing only on summarizing the advantages of the property without highlighting or bringing up expressed reservations or concerns. This is where the next technique comes in: the balance sheet approach to closing.

Using the *balance sheet* closing technique, the buyer agent summarizes advantages as well as disadvantages of the focal property, and perhaps does so in a comparative fashion (i.e., compares the "assets" and "liabilities" of those properties that have been presented to the client) to help the client make a decision. Using this technique, the buyer agent helps the client make a "rational" decision by weighing the advantages and disadvantages of the properties that are considered for selection.

The last technique is *closing on a single objection or minor details*. This closing technique prompts the buyer agent to address objections or any minor concerns expressed by the client by saying something like, "If we take care of this problem, will you sign off on the deal?" In other words, the buyer agent elicits a commitment from the client to close the deal if the objections or concerns are addressed.

Does "closing" mean only extracting a commitment from the client to seal the deal? It actually involves helping the buyer with many tasks performed after acceptance of a deal. These include:

- *appraisal* (i.e., the property will be formally appraised by the mortgage lender, who will assign this task to a professional appraiser)
- *financing* (i.e., the agent assists the buyer in dealing with the mortgage lender by explaining the various financing options, such as government financing, such as VA and FHA loans in the U.S.; conventional financing in the form of a traditional loan from a

mortgage company or a bank; and creative financing, such as having the seller assume the loan and the buyer pay the seller instead of the mortgage company; in addition, the buyer agent helps the buyer process and understand the good faith estimate provided by the mortgage company and helps the buyer resolve any noted discrepancies or problems)

- *inspections* (i.e., the property will be inspected by one or more home inspection professionals, and this may involve a house inspection, assuming the property is a residence; a pest inspection; a septic inspection, assuming that the property is not connected to the tow/city sewage system; a well inspection, assuming that the property has its own water well; and a radon test to ensure low levels of radon emissions)
- *environmental assessment* (mostly applicable for commercial real estate properties)
- *structural engineer's report* (to check the credibility of the foundation)
- *survey* (to draw specific and concrete boundaries around the property)
- *final walk-through* (to ensure that all the problems are noted and taken care of)
- *closing costs* (to help the buyer understand all the technicalities involved in closing costs—see Table 5.1 for a listing of closing costs common in most residential real estate transactions in the U.S.)[8]

Finally, the deal is sealed when the buyer signs an *earnest money contract* (referred to "purchase agreement," "purchase money offer contract of sale," "purchase binder"). This, essentially, is an agreement that states that the buyer has agreed to purchase the property for a specified amount, and the buyer pays a specified amount of money to demonstrate intent. This money is placed in an escrow account (which could be held by the buyer representative, the seller representative, or the title company). Both buyer and seller sign off on the earnest money contract, and in many cases this contract is reviewed by the attorneys of both buyer and seller. The buyer agent follows up to ensure that the process is completed by facilitating the loan process; to communicate with the buyer, seller, and the seller agent to ensure the smooth transfer of the property; and to arrange the time and place for the buyer to sign all the necessary documents at closing (normally held at either the title company or the buyer's attorney's office). See three ethical scenarios in ethics Boxes 5.1, 5.2, and 5.3 related to closing.

Table 5.1 Explanation of Buyer's Closing Costs

Closing cost item	Explanation
Loan origination fee	Fee paid to the mortgage company for processing the loan (e.g., 1 or 2 points, which amounts to 1 or 2% of the loan amount)
Appraisal fee	Fee paid to a professional appraiser for estimating the market value of the property
Recording fee	Fee for recording the transaction to make it a public record
Credit report	Fee paid to a credit reporting agency for providing credit history of the buyer
Survey fee	Fee paid to a professional surveyor for determining the exact boundaries of the property
Escrow fee	Fee paid to the escrow agent (usually the title company) for handling the transaction

(*continued*)

■ *Table 5.1* (continued)

Closing cost item	Explanation
Restrictions	This is a document that describes any restrictions on the use of the property as originally conceived by the real estate developer (e.g., homeowner cannot have farm animals on the property)
Note	The document that spells out the mortgage loan (debt) and the payment schedule
Deed of trust	The document that spells out the fact that there is a mortgage lien on the property bought by the mortgage loan shown in the note
Private Mortgage Insurance (PMI) fee	Fee for insurance (paid to a mortgage insurance company) to protect the lender in case that the buyer defaults on the loan; PMI applies only when the buyer's down payment is less than 20% of the loan amount
Transfer fee	Fee paid to the lender when the buyer assumes the existing mortgage on the property held by the seller
Attorney fee	Fee paid to the buyer's attorney for overseeing the transaction
Prorations of taxes and insurance	A document that spells out financial responsibility between buyer and seller regarding payment of current taxes and insurance (i.e., seller pays current taxes and insurance before the buyer takes over the property)
Loan application fee	Fee paid to the mortgage company for securing a mortgage to finance the purchase
Inspection fees	Fee paid to the a licensed professional who inspected the property to point out problems that need fix-ups

Source: Adapted from Rosenauer, J., & Mayfield, J. (2007). *Effective real estate sales and marketing.* Mason, OH: Thomson Higher Education (p. 181).

ETHICS BOX 5.1. UNETHICAL CLOSING TECHNIQUES?

There are many closing techniques that buyer agents commonly use; some are not ethical. For example, some agents may use a *scare technique*; that is, they make the buyer fearful of the possibility that if the buyer does not move right away to seal this deal, then the buyer stands a good chance to lose it altogether. Again, using fear to motivate the client to not only unethical, but also likely to generate dissatisfaction — the client may be forced to sign the deal, but, in the final analysis, he may not feel good about the deal. Remember, real estate business is local, and one of the most effective ways to promote real estate business is word of mouth. In other words, the client has to walk away from the deal feeling happy with the outcome. Clients happy with their outcomes create positive word-of-mouth communication and offer referrals.

Another unethical closing technique commonly used is the *alternative choice* method. Here, the buyer agent selects an undesirable alternative and compares the target property, which is highly desirable, with the undesirable alternative. Doing so induces the client to accept the deal more readily and move to closing. Again, business ethicists may frown on this closing technique because it involves psychological manipulation.

ETHICS BOX 5.2. PUT THINGS DOWN IN WRITING[i]

A buyer expresses interest in buying a home with beautiful oak bookcases. The bookcases are only attached to the wall by a single screw and bracket that keep them from tipping over. The buyer asks his agent if the bookcases come with the house. The buyer agent contacts the listing agent, who in turn confers with the seller, and the seller agrees to leave the bookcases behind.

However, when the buyer takes possession, he sees that the bookcases are missing. He asks his agent to inquire with the seller. The seller responds by saying, "I am so sorry; I forgot to instruct the movers to leave the bookcases behind." The buyer is furious and demands that the bookcases be shipped back; the seller refuses and says it would cost a small fortune to do so. He mentions the fact that he is not legally obligated to do so because the agreement was not put in writing.

In hindsight, the buyer agent should have made the agreement about the bookcases part of the formal contract. In addition, the buyer agent should have urged the seller to compensate the buyer by at least offering some financial compensation for the bookcases.

[i] Kantor A., & Hegeman, B. (2013). Do unto others. *Commonwealth: A Journal for Real Estate Professionals Published by the Virginia Association of Realtors*, April/May, 23.

ETHICS BOX 5.3. DON'T PLAY LAWYER[i]

A buyer agent is trying to finalize a deal for his client on a $900,000 home. The only remaining task is to add to the contract a contingency regarding a power-company easement. He turns to one of his colleagues within his brokerage house, another real estate agent who is good at writing technical and legal text. He asks the colleague to write the contingency into the contract. The colleague at first declines, but the buyer agent talks her into doing this by offering her football tickets.

Article 13 of the NAR Code of Ethics says: Do not engage in the unauthorized practice of law. In this case, both agents are guilty of violating this code. The buyer agent should have consulted a real estate attorney—not a fellow real estate agent.

[i] Kantor A., & Hegeman, B. (2013). Do unto others. *Commonwealth: A Journal for Real Estate Professionals Published by the Virginia Association of Realtors*, April/May, 24.

FOLLOW-UP AND SERVICING

After closing, the buyer agent becomes embroiled in three tasks: (1) supporting the buying decision, (2) dealing with dissatisfaction, and (3) enhancing the relationship.[9] With respect to *supporting the buying decision*, it is well recognized that buyers are likely to experience buyer's remorse. Given the fact that real estate acquisition is a highly emotional decision (because it represents a high-priced purchase), buyers are likely to experience cognitive dissonance and possibly feelings of regret: "Did I make the right decision?" "Maybe I should have explored other options to make sure that this is the best one?" "I am not sure whether I got the most value for the money I spent." The buyer agent has an important role to play in reducing the buyer's cognitive dissonance and possible feelings of regret. The buyer agent calls the buyer to follow up and deal with these thoughts and emotions. The agent assures the buyer that she has made the right decision and reinforces the decision by highlighting the advantages of the deal.

The second task is to *deal with dissatisfaction* arising from the use of the property. When the buyer begins to move in and use the property, there is a strong chance that he may feel dissatisfied about certain features of the property (e.g., the remote control does not work to open the garage; the security system does not work properly; the intercom system is difficult to learn). It is the buyer agent's responsibility to help the buyer get settled in his new home and to make the buyer (and his family) feel comfortable. This may mean rolling up shirtsleeves and recommending certain people and/or firms to help address the noted problems.

The last task is to *enhance the relationship*. This means that the buyer agent has to make himself available to assist the buyer to settle into the property and make the best use of it. It means continuing personal communications and becoming a resource for information about additional issues—from how to connect with utility companies to which plumber to call to fix a leak. Assisting the client after closing should enhance the relationship to the point where the agent can feel comfortable asking the client to make referrals—identify friends, colleagues, and extended family members who may be on the market to purchase real estate. To enhance the relationship long-term, buyer agents tend to use client follow-up techniques, such as sending their clients e-newsletters on a monthly basis, giving their clients a closing gift (e.g., a nice flower arrangement), and inviting them to celebratory events (e.g., social events, religious services, sports games, and educational seminars). See Case/Anecdote 5.4 for more examples of creative methods of following up.

CASE/ANECDOTE 5.4. CREATIVE FOLLOW-UP METHODS[i]

Follow-up methods are quite varied. They go from the simple to the elaborate; from the inexpensive to the expensive. Let's examine some creative ways of following up.

Closing Gifts

One real estate agent bought a portable grill for a closing gift. When the client moved in, the grill lid was open and sitting on the grill was a basket of barbecue items, such as sauces, rubs, and mesquite wood chips.

How about a home warranty for a closing gift? The client will appreciate this gift immensely in the event that something goes wrong and he has to use the warranty. This also gives the agent a chance to keep in touch with the client by calling periodically to see if everything is all right with the house.

Another creative closing gift is a contribution to the client's favorite charity. For example, donating a $100 to a dog lover's favorite animal rescue organization is likely to be viewed by the client as a very nice gesture and the client will remember this gift for years to come.

Stay in Touch Online

Many real estate agents add their clients online to social media networks such Facebook. This gives them an opportunity to stay in touch with them. They reach out to them as often as possible, not by promoting certain homes but by merely being sociable and friendly.

E-mail can be used to stay in touch, too. Many real estate agents send their clients articles of interest, YouTube videos, decent jokes, pictures, recipes of dishes they have recently discovered, favorite restaurants, shopping suggestions, and so on.

Use Snail Mail

Barbara Todaro, an agent in Massachusetts, stays in touch with her clients by keeping them abreast of new listings and sales in their neighborhood. She does this by sending them postcards with that information. Many of her clients appreciate that she remains an important source of real estate news, and she remains very memorable.

Client Appreciation Events

Wes Freas with Zephyr Realty in San Francisco holds an annual wreath decorating party every year during the holidays. It has become a tradition among his former clients. A broker from Napa Valley hosts an annual crab feed. She rents out a hall and offers salad, sourdough bread, wine, beer, and a lot of fresh crab. She gets agents to participate in the event and help out with the cooking.

[i] O'Brien, S. (2013). Real estate client follow-up: It's not just for prospects. *Realtor Magazine*, February 21, 2013. Accessed from www.marketleader.com/blog/2013/02/21/real-estate-client-follow-up, on September 15, 2013.

SUMMARY

This chapter focused on the personal selling cycle for the real estate buyer representative. The personal selling cycle involves prospecting for prospects, assessing the housing needs of prospects, qualifying the prospect, showing the prospect possible properties that are on sale, negotiating with sellers on behalf of their client, closing the deal, and following up.

Prospecting refers to the methods used to identify and contact prospective clients. These methods were discussed in terms of two groups: active and passive. Active prospecting methods are ways to identify and contact prospective clients that require the buyer representative to actively seek out prospects rather than sit back and have prospects contact him. The methods discussed in this chapter were direct inquiry, directories, referrals, canvassing, and renters. With respect to passive prospecting, these include farming, website, floor time, third-party relocation companies, buyer seminars, and community involvement.

In the needs assessment stage, the buyer representative attempts to understand and profile his client's real estate needs—the client's personal and financial characteristics (that have a direct bearing on some aspect of the real estate transaction) and property needs. The buyer agent then proceeds to qualify the prospect. That is, the buyer agent assesses the client's ability to pay the loan, motivation to repay the loan, net worth, and security for the loan.

The next stage of the buyer representative cycle is showing. Showing refers to the process in which the buyer representative takes his client on the road to show him selected properties that the buyer representative believes may fit the client's needs. Here, the buyer representative makes decisions concerning the number of showings, the times of showing, and the method of presentation. Then the buyer representative negotiates with sellers on behalf of his client. Negotiation concepts that are critical to understand include target price, reservation price, ZOPA, tradeoffs, and BATNA.

When the buyer agent is satisfied that a particular property is well suited to the client, he moves to close the deal on that particular property. There are a variety of techniques commonly used by buyer agents. These include the assumptive close, summarizing advantages, the balance sheet approach, and closing on a single objection or minor details. Closing also involves helping the buyer with many tasks performed after acceptance of a deal, such as appraisal, financing, inspection, environmental assessment, structural engineer's report, survey, final walk-through, and closing costs. Finally, the deal is sealed when the buyer signs an earnest money contract. After closing, the buyer agent helps the client with supporting the buying decision, deals with any hints of dissatisfaction, and makes an attempt to enhance the relationship.

DISCUSSION QUESTIONS

1. Briefly describe the various stages of the buyer representative's personal selling cycle.

2. Distinguish the active and passive methods of prospecting commonly used by the buyer agent.

3. How does the buyer agent use direct inquiry as a method of prospecting?

4. How does the buyer agent use directories as a method of prospecting?

5. How does the buyer agent use referrals as a method of prospecting?

6. How does the buyer agent use canvassing as a method of prospecting?

7. How does the buyer agent use renters as a method of prospecting?

8. What does farming mean, and why is it considered a method of prospecting?

9. The website is a passive method of prospecting. Please explain.

10. Buyer agents use floor time as a method of prospecting. Please comment.

11. Third-party relocation is another prospecting method. Please describe.

12. Buyer agents offer buyer seminars to community residents. Is this a prospecting method? Explain.

13. Community involvement is essential in drumming up new business. Please explain.

14. What is a needs assessment? How does the buyer agent go about conducting a needs assessment for a prospective client?

15. How does the buyer agent go about financially qualifying a prospective client?

16. Describe the decisions involved in showing properties to a client.

17. Does the buyer agent negotiate on behalf of his client? Describe some of the basic negotiation concepts, such as target price, reservation price, ZOPA, tradeoffs, and BATNA.

18. What does closing entail? Please explain.

19. How does the buyer agent go about following up with his client after closing?

NOTES

1. 47 percent of home buyers plan to use a real estate agent but do not have one yet. (2012). *Marketwatch.com*, April 30, 2012. Accessed from www.marketwatch.com/story/45-percent-of-home-buyers-plan-to-use-a-real-estate-agent-but-do-not-have-one-yet-2012–04–30, on May 5, 2012.
2. Hamilton, D. (2006). *Real estate: Marketing and essentials*. Mason, OH: Thomson Higher Education (Chapter 9).
3. Zeller, D. (2009). Capture the leads. *Realtor Magazine*, July 2009. Accessed from http://realtormag.realtor.org/for-brokers/build-your-business/article/2009/07/capture-leads, on May 15, 2013.
4. Hamilton, D. (2006). *Real estate: Marketing and essentials*. Mason, OH: Thomson Higher Education (pp. 222–224).
5. Hamilton, D. (2006). *Real estate: Marketing and essentials*. Mason, OH: Thomson Higher Education (p. 225, pp. 301–302). Rosenauer, J., & Mayfield, J. (2007). *Effective real estate sales and marketing*. Mason, OH: Thomson Higher Education (Chapter 9).
6. Bazerman, M. H., Curhan, J. R., Moore, D. A., & Valley, K. L. (2000). Negotiation. *Annual Review of Psychology*, 51, 279–314. Pruitt, D. G. (1981). *Negotiation behavior*. New York: Academic Press. Thompson, L. L. (2012). *The mind and heart of the negotiator* (3rd ed.). Upper Saddle River, NJ: Pearson Education.
7. Hamilton, D. (2006). *Real estate: Marketing and essentials*. Mason, OH: Thomson Higher Education (pp. 245–255).
8. Hamilton, D. (2006). *Real estate: Marketing and essentials*. Mason, OH: Thomson Higher Education (Chapter 15).
9. Based on Hamilton, D. (2006). *Real estate: Marketing and essentials*. Mason, OH: Thomson Higher Education (Chapter 12).

Part 3

How Real Estate Agents Negotiate

Chapter 6

The Social Psychology of Real Estate Negotiations

LEARNING OBJECTIVES

This chapter is designed to help students of real estate marketing learn:
- the rudimentary concepts of negotiation
- what negotiation effectiveness means for real estate marketing professionals
- how social scientists measure negotiation effectiveness in a real estate context
- how social influence factors influence negotiation effectiveness
- how factors related to the structural context influence negotiation effectiveness
- how factors related to the behavioral predisposition of the negotiators affect negotiation effectiveness
- how factors related to cultural context affect negotiation effectiveness

REVISITING THE RUDIMENTARY CONCEPTS OF NEGOTIATION

Before proceeding to get into the social psychology of negotiation, the student may benefit from a refresher of the rudimentary concepts. These are target price, reservation price, ZOPA, tradeoffs, and BATNA.[1] Consider the following example. The seller representative is negotiating with a buyer agent to sell a home with an appraised value of $500,000. The seller rep announces the asking price as $520,000. In other words, the *target price* (i.e., desired price) of the seller is $520,000—this is the price that the seller wishes to get for the house. The seller communicates to the seller agent a *reservation price* of $500,000—the bottom line or the price threshold that he is not willing to go below.

The buyer communicates to the buyer rep that he would like to buy the house for $480,000 (target price) but is willing to go all the way up to $510,000 (reservation price), if push comes to shove. In this case ZOPA (Zone of Potential Agreement) is the price range in which an agreement can be reached, which is the range between the two parties' reservation prices. In this case ZOPA is $10,000 (the range between the buyer's reservation price, which is $510,000, and the seller's reservation price, which is $500,000).

Tradeoffs are decisions made by the negotiating parties that help achieve a final settlement. Real estate buyers and sellers (and their agents) take into account other costs and benefits besides price in negotiating a deal. Using the preceding ZOPA example of a $10,000 range, both buyer and seller reps may make tradeoff decisions to conclude the deal. The seller may make an offer of $500,000 with additional amenities that the buyer may be interested in acquiring (e.g., the Persian rugs, the drapes, the washer and dryer, and the large tool set in the garage). The buyer may sweeten the deal by offering to work hard to expedite the closing so that the seller may be able to move on time to start his new job in another city. Note that the negotiation is not totally focused on price. It may involve price, timeliness, and other amenities. Buyers and sellers make tradeoff decisions among these various costs and benefits to strike a deal.

BATNA (Best Alternative to the Negotiated Agreement) is essentially used as leverage in real estate negotiations. In other words, negotiators are in a better position when they have alternatives they can fall back on if the negotiation falters. BATNA can influence the negotiation outcome. Consider the example of the seller rep having a BATNA in the form of a tentative offer from another prospective buyer. The tentative offer is $505,000; however, there is some uncertainty associated with the offer because that prospective buyer has to make arrangements to sell his house first before he can make a definitive offer. This BATNA provides some ammunition to the seller agent to negotiate with confidence and try to meet the seller's target price of $520,000.

NEGOTIATION EFFECTIVENESS

What is effective negotiation in real estate? Effective negotiation in real estate is the kind of negotiation that is *integrative, not distributive*. Distributive negotiation is based on the assumption that there is a fixed pie and one needs to do one's best to obtain a bigger slice of the pie (zero-sum situation). In contrast, integrative negotiation is based on the assumption that the pie can be made bigger and that each party can walk away with a slice of the pie that is fair and satisfactory (non-zero sum). That is, effective negotiation in real estate involves problem-solving and making tradeoff decisions. The goal is maximize the payoff for both buyers and sellers of real estate, not the payoff of one party at the expense of the other. The goal is reaching a deal that is satisfactory for both parties in ways that are efficient. This means that both buyers and sellers of real estate (and their representatives) end up with *positive feelings* about:

- *the instrumental outcome* (i.e., both buyer and seller and their representatives feel good about the substance of the final deal, such as the price of the property and other amenities provided by either buyer or seller or both)
- *the self* (i.e., both buyer and seller and their representatives feel good about themselves in the sense that the negotiation process did not damage their sense of self-worth but possibly enhanced their ego)
- *the process* (i.e., both buyer and seller and their representatives feel good about the concessions they made, the steps they took, and the procedure they followed in striking a deal)
- *the relationship* (i.e., both buyer and seller and their representatives feel good about the social bond that was created between the two parties during the negotiation and as a result of the negotiation)[2]

To provide the student with a better understanding of the four dimensions of negotiation effectiveness, let us examine how negotiation scientists measure the concept. Negotiation scientists have conducted many negotiation studies over the last several decades by recruiting human subjects and placing them in a specific negotiation situation in which some aspect of the negotiation situation is manipulated. The scientists then observed how the subjects act or react to the manipulated stimuli in that situation by asking the subjects how they felt about different aspects of the negotiations (feelings about the instrumental outcome, feelings about the self, feelings about the process, and feelings about the relationships) before, during, and after the negotiation session was complete.[3] Closely examining the questions asked about these feelings can give the student a better understanding of the negotiation effectiveness concept. Let us assume the role of the buyer or seller representative, not the buyer or seller. See Table 6.1.

Table 6.1 Example of Measurement Items to Capture Negotiating Effectiveness

Questions Related to Feelings about the Instrumental Outcome

- How satisfied are you with the deal or outcome—i.e., the extent to which the terms of the agreement benefit your client?
  Very dissatisfied -5 -4 -3 -2 -1 0 +1 +2 +3 +4 +5 Very satisfied
  (higher positive scores signify high negotiation effectiveness, and higher negative scores signify low negotiation effectiveness)
- How satisfied are you with the balance between the outcome for your client and your counterpart's outcome?
  Very dissatisfied -5 -4 -3 -2 -1 0 +1 +2 +3 +4 +5 Very satisfied
  (higher positive scores signify high negotiation effectiveness, and higher negative scores signify low negotiation effectiveness)
- Did you feel like your client forfeited or "lost" in this negotiation?
  Very much so -5 -4 -3 -2 -1 0 +1 +2 +3 +4 +5 Not at all
  (higher positive scores signify high negotiation effectiveness, and higher negative scores signify low negotiation effectiveness)
- Do you think the terms of the agreement are consistent with an objective standard (e.g., Comparative Market Analysis)?
  No not at all -5 -4 -3 -2 -1 0 +1 +2 +3 +4 +5 Very much so
  (higher positive scores signify high negotiation effectiveness, and higher negative scores signify low negotiation effectiveness)

Questions Related to Feelings about the Self

- Did you "lose face" (i.e., damage your sense of pride) in the negotiation?
  Very much so -5 -4 -3 -2 -1 0 +1 +2 +3 +4 +5 Not at all
  (higher positive scores signify high negotiation effectiveness, and higher negative scores signify low negotiation effectiveness)
- Did this negotiation make you feel less competent as a negotiator?
  Very much so -5 -4 -3 -2 -1 0 +1 +2 +3 +4 +5 Not at all
  (higher positive scores signify high negotiation effectiveness, and higher negative scores signify low negotiation effectiveness)
- Did you behave according to your own principles and values?
  Not at all -5 -4 -3 -2 -1 0 +1 +2 +3 +4 +5 Very much so
  (higher positive scores signify high negotiation effectiveness, and higher negative scores signify low negotiation effectiveness)
- Did this negotiation negatively impact your self-image or impression of yourself?
  Very much so -5 -4 -3 -2 -1 0 +1 +2 +3 +4 +5 Not at all
  (higher positive scores signify high negotiation effectiveness, and higher negative scores signify low negotiation effectiveness)

Questions Related to Feelings about the Process

- Do you feel your counterpart listened to your concerns?
  Not at all -5 -4 -3 -2 -1 0 +1 +2 +3 +4 +5 Very much so
  (higher positive scores signify high negotiation effectiveness, and higher negative scores signify low negotiation effectiveness)

(continued)

▨ *Table 6.1* *(continued)*

- Would you characterize the negotiation process as fair?
  Not at all -5 -4 -3 -2 -1 0 +1 +2 +3 +4 +5 Very much so
  (higher positive scores signify high negotiation effectiveness, and higher negative scores signify low negotiation effectiveness)

- How satisfied are you with the ease of reaching an agreement?
  Very dissatisfied -5 -4 -3 -2 -1 0 +1 +2 +3 +4 +5 Very satisfied
  (higher positive scores signify high negotiation effectiveness, and higher negative scores signify low negotiation effectiveness)

- Did your counterpart consider the wishes, opinions, or needs of your client?
  Not at all -5 -4 -3 -2 -1 0 +1 +2 +3 +4 +5 Very much so
  (higher positive scores signify high negotiation effectiveness, and higher negative scores signify low negotiation effectiveness)

Questions Related to Feelings about the Relationship

- What kind of "overall" impression did your counterpart make on you?
  A very negative impression -5 -4 -3 -2 -1 0 +1 +2 +3 +4 +5 A very positive impression
  (higher positive scores signify high negotiation effectiveness, and higher negative scores signify low negotiation effectiveness)

- How satisfied are you with the relationship with your counterpart as a result of this negotiation?
  Very dissatisfied -5 -4 -3 -2 -1 0 +1 +2 +3 +4 +5 Very satisfied
  (higher positive scores signify high negotiation effectiveness, and higher negative scores signify low negotiation effectiveness)

- Did the negotiation make you trust your counterpart?
  Not at all -5 -4 -3 -2 -1 0 +1 +2 +3 +4 +5 Very much so
  (higher positive scores signify high negotiation effectiveness, and higher negative scores signify low negotiation effectiveness)

- Did the negotiation build a good foundation for a future relationship with your counterpart?
  No, not at all -5 -4 -3 -2 -1 0 +1 +2 +3 +4 +5 Yes, very much so
  (higher positive scores signify high negotiation effectiveness, and higher negative scores signify low negotiation effectiveness)

In sum, the advice offered here based on our understanding of the concept of negotiation effectiveness is simple: Negotiators in real estate marketing transactions should strive to achieve negotiation effectiveness outcomes by ensuring that both parties feel positively about the instrumental outcome, the process, the self (i.e., their "egos"), and the relationship.

PREDICTORS OF NEGOTIATION EFFECTIVENESS

Social scientists traditionally classify predictors of negotiation effectiveness into four groups: (1) social influence, (2) structural context, (3) behavioral predispositions, and (4) cultural factors.[4] See Figure 6.1. Social influence predictors are factors related to the type of influence strategies the negotiation parties use in the course of negotiations. Examples of influence strategies include opening moves, countermoves, the pattern of concessions, the use of promises and threats, and the balanced use of social power. Structural context predictors are factors related to the physical environment of the negotiations (e.g., whether the negotiations take place in a neutral site or a site favorable to one party but not the other), the social environment (whether the negotiations conducted by the representatives are attended and possibly overseen by the client or clients themselves). Behavioral predisposition predictors

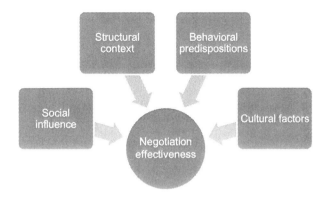

Figure 6.1
Predictors of
Negotiation
Effectiveness

are factors related to the demographic (e.g., sex, education, income), psychographic (e.g., a baby boomer, an avid cyclist, a technology buff), and personality (e.g., the extent to which the person picks up on social cues and responds to them appropriately). Cultural predictors are factors related to the type of culture in which the negotiating parties have been social-ized (e.g., two white Anglo-Saxon American real estate agents negotiating with each other versus a white Anglo-Saxon agent negotiating with an agent who emigrated from Korea five years ago). We will discuss these predictors of negotiation effectiveness in some detail in the sections that follow.

Social Influence Factors

We will discuss three different sets of social influence predictors of negotiation effectiveness. These are (1) the opening moves, (2) pattern of moves and countermoves, and (3) promises and threats. See Figure 6.2.

Opening Moves
The opening moves are very important in real estate negotiations because the moves signal cooperation versus competition. Opening moves that signal cooperation help con-tribute to a sense of trust, and trust plays a positive role in negotiation effectiveness[5] (see Figure 6.3).

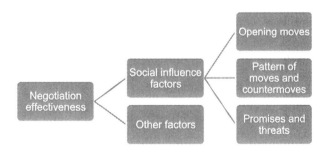

Figure 6.2
Social Influence
Factors and Their
Effect on Negotiation
Effectiveness

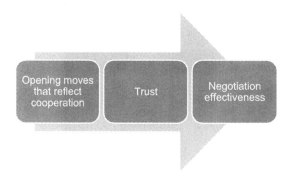

Figure 6.3
The Effect of
Opening Moves
on Negotiation
Effectiveness

Consider the following example involving two real estate agents, one representing the buyer and the other the seller. The seller is trying to sell his home for $500,000. That is, this is the target price for the seller—he would be happy if he sells the house at that price or not far below that threshold. The house is appraised for $500,000. The seller representative opens the negotiations saying that the asking price is $600,000. The buyer representative, knowing full well that the house has a market valuation of $500,000, understands that this opening move is unrealistically high. He feels that the seller representative wants to get the maximum payoff for the seller by not playing fair. That is, this opening move makes the buyer representative feel less trusting of the seller representative. However, consider a very different situation, in which the seller representative makes an opening move of $510,000. The buyer representative perceives that $510,000 is reasonable and there is room for negotiation. In other words, the $510,000 asking price signals cooperation rather than competition, and this sense of cooperation makes the buyer representative trust the seller representative to play fair and work together to arrive at a mutually satisfactory outcome. In sum, the advice based on the science of negotiation is to make an opening move most likely to signal cooperation.

However, there is some research that indicates that the more extreme the opening move, the more likely that both parties may reach a mutually satisfactory outcome because an extreme opening move allows room for making concessions. The science of negotiation has shown that the more extreme opening move leads to negotiation effectiveness when the other party does not have knowledge of the "true price," and when the extremity of the opening move leads to significant concessions by the party making the opening move.[6] Let us revisit the example of the two real estate agents negotiating the selling price of a residential house appraised for $500,000. Suppose the buyer representative has not done his homework to figure out the house's appraised value; he simply does not know. He may not think that an opening move of $600,000 is too extreme. This is then followed by the seller representative making concessions to appear to please the buyer representative. The first concession made by the seller representative is from $600,000 to $580,000, a $20,000 drop in price. The buyer representative appears reluctant to bite. The second concession is another drop by $20,000 (i.e., bringing down the asking price to $560,000). The third concession made is $10,000 (i.e., the asking price is now $550,000). The buyer representative is now feeling that the seller representative has made significant concessions, and that the $550,000 sounds like a good deal for his client. He bites. So, you see, extreme opening moves can work to lead to negotiation effectiveness under two conditions: (1) when the buyer representative has incomplete knowledge of the true price, and (2) when the seller representative makes a set of significant concessions from the initial asking price.

Further evidence also shows that extreme opening moves may backfire. The seller representative asks for $600,000 (an extreme offer). If the buyer representative knows that the appraisal was for $500,000, he thinks that the asking price is indeed extreme. The extremity of the initial offer prompts the buyer representative to "do the same" (i.e., present a counteroffer that is also "extreme"). This may be $400,000. Of course, in this case, the seller representative would reject the counter offer outright. Thus, in this situation, the extremity of the offer undermines negotiation effectiveness. Another factor that undermines the effect of an extreme offer on negotiation effectiveness is BATNA. Suppose the seller representative asks for $600,000. The buyer representative knows the appraised value of the house is $500,000. He also knows that the seller does not have offers or potential offers from other prospective buyers. In this situation, the buyer representative may find the $600,000 to be extreme and ludicrous. He rejects the offer outright and terminates the negotiation. Again, in this case when the buyer representative is not impressed with the seller's BATNA, the extremity of the opening move undermines negotiation effectiveness.[7]

In sum, the advice to real estate negotiators is to make an opening move that may be construed as high only if you expect a bunch of gradual concessions and if the other party does not have complete knowledge of the true market valuation of the property. Don't make a high opening move if you know that the other party is likely to counter your high offer with an extreme counteroffer or if you know that the other party has a strong BATNA.

Pattern of Moves and Countermoves

The overall pattern of moves and countermoves signals cooperativeness versus competitiveness, which in turn influences negotiation effectiveness.[8] Consider the following pattern of moves and countermoves. The seller representative makes an opening move of $550,000. The buyer agent makes a countermove of $450,000. The seller makes a concession by reducing his asking price from $550,000 to $530,000. The buyer agent reciprocates by jacking up his offer from $450,000 to $480,000. This move is interpreted by the seller rep as "cooperative" in the sense that he sees the buyer agent trying to cooperate by making a concession. Hence, the seller agent makes another concession: $520,000. The buyer agent responds by making another significant concession: $495,000. Note that this pattern of moves and countermoves signals cooperation—both buyer and seller agents are trying to cooperate to reach a mutually satisfactory agreement. In this case, it is very likely that a deal will be struck, and both buyer and seller reps will walk away happy with the final outcome. Now let's consider a pattern of moves and countermoves that signal competition. The seller agent makes an opening move of $550,000. The buyer agent makes a counteroffer of $450,000. The seller agent does not budge, holding at $550,000. The buyer agent attempts to make a concession: $470,000. The seller agent does not budge. In this case, the buyer agent will perceive lack of reciprocal concessions as a signal of competitiveness, which may lead the buyer agent to withdraw from further negotiations. In sum, the advice based on the science of negotiation suggests: Develop a pattern of concessions that signals cooperation and avoid patterns that signal competitiveness. See Figure 6.4.

There is also some evidence in the negotiation literature that the magnitude of the concessions can be used to signal one's reservation price (the price threshold that one is not likely to cross).[9] And doing so contributes to negotiation effectiveness. Consider the following example. The seller agent starts out with an asking price of $550,000. He then proceeds to make concessions in the following order: $525,000, $510,000, $507,000, and $505,000.

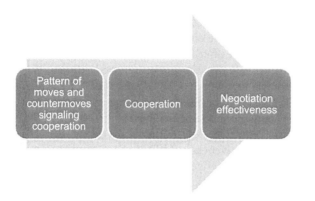

Figure 6.4
The Effect of the
Pattern of Moves
and Countermoves
on Negotiation
Effectiveness

Note that the magnitude of concessions here is declining incrementally—starting out with large concessions and moving toward smaller and smaller concessions. The buyer representative is likely to infer that the seller reservation price is around $500,000. In other words, the seller is not likely to accept an offer below $500,000. Knowing this can help the buyer representative make an offer that the seller is likely to accept. In sum, the advice to real estate negotiators is: Develop a pattern of concessions with incremental magnitudes that could signal to the other party your reservation price.

Promises and Threats

Promises and threats are commonly used in real estate negotiations. The typical threat is to terminate the negotiation and walk away without making a deal. The typical promise is to offer extra benefits to sweeten the deal (e.g., the seller rep may say that he will try to convince the seller to leave behind the Persian rugs that the buyer likes so much).

Promises and threats are used to break through obstacles in the negotiation.[10] Specifically, promises are used to help close the deal, whereas threats are used as a last resort to prevent the negotiation from falling apart. Promises and threats made by one party tend to generate compliance by the other party in the form of concessions. However, the extent to which a promise (or threat) is successful in extracting a concession is dependent on the perceived credibility of the promise (or threat). Consider the following example. A buyer agent feeling exasperated threatens to walk out (i.e., terminate the negotiation). However, the seller agent thinks that the buyer agent is using the threat as a ploy to get him (the seller agent) to make additional concessions. He thinks it is a ploy because he knows that the buyer wants the seller's property badly, and that the buyer has no BATNA. In other words, the seller agent does not believe that the threat is credible; therefore, he simply dismisses the threat and does not comply with the buyer's wish for the additional concession. In technical terms, the effectiveness of a threat (or promise) made by one party in generating compliance from the other party is dependent on the perceived credibility of the threat (or promise).

Are threats more effective than promises, or is it the opposite? It depends on what we mean by "effective." If "effective" means compliance, then the evidence points to the fact that threats by one party are more effective than promises in generating compliance from the other party. Compliance is essentially a short-term goal—trying to make sure that the negotiation does not stall. Compliance means motivating the other party to make a concession.[11] For example, if the seller representative insists on $510,000 with no further concessions, the buyer representative may make a promise (e.g., pay $505,000 and help the seller move out by the seller's desired target date to help him settle in his new job in another city) or may make a threat (e.g., threaten to walk out of the deal to take another offer). Which one would be more effective in motivating the seller (or the seller agent) to acquiesce and accept the last offer of $505,000 (instead of the desired price of $510,000)? The evidence from negotiation studies points to the fact that a threat is more effective than a promise in extracting a final concession. Negative emotions as a motivator are stronger than positive emotions. Threats tend to induce negative emotions such as fear. Positive emotions, such as joy and happiness, cannot compete with negative emotions because people, as animal organisms, are wired by a flight or fight response. In other words, human beings, like all animals, act immediately when presented with a threat from the environment—fight or run away. Hence, a threat is likely to induce immediate compliance, much more so than a promise.

However, the use of threats by one party may also cause the other party to dislike the other.[12] In other words, the seller rep may end up disliking the buyer agent for threatening to walk out and recommend his client take up an alternative deal. The bad feelings resulting from the threat would then get in the way of reaching a mutually favorable agreement. In other words, the "effectiveness" of threats versus promises depends on whether "effectiveness" means immediate compliance or reaching a mutually favorable agreement. Threats tend to be more effective than promises in inducing compliance but less effective in paving way for a mutually favorable agreement. Conversely, promises are less effective in inducing compliance but more effective in helping the two parties reach a mutually favorable agreement. In sum, the advice that real estate negotiators can use from the research findings is this: Use threats and promises as a last resort to resolve stalemates and get the negotiations moving again. If push comes to shove, threats are more effective than promises to get the other party to nudge. But use threats only as a final resort. In sum, the use of promises, although less effective than threats, should lead to better outcomes.

Structural Context

Negotiation scientists break down the structural context into three sets of characteristics: (1) physical, (2) social, and (3) issue-related.[13] In other words, the structural context refers to the environment or context surrounding the negotiation. This environment can be described in terms of its physical characteristics (e.g., location), social characteristics (e.g., the presence of an audience), and issue-related characteristics (e.g., tangible versus intangible issues). See Figure 6.5.

Figure 6.5
Factors Related to the Structural Context and Their Effect on Negotiation Effectiveness

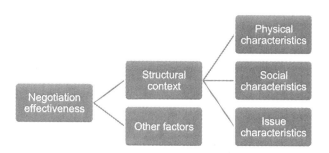

Physical Characteristics

Social scientists have studied many aspects of the physical characteristics of the structural context of negotiation. We will discuss several of these important characteristics. These are (1) site neutrality, (2) physical arrangements at the site, (3) communication channels, and (4) agenda effects.

Site neutrality refers to the extent to which the physical meeting place of the negotiation session(s) is on neutral grounds (as opposed to the home turf of either party).[14] If the negotiation takes place at the seller home, for example, the seller and possibly the seller representative are likely to have an advantage, and may behave exploitatively. However, meeting at a site that is neutral for both parties (e.g., a coffee shop located in a convenient place for both parties) may instill a feeling of trust and fair play in the negotiation process. This may lead to positive feelings about the process, which is an important dimension of negotiation effectiveness. In sum, the advice here is simple: Select a neutral site for negotiations. See Figure 6.6.

The *physical arrangements at the site* refer to the extent to which the immediate environment prompts feelings of trust and fairness.[15] Consider a buyer rep and a seller agent getting together to negotiate a home sale at a Starbucks located conveniently for both parties. The physical arrangements at Starbucks invite positive feelings of friendliness that can lead to feelings of trust and fairness. The comfortable armchairs, the coffee odor, the pastries and assortment of coffees and teas, the soft music, the soft wall colors and other wall decorations—these are environmental cues that may induce feelings of friendliness, cooperation, and trust. In sum, the advice here is to ensure that the physical arrangements at the negotiation site prompt feelings of trust and friendliness. See Figure 6.7.

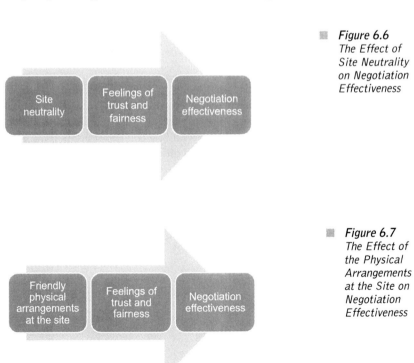

Figure 6.6
The Effect of Site Neutrality on Negotiation Effectiveness

Figure 6.7
The Effect of the Physical Arrangements at the Site on Negotiation Effectiveness

Table 6.2 Channels of Communications

Channel type	Frequency	Duration	Interactivity
Face-to-face	Moderate	High	Very high
Video conferencing	Moderate-to-high	Moderate	High
Texting	High	Low	Moderate-to-high
E-mail	High	Moderate	Moderate
Instant messaging	High	Moderate	High
Postal mail	Low	Low	Low

The availability of *communication channels* that allow the negotiating parties to communicate frequently, at length, and with a great deal of interactivity serves to enhance negotiation effectiveness. Producing positive feelings regarding the deal, the self, the process, and the relationship requires a great deal of communication. Communication channels that allow the negotiating parties to communicate frequently, at length, and interactively should bolster these positive feelings.[16] Let us consider the various channels of communications that buyers and sellers (and their representatives) use to communicate with each other regarding a possible real estate transaction. These may include face-to-face interaction, video conferencing, telephone, texting, e-mail, instant messaging, and postal mail. See Table 6.2.

Table 6.2 breaks down the various channels of communications and rates them in terms of the three dimensions of effective communication: frequency, duration, and interactivity. Based on the ratings shown in the table, one may conclude that face-to-face communication is the best channel of communication. It allows the negotiating parties to interact with moderate frequency and high duration. The worst communication channel is postal mail (i.e., writing letters and sending those letters through postal mail). In sum, the advice to real estate negotiators is to use communication devices that enhance the likelihood of interactive, frequent, and durable communications between the negotiating parties.

In every negotiation session, the negotiating parties prepare an *agenda* of issues for discussion. The agenda issues can be discussed separately and sequentially or holistically. Which is more effective in terms of negotiation effectiveness? The science of negotiation tells us that processing agenda issues holistically is more effective than processing the issues separately and sequentially.[17] This may be due to the fact that making tradeoff decisions is an important element in negotiation effectiveness, and processing agenda issues separately and sequentially does not allow for making good tradeoff decisions. In sum, the advice here is to discuss agenda items holistically, not item by item. See Figure 6.8.

Social Characteristics

There are two important social characteristics of the structural context that we will discuss here. These are (1) the presence of an audience during negotiation, and (2) direct versus representative negotiation.

The presence of an audience in real estate negotiations may take form in the presence of the actual client(s), or the client's family members, when both seller and buyer agents are negotiating. Does the presence of such an audience help or hurt negotiation effectiveness?

Figure 6.8
The Effect of the
Processing Agenda
Issues (Holistically
versus Separately/
Sequentially)
on Negotiation
Effectiveness

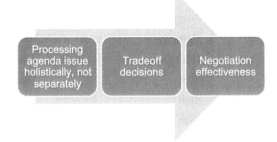

Studies in negotiations show that the presence of an audience may hamper negotiation effectiveness because it makes the negotiators (the agents) feel that every move is scrutinized by the client(s) and that they (the agents) will be held accountable for every decision.[18] Accountability and saving face become a major issue in this situation. Such a situation leads to hardening of positions taken in relation to desired and reservation prices. The agents can't change their minds, adapt to the demands of the situation, and make concessions. Why? This may be due to the fact that every concession is likely to be questioned by the client(s). Hence, the presence of an audience (client and or client's family members) serves to work against making concessions, and as such hampers negotiation effectiveness. Not having an audience allows the negotiating agents a free hand to make concessions and tradeoff decisions. In sum, the advice here is to ensure that the negotiations are conducted privately, without the clients at the scene. See Figure 6.9.

The next social characteristic that plays an important role in negotiation effectiveness is whether the negotiation is conducted by professional representatives (a listing agent representing the seller and a buyer representative) or the actual buyer and seller themselves. Much research has shown *representative negotiators* (real estate agents representing buyer and seller) are more effective in generating a mutually favorable agreement than direct negotiators (an actual buyer negotiating with an actual seller).[19] Why? This may be due to the fact that representative negotiators have expertise in real estate negotiation. They

Figure 6.9
The Effect of the
Presence/Absence
of an Audience
on Negotiation
Effectiveness

Figure 6.10
The Effect of the
Representative versus
Direct Negotiators
on Negotiation
Effectiveness

Representative negotiators (not actual buyers and sellers) → Expertise, detachment, and tactical flexibility → Negotiation effectiveness

know how to negotiate in ways to strike a good deal for all parties concerned. They are not emotionally involved in this endeavor compared to the actual buyers and sellers. Such detachment allows them not to overreact to unacceptable moves and countermoves made by the other party. Remaining cool and objective allows the representative negotiators to engage in problem-solving and make tradeoff decisions, all conducive to reaching a mutually favorable agreement. However, representative negotiators are likely to be better negotiators than direct negotiators when they (representative negotiators) have the full support of their clients. In other words, if every decision and every concession is second-guessed and micromanaged by their clients, then representative negotiators are not likely to do any better than direct negotiators. In sum, the advice based on the science is to opt for representative negotiations (not direct negotiations) with the provision that the representative have a free hand to negotiate fully without being second-guessed by their clients. See Figure 6.10.

Issue Characteristics

Issue characteristics refer to the extent an issue involved in negotiation is considered tangible or intangible. A tangible issue is directly related to the substance of the negotiation (e.g., price, timeliness, and amenities involved in tradeoff decisions). Intangible issues, in contrast, are issues related to honor, face, integrity, and status. In other words, intangible issues are ego issues. Because negotiators are human beings with egos, it is very difficult to focus strictly on the tangible issues. Intangible issues invariably infiltrate the process. Consider the following situation. Two representatives (one for the buyer and one for the seller) make an appointment to meet at a neighboring coffee shop to negotiate a deal for their respective clients. The buyer agent arrives at the coffee shop early and helps himself to a coffee and pastry. The seller agent arrives 45 minutes late—the buyer agent is almost ready to leave when he pulls up in the parking lot. The buyer agent is fuming. On top of that, the seller agent does not bother to explain why he was late. When they finally sit down to negotiate at the coffee shop, the seller agent answers five calls on his mobile phone. He doesn't bother to say, "Excuse me, but this is a call I have to take." The buyer agent is so upset by this encounter that he walks out on the fifth cell phone call. What we have here is a negotiation situation that is riddled with negative intangible issues—issues dealing with rudeness, lack of respect, bad manners, lack of etiquette, and ego deflation. Negative intangible issues influence negotiation effectiveness in negative ways. The buyer representative experiences negative feelings regarding the self (ego deflation), which spiral into more negative feelings about the relationship—two important elements of negotiation effectiveness. Furthermore,

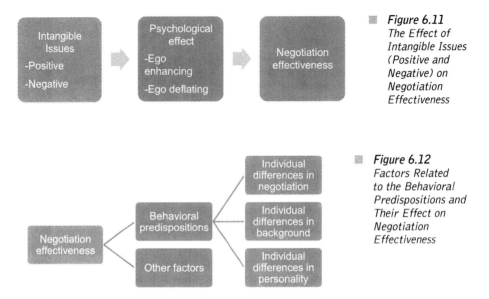

Figure 6.11
The Effect of Intangible Issues (Positive and Negative) on Negotiation Effectiveness

Figure 6.12
Factors Related to the Behavioral Predispositions and Their Effect on Negotiation Effectiveness

these bad feelings are likely to negatively influence the process by which they negotiate and the instrumental outcome, too.[20]

The converse is true about positive intangible issues. These influence negotiation effectiveness in positive ways. Consider this scenario. Both buyer and seller agents arrive on time and exchange pleasantries. They discover that they both are graduates of the same university, both having majored in business with a real estate minor. They feel a sense of camaraderie, a sense of connectedness. During the course of the negotiations, they compliment each other for "knowing their stuff." In this situation, both representatives are likely to experience positive feelings about the self and the relationship, which in turn may spill over into the process and the instrumental outcome in positive ways. In sum, the advice here is to minimize the negative intangible issues and allow for some positive ones. See Figure 6.11.

Behavioral Predispositions

Sales agents bring with them to the negotiating table a package of "behavioral predispositions" influencing negotiation effectiveness. These behavioral predispositions may be in the form of individual differences in negotiation style (e.g., cooperative versus competitive style of negotiation), individual differences in background (e.g., male versus female), and individual differences in personality (e.g., self-esteem and neuroticism). We'll address these in some detail. See Figure 6.12.

Individual Differences in Negotiation Style

The science of negotiation has revealed three behavioral predispositions in negotiation style that have a direct bearing on negotiation effectiveness: (1) interpersonal orientation, (2) motivational orientation, and (3) resistance to yielding.[21]

What is *interpersonal orientation?* Interpersonal orientation denotes the extent to which negotiators are sensitive to interpersonal aspects of their relationship with one another. A high interpersonal-orientation negotiator is one who has been sensitized, and is therefore likely to be especially reactive to variations in the other's behavior. A low interpersonal-orientation negotiator, on the other hand, is relatively insensitive to the interpersonal aspects of his relationship with the other, and is therefore likely to be less responsive to variations in the other's behavior. High interpersonal-orientation negotiators tend to make effective and proportionate concessions, contributing to negotiation effectiveness.[22] For example, consider a situation in which a buyer rep is negotiating with a home seller. The home seller has made a number of concessions, but the buyer agent is pressing for more concessions. The home seller is highly distressed and his distress is showing through facial expressions, bodily gestures, and high-pitched speech. If the buyer agent is high in interpersonal orientation, then he is very likely to pick up on these distress signals, which may indicate that the seller has reached his reservation price and demands for further concessions are likely to generate bad feelings that may sour the whole deal. If the agent is low on interpersonal orientation, he will press on because he won't pick up on the distress signals. In sum, the advice that can be deduced from the science is to select negotiation representatives who are high on interpersonal orientation. The alternative is to train real estate agents to become high on interpersonal orientation.

Let's now turn to *motivational orientation*. Motivational orientation refers most generally to one negotiator's attitudinal disposition toward another.[23] A negotiator has a *cooperative* negotiation style to the extent that he has a positive interest in the other's welfare as well as his own. A *competitive* style denotes an interest in doing better than the other, while at the same time doing as well for oneself as possible. A negotiator with an *individualistic* style is simply interested in maximizing his own outcomes, regardless of how the other fares. Much research has shown that a cooperative motivational orientation tends to lead to more effective negotiation than an individualistic, and especially than a competitive, orientation.[24] Why? This may be due to the possibility that a cooperative negotiation style imbues feelings of trust and fairness. See Figure 6.13. Let us revisit the last example in which a buyer agent is negotiating directly with a home seller and the home seller has expressed the fact that he cannot make additional concessions. If the buyer agent has a cooperative style, he will demand no more concessions. However, he may get creative by helping the home seller make tradeoff decisions. "Okay, I think I may convince my client to accept your last offer if you can leave the Persian rugs behind, because my client really likes those rugs, and, frankly, I don't think these rugs will fit in your new home." If the buyer agent has an individualistic style, he will press on for more concessions despite the fact that the home seller has expressed that he has reached his bottom line and he can make no further concessions. A buyer agent with a competitive style may try to extract further concessions from the home seller by trying to manipulate the situation, saying something like, "My client has been considering several other homes, and if you don't

Figure 6.13
The Effect of
Motivational
Orientation on
Negotiation
Effectiveness

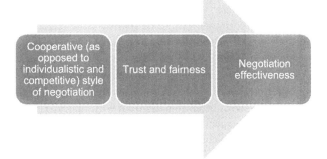

meet his terms, he will take this other offer by tomorrow," even if this is not true. In sum, the advice based on the science of negotiation is to select negotiation representatives likely to be cooperative and avoid those likely to be individualistic and competitive. Alternatively, train the real estate agents to be more cooperative and less individualistic and competitive.

Resistance to yielding refers most generally to one negotiator's resistance to give in to the demands of the other. Research has shown that resistance to yielding interacts with social motives in influencing negotiation effectiveness. Specifically, negotiators with a cooperative style engage in more problem-solving than negotiators with an egoistic motive (i.e., individualistic or competitive styles of negotiation), but only when they have high rather than low resistance to yielding.[25] This is because the combination of being cooperative and not too accommodating (high resistance to yielding) prompts a good deal of problem-solving, leading to tradeoff decisions. Ultimately, these tradeoff decisions are likely to generate a mutually satisfactory deal for both parties. Again, revisiting the example of the buyer agent pressing the home seller for additional concessions, we can predict that if the buyer agent has a cooperative style and is high in resistance to yielding, he will brainstorm with the seller a bunch of tradeoff alternatives that are likely to make both the home seller and his client equally happy. "My client may accept your final offer if you leave the Persian rugs behind. He really likes those Persian rugs. Or how about leaving the washer, dryer, and the other stuff in the utility room? He made several comments to me that he does not have a washer/dryer, and he certainly would appreciate those very much. Or how about the drapes? He likes those drapes in the living room too. They match the color of his furniture." In sum, the advice to real estate professionals is to select negotiators who are high in both cooperation and resistance to yielding. Alternatively, train real estate agents to not only be cooperative in negotiation, but also to not yield to the demands of the other party. Resistance to yielding should occur through a diligent amount of problem-solving.

Individual Differences in Background

Do individual differences in background (e.g., gender, age, education, income, marital status, ethnicity, and so on) matter in negotiation effectiveness? Yes and no. Much of the research on negotiations shows that there are two factors that influence negotiation effectiveness: age and gender.

With respect to age, the research shows that older negotiators tend to differ from their younger counterparts in terms of interpersonal and motivational orientation. Specifically, older negotiators tend to be cooperative and have high interpersonal sensitivity towards the party with whom they are negotiating, whereas younger negotiators tend to be competitive and lacking in terms of interpersonal sensitivity.[26] This finding has a direct bearing on negotiator selection. For example, for very important deals, a real estate broker (or sales manager) may choose to send older sales associates to the negotiation rather than younger ones. This may occur in commercial real estate deals when the stakes are very high.

When it comes to *gender*, the research literature shows a pattern more complex than age. In general, female negotiators tend to be slightly more effective in negotiations than male negotiators under certain conditions:

- *When the other party is also a female negotiator:* Female negotiators have higher interpersonal orientation than their male counterparts; as such, they tend to be more cooperative when negotiating with other females than with males. This may be due to the fact that female negotiators expect male negotiators to be competitive, which makes them react competitively. Conversely, they expect female negotiators to be cooperative, hence they act cooperatively.

- *When means of communications are available:* Because of their high interpersonal orientation, female negotiators tend to communicate interactively and with greater frequency with the other party. As such, female negotiators become better negotiators if and when they have communication means available to them. If so, they are likely to use these communication means to make concessions and extract concessions from the other party.
- *When the other party is attractive and cooperative:* Because female negotiators are high in interpersonal orientation, they become extra sensitive to tangible and intangible cues present during negotiations. Tangible cues may be in the form of a pattern of moves and concessions made by the other party that may signal cooperation versus competition. Thus female negotiators are likely to become effective negotiators if their negotiating counterpart acts cooperatively, not competitively. Cooperation leads to mutually satisfactory deals. Intangible cues may be in the form of physical appearance and attractiveness. Female negotiators are more sensitive to cues of physical attractiveness than males. As such, female negotiators tend to be more cooperative when negotiating with a physically attractive counterpart than with a non-attractive one.[27]

The implications of the gender findings can be helpful in ensuring successful outcomes in real estate negotiations. If the sales manager or broker expects that the other party will be represented by a female negotiator, then the manager may be better off selecting a female agent to negotiate on behalf of the organization. The sales manager or broker should also make sure that means of communications allowing the female negotiators to communicate interactively and frequently are at their disposal. Also, the sales manager may educate their agents about how attractiveness may affect their assumptions about their negotiating counterpart. Last but least, the sales manager or broker may encourage the female negotiator to be cooperative, make concessions, and expect the other party to reciprocate.

Individual Differences in Personality

There is a great deal of research in the science of negotiation dealing with personality factors. We will discuss factors related to (1) cognitive/affective processes, (2) special and abnormal populations, and (3) social motives.

With respect to *cognitive/affective processes*, we can classify negotiators as emotional versus rational.[28] The *emotional negotiator* can be described as follows:

- He prefers to talk through issues.
- He relies on his highly developed verbal persuasive skills to control the other party.
- He likes to "feel" how negotiations are progressing—relying heavily on his feelings, not liking lots of details, charts, written material.
- He gets bored quickly with too much detail.
- He feels that too much detail pins him down.
- He skips from topic to topic.
- He depends on emotion or what might refer to as "gut feeling" over logic.
- He finds it difficult to read through a complex contract and loses concentration.

In contrast, the *rational negotiator* can be described as follows:

- He prefers to listen more than talk, as long as talk is direct and not wasting his time.
- He relies on logic and proof.

- He has no need to feel anything about the process or people.
- He expects detail and written material.
- He will take time away from negotiating to read documents and prepare responses.
- He desires to get to the "bottom line" as quickly as possible.
- He values details and back-up materials with no value assigned to emotional reactions and outbursts.
- He controls the negotiating process and the end result, if he can, based on hard facts.
- He is always prepared to defend his "right" position in negotiating, based on hard fact and logical conclusion.
- He will not accept results that do not meet his specifications.

The research shows that effective negotiation cannot be predicted by knowing whether the negotiators are emotional or rational but instead can be predicted by how they deal with each other as a direct function of their emotional and rational reactions during the negotiations.[29] See Figure 6.14. Specifically, effective negotiations can be enhanced when the negotiating party can effectively deal with the rational and emotional reactions of the other party. Specifically, if a sales agent expects his counterpart to be emotional, he may be able to effectively deal with the emotional reactions of his counterpart by:

- letting him talk
- taking notes
- confirming his impressions often when he interprets the process appropriately
- asking him to summarize periodically
- asking him if he agrees after short pieces of the process have been presented, making sure that he feels good about the negotiations
- allowing him to talk until he feels secure about making a decision

If an agent negotiator expects his counterpart to be rational, he may be able to effectively deal with the rational reactions of his counterpart by:

- preparing to defend his position with good logic and hard facts
- providing clear answers and asking good questions (because rationals do not respect or trust a profusion of verbal communication)

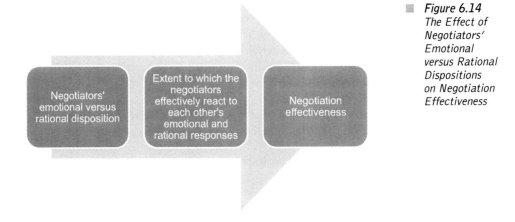

Figure 6.14
The Effect of Negotiators' Emotional versus Rational Dispositions on Negotiation Effectiveness

- making sure that the agenda is established clearly up front (unexpected changes not disclosed early in the negotiations erode trust of rationals and may derail the process)
- trying to understand his conception of the "right" end result
- not letting difference of opinion degenerate into arguments
- scheduling work sessions prior to the actual negotiating process, with the purpose of clearly defining issues, possible problems, and potential solutions
- minimizing and resolving problems during preliminary meetings so that the rational thinks and feels that all parties are in agreement as to what the "right" results of the negotiating should be

The second set of factors dealing with personality differences involves *special and abnormal populations*. Here, we would like to focus on abnormal psychology (i.e., negotiators who have mental illness). There are many instances in which sales agents have to negotiate directly with buyers and sellers of real estate who have mental health issues. Real estate professionals may also have mental health issues. Examples of typical mental illnesses include neuroses, depression, acute anxiety, Asperger syndrome, borderline personality, mania, and attention deficit hyperactivity disorder, but there are many others. Negotiators who are "mentally healthy" (relative to those who are less healthy) tend to behave like people with high interpersonal orientation; their less healthy counterparts behave more like those with low interpersonal orientation. This makes negotiators with issues of mental illness less effective as negotiators because they are less sensitive to social cues during negotiations—they fail to pick up on cues that signal the other party's reservation price, the other party's cooperativeness, the other party's BATNA, etc.[30] So how do real estate negotiators proceed if they know that their counterpart has mental health issues? The advice here is the same advice we presented in relation to dealing with people who are highly emotional (compared to people who are highly rational). To reiterate: let them talk, take notes, confirm them often when they interpret the process appropriately, ask them to summarize periodically, ask them if they agree after short pieces of the process have been presented, make sure that they feel good about the negotiations, and allow them to talk until they feel secure about making a decision.

The third set of personal factors we will discuss here involves *social motives*. Some negotiators may have strong motives such as *need for achievement* (a motive that drives people to excel in their jobs—to achieve excellence), *need for affiliation* (a motive that drives people to connect with others and be sociable and amiable), and *need for power* (a motive that drives people to control others and to demonstrate superior status). These social motives do have a direct bearing on negotiation effectiveness.[31] The research has demonstrated that negotiators motivated by a high need for achievement are likely to be most effective in negotiation, followed by those motivated by a high need for affiliation and need for power, respectively. Real estate negotiators who are high on the need for achievement are likely to be best in negotiation in the sense that they can achieve their clients' goals while also satisfying the demands of the other party. Real estate agents who are high on need for affiliation may fall short of achieving mutually satisfactory outcomes because they are likely to be too accommodating to the other party, thereby failing to realize a good payoff for their own clients. In contrast, agents with high need for power are likely to be too competitive, to the point where they may maximize the payoff of their own clients at the expense of satisfaction from the other party—thereby jeopardizing future relationships and possibly causing the other party to retaliate. In sum, the advice we can offer based on this research is to select agents for important negotiations who are high on need for achievement, moderate on need for affiliation, and low on need for power. Alternatively, train real estate agents to become more achievement-oriented (i.e., make them more goal-focused and teach them about ways to plan and implement strategies to achieve the stated goals).

Cultural Context

An important element of what real estate agents bring to the negotiating table is not only their negotiating style, their demographic background, and their personality characteristics, but also their culture. Hence the cultural context matters a great deal in negotiation effectiveness. We will discuss three major factors related to the cultural context: (1) high versus low context, (2) hierarchy versus egalitarianism, and (3) cultural adaptation. We'll address these in some detail. See Figure 6.15.

Low versus High Context Cultures

There is quite a bit of research on low versus high cultural context. A low context culture is one in which people communicate with one another through text or written materials. For example, in the world of real estate in the U.S., there are many steps in a real estate transaction, and every step involves many documents that are exchanged by the negotiating parties. Not only are they exchanged, but they are also signed every step of the way. When a client closes a deal, the attorney relegated to finalize the deal legally prepares a package of documents that both clients have to sign before the deal is sealed. For those who have experience with a real estate transaction, the documentation is indeed overwhelming. The world of real estate marketing is a world governed by many rules and regulations, and therefore is highly influenced by litigation and litigators. Litigators tend to anticipate the worst possible scenario of every situation. As a result, to reduce liability they try to ensure that every aspect of the real estate transaction is documented and agreed upon by all parties involved. This is the reality in the U.S., as well as in Canada, the U.K., Australia, New Zealand, and other Anglo-Saxon countries (Netherlands, Denmark, Norway, Sweden, etc.). In one way, we can say that the Anglo-Saxon culture is "low context" because of the heavy reliance on written communication in business transactions. This does not mean that written communications is the only mode of communication in most business transactions. Of course, oral communication is involved too, but written communication is considered paramount. In contrast, the vast majority of non-Anglo-Saxon cultures tend to be "high context." This means that communications among various parties engaged in business transactions tend to involve both oral and written communications, and that oral communications can substitute for written communications. In many high context cultures and countries, business transactions are planned and executed merely by oral communications sealed with a hand shake. Our interest here is what happens when we have two real estate negotiating parties that differ in cultural context. Of course, this can occur within the U.S., Canada, the U.K., Australia, and New

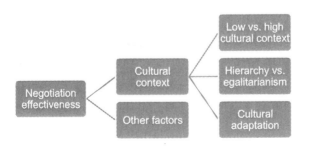

Figure 6.15
Factors Related to the Cultural Context and Their Effect on Negotiation Effectiveness

Zealand because of the high rate of immigration to these countries. In other words, there is a significant segment of the population in these countries who are not only buyers and sellers of real estate but also real estate sales professionals. Consider the situation in which a Korean immigrant in the U.S., acting in the role of a buyer representative negotiating with another agent representing the seller (a white Anglo-Saxon American, born and raised in the heartland). Negotiation scientists describe this condition as "intercultural" in the sense that the two negotiating parties come from different cultures. In contrast, "intracultural" means that the two parties have the same culture. The research has shown the following:

- Negotiators from low context cultures negotiating with others from low context cultures are likely to exchange much information *directly*, contributing to negotiation effectiveness (intracultural).
- Negotiators from high context cultures negotiating with others from high context cultures are likely to exchange much information *indirectly*, contributing to negotiation effectiveness (intracultural).
- Negotiators from different context cultures (intercultural) are likely to experience miscommunication (especially for low context culture negotiators), detracting from negotiation effectiveness.[32]

In sum, the managerial recommendation here is to ensure that communication is not hampered by cultural barriers. This can be done by selecting negotiators with similar cultural context.

Hierarchical versus Egalitarian Cultures

Scientists classify cultures along a dimension varying from highly egalitarian to highly hierarchical. Highly egalitarian societies have a belief and value system that touts equality and social justice. The vast majority of the industrialized world has laws that make it unlawful to discriminate against the poor, women, unmarried women, the aged, the disabled, homosexuals, and other vulnerable and disenfranchised groups. etc. This is of course the case in countries such as the U.S., the U.K., Canada, Australia, and New Zealand. However, we are not interested in laws. We want to focus on the intercultural negotiations among real estate buyers and sellers and their representatives. Suppose we have an immigrant male buyer (country of origin is Pakistan) negotiating directly with an American white woman representing the seller. Scientists may characterize the Pakistani buyer as coming from a hierarchical culture that regards men with higher status than women. What would we predict would happen in this situation? Research has demonstrated that negotiators from different cultures are likely to use mixed tactics (those from egalitarian cultures using cooperative tactics while those from hierarchical cultures using power tactics), detracting from negotiation effectiveness (intercultural). In contrast, negotiators from hierarchical cultures negotiating with others from hierarchical cultures are both likely to use power or influence tactics, contributing to negotiation effectiveness (intracultural). Similarly, negotiators from egalitarian cultures negotiating with others from egalitarian cultures are both likely to use cooperative tactics, again contributing to negotiation effectiveness (intracultural).[33] It is somewhat nonintuitive to think that two negotiating parties from a high hierarchical culture can negotiate effectively. The research shows that both parties are likely to use power and influence tactics. In other words, both of them are likely to use a distributive approach to negotiation, rather than an integrative approach. As long as their use of power and influence tactics is balanced, they may be able to strike a deal that is mutually satisfactory. The problem may arise when there is imbalance in the use of power and influence tactics. In this situation, the party that is more "powerful" may exploit the other, leading to an outcome that is favorable to the more powerful party and less favorable to the less powerful party. The ideal situation is when we have two negotiators with an egalitarian culture. Both are likely to use cooperative tactics, paving way for negotiation effectiveness.

In sum, the advice based on the science is to enhance negotiation effectiveness by selecting agents to negotiate whose cultures are similar—if you know that your organization has to negotiate with a party whose representative is from a hierarchical cultural, make sure that your representative also has a similar cultural background (i.e., has roots in a hierarchical culture). If you have to negotiate with a person from an egalitarian culture, make sure that your representative has cultural roots in an egalitarian culture too.

Cultural Adaptation

The science of negotiation points to the fact that negotiators who are familiar with the other party's culture are likely to adapt the other party's culturally normative behaviors, contributing to negotiation effectiveness. This is the principle of *cultural adaptation*. Consider the following example in commercial real estate. A Japanese real estate development wants to buy a huge parcel of land in a major U.S. city to develop a major shopping mall. The management of the Japanese firm selects a negotiator within the firm who is familiar with U.S. culture (i.e., he is already adapted to the U.S. culture, perhaps because he has received his graduate degree from a U.S. university and thus has resided in the U.S. for several years). In negotiating with the American land owner, the Japanese negotiator uses more direct information sharing (and less indirect information sharing and power/influence tactics). The Japanese negotiator is able to strike a good deal—the American landowner and upper management of the Japanese firm are all happy with the outcome. When asked why he negotiated the way he did (by sharing information directly with the American landowner and not by using power tactics), he responds by saying, "Because most U.S. business negotiations are conducted in that fashion." This statement reveals that the Japanese negotiator has been successful in this negotiation mostly because he culturally adapted to American style negotiations.[34] In sum, the advice is to select negotiators to represent your party who have a demonstrated record of being culturally adapted to the culture of the other party.

SUMMARY

We started out the chapter by revisiting some rudimentary concepts of real estate negotiation. These are target price, reservation price, ZOPA, tradeoffs, and BATNA. Then we focused on defining and operationalizing the concept of negotiation effectiveness. We defined it as a bargaining outcome in which both buyer and seller (and their agents) are happy with the instrumental outcome, the process, the self, and the relationship. Then we shifted our attention to identifying and describing predictors of negotiation effectiveness: We classified predictors of negotiation effectiveness in four groups: social influence, structural context, behavioral predispositions, and cultural factors.

With respect to the first set of predictors (social influence), we described five factors: the opening moves, pattern of moves and countermoves, promises and threats, and balanced use of social power. Opening moves signal cooperation versus competition. Based on the science of negotiation, real estate negotiators should make a high opening move only if the negotiator expects a series of gradual concessions and if the other party does not have complete knowledge of the true market valuation of the property. Negotiators should not make a high opening move if they know that the other party is likely to counter the opening move by an extreme counteroffer of their own and if the negotiators know that the other party has a strong BATNA. With respect to pattern of concessions, the advice that was provided is for negotiators to develop a pattern of concessions that signals cooperation and avoid patterns that signal competitiveness. Negotiators should develop a pattern of concessions with incremental magnitudes that can signal to the other party their reservation price. With respect to promises and threats, negotiators should use threats and promises as a last resort to resolve stalemates and get the negotiations moving again. But threats should be used only as a final resort. Although less effective than threats, promises should lead to better outcomes. With respect to balanced power, management should select representatives in ways to achieve balance in referent, expert, and legitimate power. Negotiators well balanced in social power are likely to negotiate more effectively than unbalanced dyads.

The structural context was identified in terms of three sets of characteristics: physical, social, and issue-related. With respect to physical characteristics, we discussed the effects of site neutrality, physical arrangements at the site, communication channels, and agenda effects. Real estate negotiators should select a neutral site for negotiations. Negotiators should also ensure that the physical arrangements at the negotiation site prompt feelings of trust and friendliness. Negotiators should use communication devices that enhance the likelihood of interactive, frequent, and durable communications between the negotiating parties. Also, negotiators should discuss agenda items holistically, not item by item. Turning to social characteristics, we discussed two important social characteristics of the structural context—namely, the presence of an audience during negotiation and direct versus representative negotiation. We concluded that negotiators should avoid having an audience during negotiations. Also, representative negotiations are more effective than direct ones, as long as the representative negotiators have the full authority and support from their clients to make decisions. Issue characteristics refer to the extent an issue involved in negotiation is considered tangible or intangible. We concluded that negotiators should minimize the negative intangible issues and allow for some positive ones.

The next of the predictors that we discussed involves behavioral predispositions of the negotiators. We described the effects of individual differences in negotiation style (interpersonal orientation, motivational orientation, and resistance to yielding), individual differences in background (age and gender), and individual differences in personality. Interpersonal orientation was defined as the extent to which negotiators are sensitive to interpersonal aspects of their relationship with one another. Effective negotiators tend to be high rather than low on interpersonal orientation. Therefore, management should recruit agents that are high on that trait. Alternatively, management can train real estate agents to become high on interpersonal orientation. Motivational orientation was defined as the negotiator's attitudinal disposition toward another—cooperative, individualistic, or competitive. Management should recruit agents likely to be cooperative and avoid those likely to be individualistic and competitive. Alternatively, management should train their agents to be more cooperative and less individualistic and competitive. Resistance to yielding refers most generally to one negotiator's resistance to give in to the demands of the counterpart in negotiation. Management should select negotiators who are high on both cooperation and resistance to yielding. Alternatively, management should train their sales associates to not only be cooperative in negotiation, but also to not yield easily to the demands of the other party. Resistance to yielding should occur through a diligent amount of problem-solving.

We then discussed individual differences in background (age and gender effects). We concluded that management should recruit older agents because older agents are likely to have high interpersonal orientation and a cooperative motivational orientation—both are necessary for negotiation effectiveness. In regards to gender effects, we concluded by arguing that if management expects that the other party will be represented by a female negotiator, then management should select a female negotiator. Management should also make sure that means of communications allowing female negotiators to communicate interactively and frequently are at their disposal. Management should also encourage female negotiators to be cooperative, make concessions, and expect the other party to reciprocate.

With respect to individual differences in personality, we discussed factors related to cognitive/affective processes, special and abnormal populations, and social motives. In relation to cognitive/affective processes, we classified negotiators as emotional versus rational. Effective negotiation cannot be predicted by knowing whether the negotiators are emotional or rational, but by how they deal with each other as a direct function of their emotional and rational reactions during the negotiations.

We shifted our attention to discussing personality differences involving special and abnormal populations by focusing on negotiators with mental health issues (neuroses, depression, acute anxiety, Asperger syndrome, borderline personality, mania, attention deficit hyperactivity disorder, etc.). It was recommended that negotiators can be trained to effectively handle those with significant mental health issues. We then addressed the effects of social motives (need for achievement, need for affiliation, and need for power) and concluded that management should select agents for important negotiations who are high on need for achievement, moderate on need for affiliation, and low on need for power. Alternatively, management should train their sales associates to become more achievement-oriented.

With respect to the cultural context, we discussed the effects of high versus low context, hierarchy versus egalitarianism, and cultural adaptation. A low context culture is one in which people communicate with one another through text or written materials. Conversely, people from a high context culture tend to use both oral and written communications in real estate transactions. We conclude intracultural negotiation (both representatives are from a low or high cultural context) is more effective than intercultural negotiation (one representative from a low context culture, the other from a high context culture). Therefore, management should train their representatives to gain the benefits of intracultural negotiation and avoid the pitfalls of intercultural negotiation. In regards to egalitarian versus hierarchical cultures, we also concluded that intracultural negotiations are more effective than intercultural ones. Finally, we addressed the effect of cultural adaptation on negotiation effectiveness. We concluded that management should select representative negotiators who have a demonstrated record of being culturally adapted to the culture of the other party.

DISCUSSION QUESTIONS

1. How do negotiation experts define the concept of "negotiation effectiveness"? Provide examples of survey items to measure this concept.

2. What advice do you recommend to real estate negotiators concerning opening moves to enhance negotiation effectiveness?

3. What advice do you recommend to real estate negotiators concerning pattern of concessions to enhance negotiation effectiveness?

4. What advice do you recommend to real estate negotiators concerning the use of promises and threats to enhance negotiation effectiveness?

5. What advice do you recommend to real estate negotiators concerning the use of balanced social power to enhance negotiation effectiveness?

6. What advice do you recommend to real estate negotiators concerning site neutrality to enhance negotiation effectiveness?

7. What advice do you recommend to real estate negotiators concerning the physical arrangements at the negotiation site to enhance negotiation effectiveness?

8. What advice do you recommend to real estate negotiators concerning the availability and use of communication devices to enhance negotiation effectiveness?

9. What advice do you recommend to real estate negotiators concerning the use of agenda to enhance negotiation effectiveness?

10. What advice do you recommend to real estate negotiators concerning the presence of the client(s) during negotiations?

11. Are direct or representative negotiators more effective? Why?

12. How about issue characteristics? What advice can you offer real estate negotiators?

13. What is interpersonal orientation? How does the research on interpersonal orientation help management in recruitment and training of sales agents?

14. What is motivational orientation? How does the research on motivational orientation help management in recruitment and training of sales agents?

15. What is resistance to yielding? How does the research on resistance to yielding help management in recruitment and training of sales agents?

16. Is age associated with negotiation effectiveness? Explain. What are the management implications?

17. Is gender associated with negotiation effectiveness? Explain. What are the management implications?

18. How should real estate negotiators deal with a highly emotional counterpart? A highly rational counterpart?

19. How should real estate negotiators deal with a counterpart who may have "mental issues"?

20. Do effective real estate negotiators have strong social motives? If so, which one? Explain. What are the management implications?

21. What does high versus low cultural context mean? What does the science of negotiations say about this concept in relation to negotiation effectiveness? Management implications?

22. How about egalitarian versus hierarchical culture? Explain. What are the management implications?

23. Explain cultural adaptation? What are the management implications?

NOTES

1. Bazerman, M. H., Curhan, J. R., Moore, D. A., & Valley, K. L. (2000). Negotiation. *Annual Review of Psychology*, 51, 279–314. Pruitt, D. G. (1981). *Negotiation behavior*. New York: Academic Press. Thompson, L. L. (2012). *The mind and heart of the negotiator* (3rd ed.). Upper Saddle River, NJ: Pearson Education.
2. Rubin, J. Z., & Brown, B. R. (1975). *The social psychology of bargaining and negotiation*. New York: Academic Press.
3. Curhan, J. R., Elfenbein, H. A., & Xu, H. (2006). What do people value when they negotiate? Mapping the domain of subjective value in negotiation. *Journal of Personality and Social Psychology*, 91(3), 493–512.
4. Rubin, J. Z., & Brown, B. R. (1975). *The social psychology of bargaining and negotiation*. New York: Academic Press.
5. Rubin, J. Z., & Brown, B. R. (1975). *The social psychology of bargaining and negotiation*. New York: Academic Press. Yukl, G. A. (1974). Effects of opponent's initial offer, concession magnitude, and concession frequency on bargaining behavior. *Journal of Personality and Social Psychology*, 30(3), 323–335.
6. Galinsky, A. D., & Mussweiler, T. (2001). First offers as anchors: The role of perspective-taking and negotiator focus. *Journal of Personality and Social Psychology*, 81(4), 657–669.
7. Galinsky, A. D., & Mussweiler, T. (2001). First offers as anchors: The role of perspective-taking and negotiator focus. *Journal of Personality and Social Psychology*, 81(4), 657–669.
8. Rubin, J. Z., & Brown, B. R. (1975). *The social psychology of bargaining and negotiation*. New York: Academic Press.
9. Rubin, J. Z., & Brown, B. R. (1975). *The social psychology of bargaining and negotiation*. New York: Academic Press.
10. Rubin, J. Z., & Brown, B. R. (1975). *The social psychology of bargaining and negotiation*. New York: Academic Press.
11. Rubin, J. Z., & Brown, B. R. (1975). *The social psychology of bargaining and negotiation*. New York: Academic Press. Van Kleef, G. A., & Cote, S. (2007). Expressing anger in conflict: When it helps and when it hurts. *Journal of Applied Psychology*, 92(6), 1557–1569. Van Kleef, G. A., De Dreu, C. K. W., & Manstead, A. S. R. (2004). The interpersonal effects of anger and happiness in negotiations. *Journal of Personality and Social Psychology*, 86(1), 57–76.
12. White, J. B., Tynan, R., Galinsky, A. D., & Thompson, L. (2004). Face threat sensitivity in negotiation: Roadblock to agreement and joint gain. *Organizational Behavior and Human Decision Processes*, 94(2), 102–124.
13. Rubin, J. Z., & Brown, B. R. (1975). *The social psychology of bargaining and negotiation*. New York: Academic Press.
14. Rubin, J. Z., & Brown, B. R. (1975). *The social psychology of bargaining and negotiation*. New York: Academic Press.
15. Rubin, J. Z., & Brown, B. R. (1975). *The social psychology of bargaining and negotiation*. New York: Academic Press.
16. Rubin, J. Z., & Brown, B. R. (1975). *The social psychology of bargaining and negotiation*. New York: Academic Press.
17. Cohen, G. L., Sherman, D. K., Bastradi, A., Hsu, L., & McGoey, M. (2007). Bridging the partisan divide: Self-affirmation reduces ideological closed-mindedness and inflexibility in negotiation. *Journal of Personality and Social Psychology*, 93(3), 415–430. Rubin, J. Z., & Brown, B. R. (1975). *The social psychology of bargaining and negotiation*. New York: Academic Press.
18. Kramer, R. M., Pommerenke, P., & Newton, E. (1993). The social context of negotiations: Effects of social identity and interpersonal accountability on negotiator decision making. *Journal of Conflict Resolution*, 37(4), 633–654. Rubin, J. Z., & Brown, B. R. (1975). *The social psychology of bargaining and negotiation*. New York: Academic Press.
19. Rubin, J. Z. (1980). Experimental research on third-party intervention in conflict: Toward some generalizations. *Psychological Bulletin*, 87(2), 379–391. Rubin, J. Z., & Brown, B. R. (1975). *The social psychology of bargaining and negotiation*. New York: Academic Press.
20. Rubin, J. Z., & Brown, B. R. (1975). *The social psychology of bargaining and negotiation*. New York: Academic Press. Cohen, G. L., Sherman, D. K., Bastradi, A., Hsu, L., & McGoey, M. (2007).

Bridging the partisan divide: Self-affirmation reduces ideological closed-mindedness and inflexibility in negotiation. *Journal of Personality and Social Psychology, 93*(3), 415–430.

21. Rubin, J. Z., & Brown, B. R. (1975). *The social psychology of bargaining and negotiation*. New York: Academic Press.

22. Amanatullah, E. T., Morris, M. W., & Curhan, J. R. (2008). Negotiators who give too much: Unmitigated communion, relational anxieties, and economic costs in distributive and integrative bargaining. *Journal of Personality and Social Psychology, 95*(3), 723–738. Barry, B. & Friedman, R. A. (1998). Bargainer characteristics in distributive and integrative negotiation. *Journal of Personality and Social Psychology, 74*(2), 345–359. Rubin, J. Z., & Brown, B. R. (1975). *The social psychology of bargaining and negotiation*. New York: Academic Press.

23. Weingart, L. R., Bennett, R. J., & Brett, J. M. (1993). The impact consideration of issues and motivational orientation on group negotiation process and outcome. *Journal of Applied Psychology, 78*(3), 504–517.

24. Amanatullah, E. T., Morris, M. W., & Curhan, J. R. (2008). Negotiators who give too much: Unmitigated communion, relational anxieties, and economic costs in distributive and integrative bargaining. *Journal of Personality and Social Psychology, 95*(3), 723–738. Barry, B., & Friedman, R. A. (1998). Bargainer characteristics in distributive and integrative negotiation. *Journal of Personality and Social Psychology, 74*(2), 345–359. Rubin, J. Z., & Brown, B. R. (1975). *The social psychology of bargaining and negotiation*. New York: Academic Press.

25. De Dreu, C. K. W., Weingart, L. R., & Kwoon, S. (2000). Influence of social motives on integrative negotiation: A meta-analytic review and test of two theories. *Journal of Personality and Social Psychology, 78*(5), 889–905.

26. Rubin, J. Z., & Brown, B. R. (1975). *The social psychology of bargaining and negotiation*. New York: Academic Press.

27. Bowles, H. R., Babcock, L., & McGinn, K. L. (2005). Constraints and triggers: Situational mechanics of gender in negotiation. *Journal of Personality and Social Psychology, 89*(6), 951–965. Kray, L. J., Thompson, L., & Galinsky, A. (2001). Battle of the sexes: Gender stereotypes confirmation and reactance in negotiations. *Journal of Personality and Social Psychology, 80*(6), 942–958. Miles, E. W., & Clenney, E. F. (2010). Gender differences in negotiation: A status characteristics theory view. *Negotiation and Conflict Management Research, 3*(2), 130–144. Rubin, J. Z., & Brown, B. R. (1975). *The social psychology of bargaining and negotiation*. New York: Academic Press. Small, D. A., Gelfand, M., Babcock, L., & Gettman, H. (2007). Who goes to the bargaining table? The influence of gender and framing on the initiation of negotiation. *Journal of Personality and Social Psychology, 93*(4), 600–613. Walters, A. E., Stuhlmacher, A. F., & Meyer, L. L. (1998). Gender and negotiator competitiveness: A meta-analysis. *Organizational Behavior and Human Decision Processes, 76,* 1–29.

28. Fulmer, I. S., & Barry, B. (2004). The smart negotiator: Cognitive ability and emotional intelligence in negotiation. *International Journal of Conflict Management, 15*(3), 245–272. Ritchie, J. C., Jr. (2001). *The 3 Ps of negotiating: Exploring the dimensions*. Upper Saddle River, NJ: Prentice Hall. Thompson, L. Nadler, J., & Kim, P. H. (1999). Some like it hot: The case for the emotional negotiator. In L. Thompson, J. Levine, & D. M. Messick (Eds.), *Shared cognition in organizations: The management of knowledge* (pp. 139–162). Mahwah, NJ: Erlbaum.

29. Fulmer, I. S., & Barry, B. (2004). The smart negotiator: Cognitive ability and emotional intelligence in negotiation. *International Journal of Conflict Management, 15*(3), 245–272. Ritchie, J. C., Jr. (2001). *The 3 Ps of negotiating: Exploring the dimensions*. Upper Saddle River, NJ: Prentice Hall. Thompson, L., Nadler, J., & Kim, P. H. (1999). Some like it hot: The case for the emotional negotiator. In L. Thompson, J. Levine, & D. M. Messick (Eds.), *Shared cognition in organizations: The management of knowledge* (pp. 139–162). Mahwah, NJ: Erlbaum.

30. Rubin, J. Z., & Brown, B. R. (1975). *The social psychology of bargaining and negotiation*. New York: Academic Press.

31. Bereby-Meyer, Y., Moran, S., & Sattler, L. (2010). The effects of achievement motivational goals and of debriefing on the transfer skills in integrative negotiations. *Negotiation and Conflict Management Research, 3*(1), 64–86. Langner, C. A., & Winter, D. G. (2001). The motivational basis of concessions and compromise: Archival and laboratory studies. *Journal of Personality and Social Psychology, 81*(4), 711–727.

32. Adair, W.L., Okumura, T., & Brett, J.M. (2001). Negotiation behavior when cultures collide: The United States and Japan. *Journal of Applied Psychology*, 86(3), 371–385. Adair, W.L., Weingart, L. & Brett, J.M. (2007). The timing and function of offers in the U.S. and Japanese negotiations. *Journal of Applied Psychology*, 92(4), 1056–1068.
33. Adair, W.L., Okumura, T., & Brett, J.M. (2001). Negotiation behavior when cultures collide: The United States and Japan. *Journal of Applied Psychology*, 86(3), 371–385.
34. Adair, W.L., Okumura, T., & Brett, J.M. (2001). Negotiation behavior when cultures collide: The United States and Japan. *Journal of Applied Psychology*, 86(3), 371–385.

Negotiation Strategies and Tactics

NEGOTIATION EFFECTIVENESS REVISITED

In the previous chapter on the social psychology of negotiation, we viewed negotiation effectiveness in real estate as *integrative, not distributive*. Distributive negotiation is based on the assumption that there is a fixed pie and one needs to do one's best to obtain a bigger slice of the pie (zero-sum situation). In contrast, integrative negotiation is based on the assumption that the pie can be made bigger and that both parties can walk away with their slice of the pie that is fair and satisfactory (non-zero sum). That is, effective negotiation in real estate involves problem-solving and making tradeoff decisions. The goal is maximize the payoff for both buyers and sellers of real estate, not the payoff of one party at the expense of the other, so that the deal is satisfactory for both parties in ways that are efficient.[1] This means that both buyers and sellers of real estate (and their representatives) end up with *positive feelings* about:

- *the instrumental outcome* (i.e., both buyer and seller and their representatives feel good about the substance of the final deal, such as the price of the property and other amenities provided by either buyer or seller or both)
- *the self* (i.e., both buyer and seller and their representatives feel good about themselves, in the sense that the negotiation process did not damage their sense of self-worth, but possibly enhanced their ego)
- *the process* (i.e., both buyer and seller and their representatives feel good about the concessions they made, the steps they took, and the procedure they followed in striking a deal)
- *the relationship* (i.e., both buyer and seller and their representatives feel good about the social bond that was created between the two parties during the negotiation and as a result of the negotiation)[2]

POSITIONAL NEGOTIATION

Hard positional negotiation is the most common method of negotiation.[3] Economists refer to it as "distributive negotiation," in the sense that the object of negotiation can be construed as a "pie" and the goal for each negotiator is to end up with a bigger slice of the pie. In other words, it is a competitive game in which one party wins at the expense of the other party (i.e., the other party loses). Of course, like all competitive games, there are "ties" too (i.e., there is no winner or loser).[4] Nevertheless, hard positional or distributive negotiation is viewed as a competitive game, and the goal of the game is to win.[5] For example, if you have a home seller who wants to sell his house for $520,000 (with a reservation price of $500,000) and a prospective buyer who wants to buy it for $480,000 (with a reservation price of $500,000), then the negotiators (e.g., real estate agents representing the home seller and buyer) will negotiate with the clients' positions in mind. The seller agent will try his best to bring down the price from $520,000 to $480,000, whereas the seller agent will try to get the buyer to climb up from $480,000 to $520,000. This is essentially a contest of wills, and the final outcome will reflect who won and who lost. If the final deal turns out to be $490,000, this means that the buyer has won and the seller has lost. If the final agreed-upon price is $510,000, then one can say that the seller won and the buyer lost. If the final price was $500,000, then one can view this as a "tie."

POSITIONAL NEGOTIATION IS CONSIDERED INEFFECTIVE

Positional (hard) negotiation is ineffective because arguing over positions produces unwise agreements, is inefficient, and endangers an ongoing and possibly a future relationship.[6] Consider the previous examples of home buyer and home seller battling each other—the home seller agent wants to maximize his clients' interest by striking a deal as close as possible to $520,000, whereas the home buyer agent does the same by ensuring that the price is as close as possible to $480,000. The result of the negotiation may only reflect who won and who lost, not the true interests of the clients. One client may be happy with the outcome, the other not. Or perhaps both may not be happy with the outcome, period. Thus, we can say hard positional negotiation produces *unwise agreements* because the agreement may not truly reflect the interest of the clients.

The agreement reached may be *inefficient*. The inefficiency reflects the process, not the outcome. A process that is inefficient is one that has high costs. Costs in this case can be viewed in terms of time, energy, and money. Suppose in reaching a final price of $510,000, the seller representative had to meet with the buyer representative numerous times. He had to ingratiate the buyer, his family, and his representative by hosting a special social get-together that was expensive. The countless meetings and the preparation for the social event was highly taxing. The seller agent was physically and emotionally exhausted by the time the deal was closed. The process was inefficient.

In addition, hard positional negotiations are likely to breed bad feelings. This is because this type of negotiation involves conflict and a battle of wills. When one party ends up winning, the winner may be somewhat happy, but the loser is likely to feel anguished by the defeat. The anguish is very likely to sour the current *relationship* between the two parties. The anguished party is likely to avoid doing repeat business with the other party in the foreseeable future.

Remember the discussion concerning negotiation effectiveness and its conceptualization in terms of four dimensions: positive feelings about the instrumental outcome, the process, the self, and the relationship. Hard positional negotiation violates these guidelines because it produces unwise agreement (bad feelings about the instrumental outcome) and an inefficient agreement (bad feelings about the process), while jeopardizing the relationship (bad feelings about the self and relationship). See Ethics Box 7.1.

ETHICS BOX 7.1. A QUESTIONABLE NEGOTIATION TACTIC[i]

Consider the following complaint. "I was in the middle of negotiating an offer on behalf of buyers I was representing. We had submitted a counteroffer in writing to the listing agent and we were waiting for a response. Suddenly, the listing agent contacted me and told me that there were multiple offers on the table. She said her seller instructed her to tell us that we needed to make our 'best and final offer' by a certain time that day. This doesn't seem fair to me. I feel the sellers should have completed negotiations with my buyers before considering another offer. Is the listing agent in violation of the Code of Ethics?"[ii]

Although this situation is not unethical per se, it is not consistent with the principles of integrative negotiation. Some may say that this is perfectly legitimate in the context of distributive negotiation (hard positional negotiation). This can be viewed as a ploy the listing agent has used to gain leverage by putting the buyer under time pressure and making the buyer believe that the seller has a strong BATNA, forcing the buyer to make an offer that meets or exceeds the seller's desired price.

However, Standard of Practice 1–6 of the Realtors' Code of Ethics speaks directly to the timing of the presentation of an offer: Real estate agents and brokers shall submit offers and counteroffers objectively and as quickly as possible. The question here is whether the listing agent communicated that offer to his client as quickly as possible. Did he expect other offers to come, and did he deliberately sit on the first offer until other offers arrived? If so, then the listing agent may be accused of unethical conduct.

[i] Aydt, B. (2009). No special treatment for first offer. *Realtor Magazine*, March 2009. Accessed from http://realtormag.realtor.org/law-and-ethics/ethics/article/2009/03/no-special-treatment-for-first-offer, on October 20, 2013.

[ii] Aydt, B. (2013). Speak up when offer arrives. *Realtor Magazine*, July 2013. Accessed from http://realtormag.realtor.org/law-and-ethics/ethics/article/2013/07/speak-up-when-offer-arrives, on October 20, 2013.

SOFT VERSUS HARD POSITIONAL NEGOTIATION

Distributive negotiation is essentially hard positional negotiation. There is a counterpart to hard positional negotiation, too: *soft positional negotiation*.[7] Soft positional negotiation is essentially negotiation with the goal of accommodating the other party in the best way possible. Both negotiators are "friends." The goal is agreement. Negotiations are conducted by making concessions to cultivate the relationship. Negotiators are soft on the people and on the problem. They are trusting of each other. They change their position easily to accommodate the other party. They may disclose the reservation price to the other party, and they tend to accept one-sided losses to reach an agreement. They search for the single answer: the one that the other party will accept. They insist on agreement. They try to avoid a contest of will, and they yield easily to pressure. Soft positional negotiation is common when both buyer and seller agents come from the same brokerage firm. It is very likely that the agents know each other and they may be good friends. Such a situation may not best represent the interests of the actual buyer and seller. In sum, we recommend that real estate salespeople avoid soft positional negotiation in most circumstances.

In contrast, hard positional is essentially highly competitive. The goal is to win, the other party should lose. Negotiators in this case are adversaries, not friends. The goal is victory, the other party should be defeated. In contrast to soft positional negotiation, in which people use concessions to cultivate the relationship, hard positional negotiation is essentially the opposite: negotiators demand concessions as a condition of the relationship. In soft positional negotiation, negotiators are soft on the people and the problem. In contrast, in hard positional negotiations negotiators are hard on the problem and the people. They distrust one another. They make threats. They may act to mislead the other party regarding their bottom line (or reservation price). They demand one-sided gains as the price of agreement. They search for the single answer: the one they will accept. They insist on their own position. They try to win a contest of will, and they apply much pressure on the other party to yield. Hard positional negotiation is commonly practiced in real estate, particularly among agents who are not familiar with the science of business negotiations and in situations involving competitor firms. When buyer and seller agents come together from brokerage firms that have a history of being highly competitive, it is likely that the real estate agents are prompted into this mode of negotiation—competitive, combative, and adversarial. Of course, such situations do not serve the best interest of the clients, and should be avoided.

WHAT IS PRINCIPLED (OR INTEGRATIVE) NEGOTIATION?

Principled (or integrative) negotiation is essentially negotiation based on the merits of the issues.[8] Principled negotiators are problem solvers, not friends (as in soft positional), nor adversaries (as in hard positional). The goal of principled negotiation is a wise outcome reached efficiently and amicably, not accommodating the other party (soft positional), nor beating them (hard positional). Principled negotiators try their best to separate the people from the problem, not being soft on the people and the problem (as in soft positional), nor hard on the people and the problem (as in hard positional). Principled negotiators proceed independent of trust, not being too trusting of the other party (as in soft positional), nor too distrusting (as in hard positional). Principled negotiators focus on interests, not positions. That is, they explore the interests underlying the positions, and address those interests, not the positions. Both soft positional and hard positional negotiators focus on their positions. Principled negotiators try to avoid setting in stone a bottom line. Soft positional negotiators tend to share their bottom line with each other, whereas hard positional negotiators try to mislead the other party about their own bottom line. Principled negotiators invent options for mutual gain, whereas soft positional negotiators search for the single answer that the other party will accept, and hard positional negotiators search for the single answer that is acceptable to them, irrespective of the other party. In doing so, principled negotiators develop multiple options first, and make a decision second. Principled negotiators make every attempt possible to use objective criteria to assess value. In other words, principled negotiators try to reach a result based on objective standards, independent of will. In contrast, soft positional negotiators try to avoid a contest of will, whereas hard positional negotiators try to win a contest of will. Finally, principled negotiators try to reason and are open to reason; they yield to principle, not pressure. Soft positional negotiators yield easily to pressure, whereas hard positional negotiators do their best to apply pressure to get their way.[9]

In sum, principled negotiation can best be captured in terms of four basic strategies: (1) separate the people from the problem; (2) focus on interest, not positions; (3) invent options for mutual gain; and (4) insist on using objective criteria.[10]

STRATEGIES AND TACTICS OF PRINCIPLED (OR INTEGRATIVE) NEGOTIATION

To reiterate, there are four basic strategies in principled negotiation. These are (1) separate the people from the problem; (2) focus on interests, not positions; (3) invent options for mutual gain; and (4) insist on using objective criteria. We will address these in some detail in the following sections. See Figure 7.1.

Separate the People from the Problem

Principled negotiators try their best to "separate the people from the problem."[11] In other words, they recognize that negotiators are people too. This means that people have feelings and egos. Every negotiation involves a certain degree of conflict: one party has a goal that is not easily reconcilable with the goal of the other party. Hence, the vast majority of situations reflect goal conflict. It is easy in those situations to have one's feelings hurt or one's ego bashed. This is due to the inherent nature of negotiable situations.

To separate the people from the problem, the negotiator has to realize that there are tangible and intangible issues in all negotiations. The tangible issues are directly related to the substance of the negotiations. Intangible issues are related to the negotiators' egos and the relationship between the two parties. Therefore, the challenge is separate the tangible issues (i.e., the problem) from the intangible ones (the people).

Misperceptions, emotional reactions and overreactions, and miscommunication are the hallmarks of intangible issues spoiling negotiations.[12] Consider the following example related to *misperceptions*. Two representative agents (one for the buyer and the other for the seller) are scheduled to meet at a coffee shop to negotiate the sale of a house. They

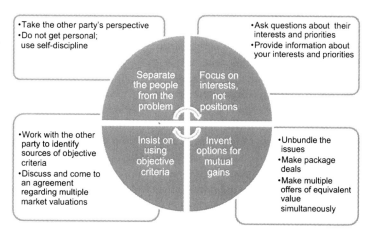

Figure 7.1
Strategies and Tactics of Principled (or Integrative) Negotiation

- Take the other party's perspective
- Do not get personal; use self-discipline

- Ask questions about their interests and priorities
- Provide information about your interests and priorities

Separate the people from the problem

Focus on interests, not positions

Insist on using objective criteria

Invent options for mutual gains

- Work with the other party to identify sources of objective criteria
- Discuss and come to an agreement regarding multiple market valuations

- Unbundle the issues
- Make package deals
- Make multiple offers of equivalent value simultaneously

agree to meet at 10:00 a.m. They don't know much about each other; this is their first meeting. The buyer representative arrives at the coffee shop a few minutes before 10:00 a.m. The seller representative arrives at 10:30 a.m. By the time the seller representative arrives for the meeting, the buyer representative is fuming. The seller representative apologizes for being late without offering an explanation. Because of this incident, the buyer representative perceives the seller representative to be disrespectful and unprofessional. The meeting is off to a bad start. Why has the meeting gotten off on a bad start? The buyer representative has used the tardiness incident (a situational characteristic) to infer that the seller representative is disrespectful and unprofessional (personal characteristics). The true reasons for the seller representative's tardiness are multiple. He had to make urgent calls that morning, and it was impossible to cut those calls short. When he finally got off the phone, it was 9:50 a.m., and then he got caught in traffic and had a hard time finding parking. Already late for the meeting, he didn't want to take the additional time to explain his tardiness to the buyer representative, so he apologized for being late and started discussing the deal. Note that his reasons for being late are due to situational factors (the phone calls, the congested traffic, the parking problem), not lack of professionalism (a personal characteristic). Psychologists call this type of misperception a *fundamental attribution error*.[13] People have a tendency to infer a personal characteristic to explain the behavior of another person. Conversely, they use situational characteristics to explain their own behavior.[14] If the seller representative had made an effort to explain his tardiness to the buyer representative, then the buyer representative might have corrected his misperception of him as being unprofessional. If the buyer representative had taken a few minutes to get a better feel for the circumstances leading to the tardiness, then his anger could have been tempered, and the meeting would not have gotten off to bad start.

One important tactic that principled negotiators use to decrease the likelihood of misperceptions, emotional overreactions, and miscommunications is *perspective taking*. Negotiators take the perspective of the other party by trying to see the circumstances surrounding the negotiation from that party's eyes.[15] Perspective taking is somewhat different from empathy; it is a cognitive ability to understand the world from the other party's point of view, while empathy is the ability to emotionally connect with the other party. Research has shown that negotiators who are good at perspective taking are more successful in identifying and reaching integrative solutions.

Another tactic commonly used to separate the people from the problem is to *not get personal; use self-discipline*. Consider the following study about marriage and divorce. The study involved tracking couples over a 14-year period. The study found divorce was predicted by not the amount of arguing, nor the amount of anger, but by the use of personal attacks.[16] Negotiation experts recommend self-discipline in controlling one's emotions; avoid reacting to the other party's emotional outbursts, inappropriate actions, or personal attacks. Don't take things personally. Be self-disciplined and professional, remain objective, and take time to cool off if needed. Don't get defensive and don't reciprocate by counterattacking.[17] Research from the social psychology of negotiations has revealed that representative negotiations (e.g., negotiations conducted by real estate agents representing their clients) tend to lead to higher levels of negotiation effectiveness than direct negotiations (e.g., negotiations conducted by actual clients—buyers and sellers of real estate). A primary reason accounting for this difference is the fact that representative negotiators are less likely to get personal, and they are more self-disciplined as professionals, compared to clients, who have a significant personal stake in the negotiation.[18] See additional tips related to separating people from the problem in Zooming-In 7.1. Also see Case/Anecdote 7.1.

ZOOMING-IN BOX 7.1. TIPS RELATED TO SEPARATING THE PEOPLE FROM THE PROBLEM[i]

Here are tips about how to separate the people from the problem, offered by experienced real estate professionals. The tips are:

Be calm, yet assertive. Try to stay calm; don't get flustered; getting upset does not accomplish anything positive; remember that anger begets anger and negotiations can spiral out of control.

Listen closely. Try to understand where the other party is coming from. Why did they express themselves the way they did? Repeat what they just said, and perhaps echo their sentiment by saying, "I hear you say so-and-so. Right?" Get a confirmation to make sure that you are interpreting them correctly.

Put yourself in their shoes. Try to express empathy. In other words, don't try to simply understand how they feel; try to experience these feelings by putting yourself in their shoes.

Stay positive. It is easy to slip into negativity when the other party is negative. Don't succumb to the negativity; stay positive. Don't trade insults. Frame your thoughts positively and communicate your ideas accordingly.

Visualize. Visualize the outcome of successful negotiation. What will be life like given the successful completion of this deal? In other words, try to feel what success means. Doing so helps you to commit to goal attainment, and to make every attempt possible to persevere and overcome any obstacle.

[i] Dittman Tracey, M. (2011). My best negotiating secrets. *Realtor Magazine*, September 2011. Accessed from http://realtormag.realtor.org/sales-and-marketing/feature/article/2011/09/my-best-negotiating-secrets, on October 22, 2013.

CASE/ANECDOTE 7.1. DEALING WITH EMOTIONAL OUTBURSTS IN REAL ESTATE NEGOTIATIONS[i]

Martha Erdem, with Prudential Howe & Doherty Real Estate in Andover, Massachusetts, tells this story. She is a listing broker for a client who reacted with an angry outburst when a counteroffer came in just $200 below the asking price. The client barked at her, "Absolutely not! The deal is off!" Instead of relaying this decision to the prospective buyer, Erdem decided to wait and allow a cooling off period. However, she contacted the buyer agent and told her that her client will decide in a day or two. The next day Erdem was able to get the seller to agree to the $200 lower offer. The sale proceeded to closing.

Erdem believes that third party negotiations are better than direct negotiations. This is why real estate agents play a vital role in this business. They negotiate with some degree of detachment and objectivity. Their egos don't get easily bruised.

Anne Meczywor, with Roberts & Associates Realty Inc. in Lenox, Massachusetts, admits, "Without us, there would likely be many more physical assaults on the books and far fewer closed transactions. The most important thing real estate professionals can do is help to take the emotion out of the deal. While we empathize with the party directly in front of us, we do not need to repeat every nasty thing said or every overdone point made by the other party's agent. We present facts. We see the big picture, the overall goal, and all the potential obstacles in the way. Most of all, we should be able to see the common ground, and must use that as the building blocks for a sound agreement."

[i] Dittmann Tracey, M. (2011). Finding the peace: Strategies for negotiating in today's marketplace. *Realtor Magazine*, September 2011, accessed from http://realtormag.realtor.org/sales-and-marketing/feature/article/2011/09/finding-peace-strategies-for-negotiating-in-todays-market, on September 15, 2013.

Focus on Interests, Not Positions

To reiterate, hard positional (or distributive) negotiation is all about positions—staking out one's position (desired and bottom line), figuring out the other party's position (desired and bottom line), and making an attempt to strike a deal to capture one's desired position (or a position that is better than one's bottom line). Principled (or integrative) negotiation involves interests—the interests underlying the positions.[19] If negotiators address interests rather than positions, they can strike a deal faster and more amicably.[20] Consider the following example. A home is up for sale for $500,000. The seller representative knows that his client has accepted a position in another city and that he and his family have already purchased another house with move-in date of two-and-a-half weeks. Therefore, it is imperative for the home seller to complete the transaction in less than three weeks. The same home seller also has three other residential units that he rents out, and he needs to find a person or property management company to manage these rental units. A prospective buyer and his family fall in love with the house; however, the buyer agent informs the seller agent that the prospective buyer cannot qualify for a mortgage exceeding $450,000. The prospective buyer also does not have the financial resources to make a significant down payment. If the agents focus on the positions of the home seller and home buyer, then they focus on $500,000 (home seller's position) and $450,000 (home buyer's position). Can they strike a deal? No, if they emphasize their positions. However, they *can* strike a deal if they focus on the *interests* underlying their positions. How? The home seller's interests are (1) timeliness (he doesn't have time to wait and entertain other offers from prospective buyers) and (2) property management (finding a person or a property management firm to manage his other rental units). The home buyer's interest is financial resources. Perhaps an arrangement can be made as follows:

- The price would be $450,000.
- The home seller would assume the loan (i.e., mortgage payments can be made to him directly in lieu of the mortgage company).
- The home buyer would help manage the three other rental units for five years.
- The transaction can be completed in two weeks.

Tactics involved in the implementation of this strategy (focus on interests, not positions) include *asking questions about interests and priorities* and providing information about the same.[21] Questions can be asked concerning underlying interests for the purpose of expanding the "size of the pie." It is interesting to note that research shows that only 7% of negotiators seek information about the other party's preferences during negotiation. Such information should clearly help in integrative negotiations.[22] Answers to these questions can help a negotiator discover what the other party values; such understanding should help negotiators make tradeoff decisions for mutual gains. Also, asking the other party about his BATNA or reservation price may motivate the other party to misrepresent his situation; however, asking him about his interests may be less likely to. Such questions signal to the other party that the negotiator means well, that he is interested in cooperation, by addressing the other party's needs and concerns. Research has also found that asking questions concerning the other party's interests prompts the other party to be more open, to trust, to make positive attributions about the negotiator's intentions, and to seek a relationship that involves future repeat business.[23] Furthermore, research has uncovered the notion that negotiators who have high epistemic motivation (a personal need for structure) are more likely to produce an outcome characterized by mutual gains, compared to negotiators with low epistemic motivation.[24]

The second tactic is *providing information about one's interests and priorities*. Research has shown that negotiators who are straightforward in the way they share information about their interests and priorities tend to show greater concern for the other party and make more concessions. This is advantageous in integrative negotiations but disadvantageous in distributive ones (in the sense that straightforward negotiators end up with less share of the payoff).[25] Because our focus here is integrative negotiation, not distributive, we recommend real estate negotiators be straightforward in revealing their interests and priorities to the other party. Doing so signals trust, paving the way for cooperation; this, in turn, paves the way to mutual gains. Also, volunteering information about interests and priorities serves to induce the other party to reciprocate (i.e., if the negotiator shares information, it is more likely that the other party will do the same). Thus, information exchange is very helpful in reaching mutually satisfactory agreements.[26] See Ethics Box 7.2 for an interesting discussion related to serving clients' interests and dual agency.

ETHICS BOX 7.2. CONFLICTING INTERESTS[i]

Mike Crowley, a successful real estate broker for 18 years in Spokane, Washington, has recently made a decision to serve real estate buyers only. That is, he is getting out of the business of allowing dual agency (allowing representation of both buyers and sellers of the same property). While dual agency is a widely accepted practice, it is replete with ethical problems. Some real estate professionals assert that dual agency poses no ethical dilemmas as long as it's fully disclosed in advance of any transaction. But how do you negotiate in this situation? How can you best understand and represent the interest of both buyers and sellers at the same time?

In addition to Mike Crowley, many other real estate brokers are specializing as buyer representatives: Mike Rygiol has been an exclusive buyer's agent for more than 20 years. Chris Whitehead, ABR, broker-owner of Homebuyer Experts in Akron, Ohio, has made the switch, too. "Traditional agents have to shift their mind-set back and forth depending on whom they're negotiating for. There's no conflict in my thought patterns of whether I'm trying to get the highest price in the shortest possible time on behalf of a seller or the lowest price on behalf of a buyer," he said.

Buyers get the reassurance that their agents are ultra-focused on their needs. Crowley says, "I take on no more than five or six clients at a time. They often feel like I am working for no one else but them."

> However, Curtis Hall, ABR, CRS, broker associate with Metro Realty Professionals in Phoenix, warned that even exclusive agents can run into potential conflicts. "If an exclusive buyer agent takes on two buyers who are looking for the same type of house in the same location, they'll find themselves in the same boat as someone working as a dual agent," he says.
>
> And while it is true that negotiating a lower price for a buyer client results in a smaller commission for the buyer agent, Whitehead believes that this is not much of a disincentive for most buyer representatives. After all, a more satisfied client is likely to make referrals that boost future business. "The integrity piece is what matters. That's our leverage," Whitehead says.
>
> [i] Cole, W. (2011). Make it all about the buyers. *Realtor Magazine*, June 2011. Accessed from http://realtormag.realtor.org/for-brokers/feature/article/2011/06/make-it-all-about-buyers, on October 20, 2013.

Invent Options for Mutual Gains

Part of the challenge of principled (integrative) negotiations is brainstorming with the other party to develop alternative solutions to the "problem."[27] In other words, principled negotiators approach the negotiation as a "problem" that can be solved jointly, by both parties. As such, developing alternative solutions and selecting the best solution for both parties is the recommended course of action.[28]

Before alternative solutions can be identified, the negotiators have to first and foremost diagnose the problem. This means that both negotiators should refrain from making premature judgments. They should diagnose the problem first and, based on the diagnosis, find a prescription—or several alternative prescriptions. In thinking about the problem, principled negotiators have to realize that the problem is essentially joint (i.e., "It is not my problem; it is not their problem; it is, however, our problem"). There are no losers and winners in this situation; instead, both parties will work jointly to solve the problem. The size of the pie is not fixed. If the parties work together, the size of the pie can be expanded through tradeoff decisions. Hence, the goal is to understand the interests of both parties and generate alternative solutions to solve the problem that can be acceptable to both parties.

Let us revisit the previous example involving the home seller who is pressed for time and the home buyer who does not have the financial resources to qualify for the mortgage. We identified one possible solution that is acceptable to both parties: the price would be $450,000; the home seller would assume the loan; the home buyer would help manage the three other rental units for five years; the transaction would be completed in two weeks. Are there other possible solutions? Let's brainstorm. Here is another:

- The price would be $500,000.
- The home seller would loan the buyer $50,000 from his equity line (the home buyer would make payments to the seller directly with interest).
- The transaction would be completed in two weeks.

Presenting these two alternative solutions to both buyer and seller may be more attractive than presenting them with only one solution. Both buyer and seller and their agents may deliberate between the two solutions to see which one is the most palatable. See Case/Anecdote 7.2.

CASE/ANECDOTE 7.2. TRADE-OFF DECISIONS IN REAL ESTATE NEGOTIATIONS[i]

Mehrad Nazari is president of Nazari Inc., a real estate investment brokerage and consulting firm in La Jolla, California. Nazari, has 30 years experience and also holds a Ph.D. in leadership and human behavior. Nazari says, "Don't get stuck in one way of reaching the goal. Creativity is essential." In other words, develop alternatives that can satisfy your clients' needs. Doing so can help reach the goal of selling the home.

Nazari recalls a situation where a seller refused to reduce the price. The client said that she can't afford to do so because she was going to have to vacate earlier than expected and find a short-term rental until she could move into her new home. The problem was the buyer was qualified for a certain loan amount and couldn't pay the asking price. Stalemate!

To solve this problem Nazari came up with the following solution: the seller would stay in the property for up to two months rent-free. Doing so would help her save nearly $5,000 in moving costs and allow her to meet the buyer's offer. This plan saved the deal.

[i] Dittmann Tracey, M. (2011). Finding the peace: Strategies for negotiating in today's marketplace. *Realtor Magazine*, September 2011, accessed from http://realtormag.realtor.org/sales-and-marketing/feature/article/2011/09/finding-peace-strategies-for-negotiating-in-todays-market, on September 15, 2013.

What are some tactics designed to facilitate the invent-options-for-mutual-gains strategy? Here are three tactics: (1) unbundle the issues, (2) make package deals, not single-issue offers, and (3) make multiple offers of equivalent value simultaneously. What do we mean by *unbundling the issues*? Research in business negotiations has long recognized that negotiators tend to fail when they focus exclusively on price. By focusing on price (single issue), negotiation becomes distributive, in the sense that the buyer will try to reduce the price as much as possible while the seller will do the opposite (i.e., try to take more of the fixed pie). Unbundling the issues means bringing new issues to the negotiating table with the goal of transforming a single-issue/fixed-pie negotiation into an integrative, multi-issue/expanding-pie negotiation.[29] The example we used above epitomizes the unbundling-the-issues tactic by bringing to the negotiating table other issues besides price. In this case, the issues involve price (an issue important to both buyer and seller), timeliness (an issue important to the seller), financial resources and qualifying for a mortgage loan (an issue important for the buyer), and the management of other rental units (an issue important for the seller).

Once the issues are unbundled, the challenge is to configure alternative combinations of solutions addressing the unbundled issues. This is where the second tactic comes in: *make package deals, not single-issue offers*.[30] Negotiating issue-by-issue is a hallmark of distributive negotiation. The hallmark of integrative negotiation is to bundle the issues in alternative "packages" that reflect tradeoff decisions—tradeoffs among the issues. Thus, effective negotiators compare and contrast the issues and trade them off in ways to reflect the underlying interests of both parties. These tradeoffs allow both parties to create bridges in ZOPA (Zone of Possible Agreement) and, of course, doing so may break an impasse, resulting in a final deal that is a mutually profitable outcome. Single-issue offers force negotiators into accepting a compromise that may not be acceptable to both parties. Making package deals by bundling the issues allows negotiators to do better than compromise. Again, the examples we used above reflect this tactic well.

The third tactic is to make *multiple offers of equivalent value simultaneously*.[31] This tactic involves a negotiator presenting the other party with at least two proposals that are equally valuable to the client. Consider the two preceding examples we used. The first proposal was:

- The price would be $450,000.
- The home seller would assume the loan (i.e., mortgage payments can be made to him directly, in lieu of the mortgage company).
- The home buyer would help manage the three other rental units for five years.
- The transaction would be completed in two weeks.

The other proposal was:

- The price would be $500,000.
- The home seller would loan the buyer $50,000 from his equity line (the home buyer would make payments to the seller directly with interest).
- The transaction would be completed in two weeks.

These two options, proposed by the seller's agent, are of equal value to the seller. Research has shown that negotiators who make multiple equivalent offers are likely to be rated highly on measures of negotiation effectiveness (both parties express positive feelings about the instrumental outcome, process, self, and relationship).[32]

Insist on Using Objective Criteria

One of the important elements of principled negotiation is fairness. If negotiators develop a solution to the problem that is "fair," then both parties are likely feel happy about the outcome as well as the process that led them to that outcome.[33] This is where the use of objective criteria comes in. What is a fair deal? A fair deal in real estate transaction is a deal that has fair market value. Fair market value can be assessed through objective criteria. Remember the pricing methods covered in Part 1 of this book. The pricing methods we discussed include comparative market analysis (commonly used in residential real estate marketing), the costing method (commonly used in commercial real estate and by builders of new homes), and the income method (commonly used in rental and property management). These pricing methods are objective standards considered to be fair by the vast majority of real estate professionals.

Let us revisit the example of the home seller with an asking price of $520,000. The buyer would like to acquire the home for $480,000. The bank appraisal of the house is $500,000. The tax appraisal is $490,000. Here we have two objective criteria that reflect the market valuation of the property: the bank appraisal and the tax appraisal. One can stipulate that a fair price is somewhere within the range of $490,000 and $500,000. If both negotiators use objective criteria as their guide, they can easily converge on a final price. It is also very likely that both parties will be happy about the final price, given that it is based on objective criteria.

There are two main tactics related to the use of objective criteria. The first tactic is to *work together with the other party to identify sources of objective criteria* and reach an agreement about the applicability of the objective criteria in current negotiations. It is important that the two negotiating parties discuss which objective criteria establish a fair market value and come to an agreement about their use in the situation in question.

The second tactic is to *discuss and come to an agreement regarding multiple market valuations* by applying the agreed-upon objective criteria. For example, if the case is residential real estate, then it is important to come to an agreement that there are two different sources of market valuations based on the comparative market analysis: one is a bank appraisal and the other is a tax assessment. The question is which one to use? Perhaps both! The discussion should be focused on the credibility of the sources.

See Ethics Box 7.3 for a discussion on the ethics of manipulation versus negotiation in real estate.

ETHICS BOX 7.3. MANIPULATING VERSUS NEGOTIATING IN REAL ESTATE[i]

Bob Corcoran, founder and president of Corcoran Consulting Inc. and a specialist in the implementation of sound business practices in residential and commercial real estate, says that "manipulation while negotiating can kill your business." Many real estate agents engage in manipulation, yet they don't think they are doing anything wrong. They don't realize that it is unethical and, in addition, it can destroy an agent's reputation as a professional with high integrity. If an agent becomes known in the community as not having integrity in negotiations (i.e., she doesn't play fair), this bad reputation will sooner or later kill her business. Agents who manipulate shoot themselves in the foot because it is very likely that those they manipulate know they are being manipulated, and those clients will not do business with them again—and will eventually ruin their reputation through word of mouth.

Corcoran advises real estate brokers and agents to give clients pleasant and memorable experiences by:

- *letting both sides win* (i.e., don't negotiate from an adversarial winner-take-all perspective; stay focused on your top priorities; keep your emotions in check; and try to achieve win-win outcomes)
- *respecting the other side's priorities* (i.e., learn about the other party's interests and priorities and use this information to reconcile the interests of the two parties)
- *building a bridge, not a dam* (i.e., don't make low-ball offers; instead provide an offer based on a recent comparative market analysis—a fair market valuation and a fair offer)
- *build a good impression* (i.e., people remember how you made them feel; you need to leave a positive and memorable impression; real estate business is local; and local business is very much influenced by word of mouth)

[i] Corcoran, B. (2012). Manipulating vs. negotiating: One can kill your business. *Real Town*. Accessed from http://articles.realtown.com/blog/2012/03/manipulating-vs-negotiating-one-can-kill-your-business/, on October 20, 2013.

SUMMARY

This chapter focused on negotiation strategies and tactics. We started out the chapter by reiterating the importance of integrative negotiation and making a case that integrative negotiation in real estate is much more effective and ethical than distributive negotiation. The goal in integrative negotiation is efficiently reaching a deal that is satisfactory for both parties. We discussed distributive negotiation and how it can be either hard positional or soft positional. We argued that positional negotiation (both hard and soft) is ineffective vis-à-vis integrative negotiation. We referred to integrative negotiation as principled, which is defined as based on the merits of the parties' interests. We discussed principled negotiation in terms of four basic strategies: separate the people from the problem; focus on interest, not positions; invent options for mutual gain; and insist on using objective criteria.

Principled negotiators try their best to "separate the people from the problem." We described two important tactics for this strategy. One involves taking the other party's perspective. The other is to avoid getting personal, and instead use self-discipline. We then described the second strategy of principled negotiation: focus on interests, not positions. In that vein, we discussed two tactics: asking questions about interests and priorities and providing information about the same. The third strategy of principled negotiations, invent options for mutual gains, was described as brainstorming with the other party to develop alternative solutions to the problem. The tactics involved in implementing this strategy include unbundling the issues; making package deals, not single-issue offers; and making multiple offers of equivalent value simultaneously. The fourth and final strategy is to insist on using objective criteria. A fair deal is a deal that involves fair market value and fair market value is best assessed through objective criteria. There are two main tactics involved in the use of objective criteria. The first tactic is to work together with the other party to identify sources of objective criteria and reach an agreement about the applicability of the objective criteria in current negotiations. The second tactic is to discuss and come to an agreement regarding multiple market valuations, applying the agreed-upon objective criterion.

DISCUSSION QUESTIONS

1. What is the concept of negotiation effectiveness again? Please revisit.

2. What is positional negotiation? Examples, please.

3. Why is positional negotiation is considered ineffective? Please explain.

4. What is the distinction between soft and hard positional (or distributive) negotiation? Please explain.

5. What is principled (integrative) negotiation? Please describe.

6. Compare and contrast positional and principled negotiations.

7. Describe the strategy of "separate the people from the problem" in principled negotiation.

8. What are the tactics related to the strategy of separating the people from the problem? Please describe.

9. Describe the strategy of "focus on interests, not positions" of principled negotiation.

10. What are the tactics related to the strategy of "focus on interests, not positions"? Please describe.

11. Describe the strategy of "invent options for mutual gains" of principled negotiation.

12. What are the tactics related to the strategy of "invent options for mutual gains"? Please describe.

13. Describe the strategy of "insist on using objective criteria" of principled negotiation.

14. What are the tactics related to the strategy of "insist on using objective criteria"? Please describe.

NOTES

1. Fry, R., & Ury, W. (1991). *Getting to YES: Negotiating agreements without giving in.* New York: Penguin Books (Chapter 1).
2. Curhan, J., Elfenbein, H., & Xu, H. (2006). What do people value when they negotiate? Mapping the domain of subjective value in negotiation. *Journal of Personality and Social Psychology,* 91(3), 493–512.
3. Thompson, L. L., & Hastie, R. (1990). Social perception in negotiation. *Organizational Behavior and Human Decision Processes,* 47(1), 98–123.
4. Thompson, L. L. (2012). *The mind and heart of the negotiator.* Upper Saddle River, NJ: Pearson Education (Chapter 3).
5. Thompson, L. L. (2012). *The mind and heart of the negotiator.* Upper Saddle River, NJ: Pearson Education (p. 8).
6. Fry, R., & Ury, W. (1991). *Getting to YES: Negotiating agreements without giving in.* New York: Penguin Books (Chapter 1).
7. Thompson, L. L. (2012). *The mind and heart of the negotiator.* Upper Saddle River, NJ: Pearson Education (p. 8).
8. Thompson, L. L. (2012). *The mind and heart of the negotiator.* Upper Saddle River, NJ: Pearson Education (Chapter 4).
9. Fry, R., & Ury, W. (1991). *Getting to YES: Negotiating agreements without giving in.* New York: Penguin Books (Chapter 1).
10. Fry, R., & Ury, W. (1991). *Getting to YES: Negotiating agreements without giving in.* New York: Penguin Books (Chapter 1).
11. Fry, R., & Ury, W. (1991). *Getting to YES: Negotiating agreements without giving in.* New York: Penguin Books (Chapter 2).
12. Kumar, R. (1997). The role of affect in negotiations: An integrative overview. *Journal of Applied Behavioral Science,* 33(1), 84–100.
13. Morris, M. W., Larrick, R. P., & Su, S. K. (1999). Misperceiving negotiation counterparts: When situationally determined bargaining behaviors are attributed to personality traits. *Journal of Personality and Social Psychology,* 77(1), 52–67.
14. Thompson, L. L., & Loewenstein, G. F. (1992). Egocentric interpretations of fairness and interpersonal conflict. *Organizational Behavior and Human Decision Processes,* 51(2), 176–196.
15. Galinsky, A., Ku, G., & Wang, C. (2005). Perspective-taking: Fostering social bonds and facilitating coordination. *Group Processes and Intergroup Relations,* 8, 107–125. Galinsky, A., Ku, G., & Wang, C. (2008). Perspective-takers behave more stereotypically. *Journal of Personality and Social*

Psychology, 95(2), 404–419. Galinsky, A. D., Maddux, W. W., Gilin, D., & White, J. B. (2008). Why it pays to get inside the head of your opponent: The differential effects of perspective-taking and empathy in strategic interactions. *Psychological Science*, 19(4), 378–384.

16. Gottman, J. M., & Levenson, R. W. (2000). The timing of divorce: Predicting when a couple will divorce over a 14-year period. *Journal of Marriage & the Family*, 62(3), 737–745.

17. Fry, R., & Ury, W. (1991). *Getting to YES: Negotiating agreements without giving in.* New York: Penguin Books (Chapter 3).

18. Rubin, J. Z. (1980). Experimental research on third-party intervention in conflict: Toward some generalizations. *Psychological Bulletin*, 87(2), 377–391. Rubin, J. Z., & Brown, B. R. (1975). *The social psychology of bargaining and negotiation.* New York: Academic Press.

19. Fry, R., & Ury, W. (1991). *Getting to YES: Negotiating agreements without giving in.* New York: Penguin Books (Chapter 4).

20. Fry, R., & Ury, W. (1991). *Getting to YES: Negotiating agreements without giving in.* New York: Penguin Books (Chapter 5).

21. Thompson, L. L. (2012). *The mind and heart of the negotiator.* Upper Saddle River, NJ: Pearson Education (pp. 77–79).

22. Thompson, L. L. (1991). Information exchange in negotiation. *Journal of Experimental Social Psychology*, 27(2), 161–179.

23. Chen, F. S., Minson, J. A., & Tormala, Z. L. (2010). Tell me more: The effects of expressed interest on receptiveness during dialogue. *Journal of Experimental Social Psychology*, 46(5), 850–853.

24. Ten Velden, F. S., Beersma, B., & De Dreu, C. K. W. (2010). It takes two to tango: The effects of dyads' epistemic motivation composition in negotiation. *Personality and Social Psychology*, 36(11), 1454–1466.

25. DeRue, D. S., Conlon, D. E., Moon, H., & Willaby, H. W. (2009). When is straightforwardness is a liability in negotiations? The role of integrative potential and structural power. *Journal of Applied Psychology*, 94(4), 1032–1047.

26. Bazerman, M. H., & Neale, M. A. (1992). *Negotiating rationally.* New York: Free Press. Thompson, L. L. (2012). *The mind and heart of the negotiator.* Upper Saddle River, NJ: Pearson Education (p. 79). Thompson, L. L. (1991). Information exchange in negotiation. *Journal of Experimental Social Psychology*, 27(2), 161–179.

27. Fry, R., & Ury, W. (1991). *Getting to YES: Negotiating agreements without giving in.* New York: Penguin Books (Chapter 6).

28. Thompson, L. L. (2012). *The mind and heart of the negotiator.* Upper Saddle River, NJ: Pearson Education (p. 20).

29. Lax, D. A., & Sebenius, J. K. (1986). *The manager as negotiator.* New York: Free Press. Thompson, L. L. (2012). *The mind and heart of the negotiator.* Upper Saddle River, NJ: Pearson Education (p. 81).

30. Thompson, L. L. (2012). *The mind and heart of the negotiator.* Upper Saddle River, NJ: Pearson Education (p. 81).

31. Thompson, L. L. (2012). *The mind and heart of the negotiator.* Upper Saddle River, NJ: Pearson Education (p. 82).

32. Hyder, E. B., Prietula, M. J., & Weingart, L. R. (2000). Getting to best: Efficiency versus optimality in negotiation. *Cognitive Science*, 24(2), 167–204. Leonardelli, G. J., Medvec, V., Galinsky, A. D., & Claussen-Schultz, A. (2008). Building interpersonal and economic capital by negotiating with multiple equivalent offers. Unpublished manuscript.

33. Deutsch, M. (1985). *Distributive justice: A social-psychological perspective.* New Haven, CT: Yale University Press. Loewenstein, G. F., Thompson, L. L., & Bazerman, M. H. (1989). Social utility and decision making in interpersonal contexts. *Journal of Personality and Social Psychology*, 57(3), 426–441.

Part 4

Sales Management—the Real Estate Broker

Chapter 8

Recruitment and Training of Real Estate Salespeople

LEARNING OBJECTIVES

This chapter is designed to help students of real estate marketing learn:
- why recruiting and selection of real estate salespeople is important
- the recruitment process
- how to analyze the various sales-related jobs and develop a job sheet for each
- how to determine qualifications for the various sales-related positions
- how to recruit candidates from a variety of recruitment sources
- how to select qualified prospects
- how to validate the recruitment process
- why training of incoming real estate agents is important
- how to assess training needs of salespeople
- how to budget for training needs
- how to develop training programs
- how to evaluate the effectiveness of the training program

WHY IS RECRUITING AND SELECTION OF REAL ESTATE SALESPEOPLE IMPORTANT?

Recruiting and selection in real estate firms is important because these tasks serve to accomplish important goals, both primary and secondary. The primary goal of recruitment and selection is to fill vacant positions and to ensure that those who assume the positions have the right skills, experience, and personal characteristics to be successful on the job. Secondary goals of recruitment and selection are numerous. Effective recruiting and selection leads to high levels of job performance, job satisfaction, and organizational commitment, as well as lower levels of absenteeism, presenteeism, and turnover.[1] Many real estate sales organizations count on their sales force to generate a great deal of revenue. It is estimated that adding a new recruit to the sales organization amounts to $25,000 added net revenue to the company per year.[2]

The real estate industry suffers from an incredibly high turnover rate. The estimate of turnover is approximately 55% each year. Real estate salespeople leave their companies, either to go with other real estate companies or to leave the industry entirely. High turnover is highly detrimental to the financial health of a real estate sales organization. Turnover interferes with creating long-term relationships with clients. This high rate of turnover in the real estate marketing industry is related to poor recruiting. Many real estate brokers or sales managers take a "warm-body approach" to hiring. That is, their goal is simply to make sure that every vacant salesperson position is filled, instead of making sure that every vacant position is filled by the right person with the right qualifications for the right circumstances.[3]

THE RECRUITMENT PROCESS

The recruitment process can be described in terms of five steps: (1) analyzing the job, (2) determining qualifications, (3) prospecting sources for recruiting candidates, (4) selecting prospects, and (5) validating the process (see Figure 8.1).

Analyzing the Job

Job analysis refers to the process by which the sales manager develops a carefully crafted description for each job in the sales organization that comes under his umbrella. A job description is essentially a description of activities and responsibilities inherent to a specified job. This may include position title, job objectives, who the person assuming this position reports to, the market territory or segment that the position is related to, the principal activities related to the job, and how the performance of the person in that position is evaluated and financially compensated. See Table 8.1 for a hypothetical example.

In Table 8.1, the sales organization is ABC Real Estate. It is a brokerage firm located in Blacksburg, Virginia (U.S.A.), which is a university town, home to Virginia Polytechnic Institute & State University. The table provides a job sheet for a senior sales associate position in the realty company. Notice that the job sheet spells out the job objectives, who the person assuming this position will report to, the market territory or segment assigned, the principal activities associated with the position, and the criteria related to performance evaluation and compensation.

Two *job objectives* are spelled out on the job description sheet: (1) "to assist home buyers in buying homes that meet their housing needs," and (2) "to assist home sellers in selling their homes in a timely and effective manner." Job objectives identify the primary goals

Figure 8.1
The Recruitment Process in
Real Estate Marketing

Table 8.1 An Example of a Job Description of a Senior Sales Associate Position

Position Title	Senior Sales Associate
Job Objectives	To assist home buyers in buying homes that meet their housing needs; to assist home sellers in selling their homes in a timely and effective manner
Report To	The ABC Real Estate broker
Territory/Segment	Focus on the Blacksburg market, with particular emphasis on serving the housing needs of Virginia Tech faculty and staff
Principal Activities	Seller representative activities: prospecting, information gathering, needs analysis, the presentation, negotiating the type of listing, post-agreement discussion, seller servicing, client follow-up. Buyer representative activities: prospecting, needs assessment, qualifying prospects, showing, negotiating and responding, closing, and follow-up servicing.
Performance Evaluation and Compensation	Generate at least $300,000 in commissions for the firm in 2013 and 2014. Compensation is based on 70% of the generated commissions

associated with the stated position. The example here is directly related to residential real estate and focuses on representing both home buyers and sellers in the role of a real estate agent. Of course, the job objectives on the job sheet vary considerably as a direct function of the job. In a real estate brokerage firm, the jobs may include junior sales associate, senior sales associate, partner, associate sales manager, manager, broker, administrative assistant, office manager, bookkeeper, accountant, webmaster, and receptionist. All these positions entail different job objectives.

Every position requires a supervisor (or supervising entity). Typically, most sales associates report to the sales manager or broker in charge. In some cases, junior sales associates who are assigned to assist senior sales associates report to the senior sales associates whom they assist. The supervisor is the person who is in charge of evaluating the performance of the person who is assuming the stated position. The example in Table 8.1 shows that the person assuming the senior sales associate position reports directly to the broker of ABC Real Estate.

The job sheet specifies a *market territory* or *segment* that the position is related to. This means that each sales associate may be assigned to a specific geographic territory (e.g., certain neighborhoods) or perhaps one or more market segments (e.g., retired people, faculty and staff directly affiliated with the university, or college students). Of course, this part of the job sheet is unique to sales position and may not apply to other positions in a realty company, such as associate sales manager, manager, broker, administrative assistant, office manager, bookkeeper, accountant, webmaster, and receptionist. That is not to say that every sales associate position has an assigned territory or segment. There are many instances in which the sales position is not restricted to a specific geographic territory or market segment. However, this may not be strategically optimal. A realty firm can do better to assign their sales staff to focus on specific territories or market segments. Of course, this decision is very much based on the company's overall marketing strategy (see Chapter 1 for a refresher).

In the example shown in Table 8.1, the senior sales position in question is assigned to "the Blacksburg market, with particular emphasis on serving the housing needs of Virginia Tech faculty and staff."

Part of the job sheet is *principal activities*. That is, each position is associated with a set of tasks and activities that the person assuming that position is expected to undertake. The example of the senior sales associate shown in Table 8.1 identifies the principal activities of seller/buyer representatives. Seller representative activities include prospecting, information gathering, needs analysis, the presentation, negotiating the type of listing, post-agreement discussion, seller servicing, client follow-up. Buyer representative activities include prospecting, needs assessment, qualifying prospects, showing, negotiating and responding, closing, and follow-up servicing.

Lastly, there is *performance evaluation/compensation* explicitly stated in the job sheet. It is imperative to spell out how the job performance of the person who assumes the stated position will be evaluated. The example shown in the table is "Generate at least $300,000 in commissions for the firm for 2013 and 2014." In other words, a sales quota is specified that is the basis for performance evaluation. Furthermore, the method of compensation is also specified in the job sheet. In the example shown in the table, the person assuming the senior sales associate position is to be compensated based on 70% of the generated commissions.

Determining Qualifications

Associated with each position in the sales organization are a set of qualifications. Job qualifications are customarily grouped into four major categories: (1) knowledge, (2) experience, (3) personal characteristics, and (4) legal considerations. See Table 8.2 for details.

Table 8.2 Factors Involved in Determining Qualifications

Qualifications	Description
Aptitude	Taking educational courses in real estate marketing, passing the real estate marketing exam, and obtaining a license to practice real estate marketing
Experience	Experience in personal selling in real estate or related industries (e.g., retail)
Personal Characteristics	*Psychological* (e.g., dominance, influence, steadiness, and compliance) *Physical* (e.g., good physical and mental health, active lifestyle, business attire, professional manner and speech, reliable means of transportation, information technology skills) *Social* (e.g., stable relationships, stable finances, connectedness in the community)
Legal Considerations	Civil Rights Act Age Discrimination Act American with Disabilities Act Equal Pay Act Pregnancy Discrimination Act Equal Employment Opportunity Commission Affirmative Action

Knowledge

Real estate sales positions call for a certain *knowledge* or skill level. In the U.S., skill for real estate sales is typically reflected in a real estate sales license. To obtain that license, a person has to demonstrate knowledge of real estate marketing through a knowledge test. This knowledge is typically acquired through formal or informal education and training. A person who wants to become licensed has to pass the real estate sales exam administered by the state in which the person hopes to practice. Also, applying to take the real estate sales exam requires the sponsorship of a real estate organization operating in that state.

Typically, a person interested in becoming a real estate salesperson approaches a real estate sales organization with the prospect of joining it. If the sales manager feels that this person has the knowledge, experience, and personal characteristics to become successful in real estate sales, he sponsors that person, and only through this sponsorship can that person apply to take the real estate sales exam. The assumption is that upon successful completion of the exam, the person is to join the sales organization. Most universities, colleges, and community colleges in the U.S. offer courses in real estate sales to help people interested in becoming real estate agents acquire the knowledge sufficient to pass the licensing exam.

Experience

A qualified real estate salesperson has to have some degree of *experience* in personal selling. It is best to have experience directly related to real estate sales. However, many real estate sales organizations accept related sales experience (e.g., retail sales). Experience in sales is important because many of the skills involved in personal selling in other settings (e.g., prospecting skills, communication skills, organizational skills, and negotiation skills) can be easily transferred to real estate selling.

A major decision a broker or sales manager has to make is whether to hire agents with little or no experience or seasoned agents. There are advantages and disadvantages to both. There are two significant advantages in hiring new agents. Hiring new agents means that the commission split would be much larger for the firm and much smaller for the new recruit (e.g., 50–50 split). This of course translates into more money for the firm per unit transaction. The second advantage is that the broker can easily mold the new agents. Seasoned agents, in contrast, bring in lots of baggage. They may be set in the way they conduct their business, which may not be highly compatible with the firm's culture. The molding of the new agents occurs through training and other socialization methods. Of course, the biggest disadvantage of hiring new agents is that they may not know what they're doing and can easily get themselves into trouble. Another disadvantage is that they require the manager or broker to devote much time, energy, and money to their training. New agents are typically recruited through local newspaper advertising, past clients, and referrals from instructors who teach at license training schools and at colleges and universities with real estate programs and degrees. Experienced agents are traditionally recruited through other sales agents (e.g., asking a sales agent at one's firm to identify good sales agents in other firms), through periodic meetings held by the local chapters of the real estate marketing association, by direct mail campaigns, and at community activities.[4]

Personal Characteristics

Qualified real estate salespeople have to have a set of *personal characteristics*. These can be discussed in terms of three major groups: (1) psychological characteristics, (2) physical characteristics, and (3) social characteristics. Let's first start out with *psychological characteristics*. Successful real estate agents have to have "people skills" and "drive." "People skills" can be viewed in terms of personality traits such as dominance and influence. People who are dominant tend to exert influence over others. They are persuasive and are successful in swaying people to their way of thinking. Their presence is felt in any interpersonal encounter. It is not easy to ignore them. Successful real estate salespeople also have drive. That is, they are self-motivated. They don't expect to be told what to do to get things done. They get things accomplished on their own accord. Real estate sales managers typically use a personality inventory that is administered to potential recruits called the DISC. The DISC stands for Dominance, Influence, Steadiness, and Compliance.[5]

"Dominance" is a personality trait that reflects how a person responds to problems and challenges. A high-dominance person is usually self-motivated, decisive, a problem solver, and a risk taker. He has the tendency to welcome new situations instead of avoiding them. High-drive salespeople tend to seek information from the sales manager and use the information to make their own decisions. They are optimistic and results-oriented. They tend to be direct and to the point in communicating with others and are goal-oriented (i.e., they focus on the bottom line of the transaction). An example item on the DISC Personality Profile is "Have you been told you are demanding? [Yes/No]". A "yes" response reflects high dominance.[6]

"Influence" is a personality trait that reflects how a person influences others to their point of view. A high-influence person is likely to be persuasive, enthusiastic, trusting, and optimistic. An example item on the DISC Personality Profile is "Do people tell you that you are the life of the party and gregarious? [Yes/No]". A "yes" response reflects high influence.[7]

"Steadiness" is a personality trait that reflects how a person responds to pace of change in the environment. A high-steadiness person is likely to be a good listener, a team player, predictable, understanding, friendly, emotionally stable, and loyal. A high-steadiness salesperson is motivated by loyalty and dependability, safety and security, and activities that can be started and completed on time. An example item on the DISC Personality Profile is "Have you been told you are generous almost to a fault? [Yes/No]". A "yes" response reflects high steadiness.[8]

"Compliance" refers to a personality trait that reflects how a person responds to the rules and procedures set by others. A high-compliance person accepts rules and procedures set by others. He is easy to get along with and does not fight the sales manager over decisions. He accepts the sales manager's judgment, and goes along with the company rules and policies. An example item on the DISC Personality Profile is "Do you like to follow procedures and go by the book? [Yes/No]". A "yes" response reflects high compliance.[9]

Experts who have a long history of using the DISC inventory to identify good recruits swear that the best agents tend to score high on Steadiness and Influence. Too much Compliance is not necessarily a good thing, but a moderate amount helps guarantee that the new recruits will follow the rules of the business. Being too high on Dominance is not a good thing. It may signal that the person is "too independent" to remain loyal to the company. These people are risk takers and more entrepreneurial—sometimes to the point of helping themselves at the expense of the company.[10]

See Zooming-In Box 8.1 for additional information on the use of personality inventories to screen and identify recruits in real estate marketing.

ZOOMING-IN BOX 8.1. THE PERSONALITY OF A GOOD REAL ESTATE AGENT[i]

Good salespeople can be identified by their personality profile. Research has shown that there are five personality traits associated with top salespeople. These are empathy, ego-drive, service motivation, conscientiousness, and ego-strength.

Empathy

This personality trait refers to the salesperson's ability to sense and understand other people's emotional reactions. It involves the ability to view a situation from the client's perspective. Empathetic real estate salespeople are sensitive in picking up on subtle body language and verbal cues. In other words, they are highly skilled in reading and understanding others.

Ego-drive

Ego-drive is essentially internal drive. Internal drive motivates a salesperson to succeed. Real estate salespeople with high ego-drive tend to thrive in situations in which they persuade and influence clients to their own point of view. Influence and persuasion provide a significant boost to their self-esteem. In other words, they feel that they need to make a sale to maintain their own sense of positive regard.

Service motivation

Service motivation refers to the tendency to enjoy helping others. Real estate salespeople with high service motivation feel a sense of gratification from helping others. That gratification comes from praise, gratitude, and recognition from others, including sales managers and other sales associates, but particularly from the clients who have been assisted.

Conscientiousness

Conscientiousness is a trait that reflects a sense of responsibility, a sense of obligation to follow through on an assigned task and bring it to completion. Salespeople with a high level of conscientiousness also tend to pay attention to detail. Once a conscientious agent signs up a client, you can rest assured that the agent will do his utmost best to service this client to the fullest, with a great deal of attention to detail.

Ego-strength

Ego-strength refers to having an unshakable sense of self. People with a vulnerable sense of self are easily affected by praise and rejection from others. Their sense of self is significantly bolstered by praise and they go to pieces when rejected. In contrast, real estate salespeople with high ego-strength have the mental toughness to safeguard against rejection—rejection from potential clients, rejection from other sales associates, and rejection from their sales manager. In other words, a sales associate with high ego-strength doesn't take rejection personally; he accepts the fact that rejection is part of the job.

Sales in real estate should be treated as a science, not an art. Understanding the personality traits of successful sales associates is part of this science. Real estate sales managers should use personality inventories in carefully screening applicants. This is a science-based approach to successful recruiting.

[i] Leporini, C. M. (2001). Selling is a state of mind. *Realtor Magazine*. July 1, 2001. Accessed from http://theweeklybookscan.blogs.realtor.org/2001/07/, on January 30, 2013. Greenberg, H., Weinstein, H., & Sweeney, P. (2000). *How to hire and develop your next top performer: The five qualities that make salespeople great.* New York: McGraw-Hill Professional Publishing.

Besides personality, sales managers in real estate have specific expectations regarding *physical and social characteristics* of salespeople. Effective salespeople have a healthy lifestyle. They have to be healthy, medically speaking (both physically and mentally). They have to be physically active because the job is physically demanding. This means they don't drink excessively; they don't smoke; they don't take unnecessary risks with their bodies and minds. Their dress code is businesslike, and their manner and speech are professional. They also have reliable means of transportation. Also, in today's business world it is very difficult to stay on top of things without being well-versed in information technology (the use of computers and mobile phones, word processing, spread sheets, data base, financial bookkeeping, web design and maintenance, Internet searches, video production, graphic design, etc.). See Zooming-In Box 8.2 for evidence concerning the role of physical attractiveness in real estate selling.

ZOOMING-IN BOX 8.2. THE EFFECT OF PHYSICAL ATTRACTIVENESS IN REAL ESTATE SELLING[i]

Recent studies in real estate have suggested that attractive professionals (more specifically, women) earn between 10–15% more on average than unattractive professionals. Underweight women earn significantly more money than overweight women. Does this translate into real estate sales? The evidence suggests that attractive agents are often more successful than their less attractive colleagues. Specifically, recent studies indicate:

- Male agents (both buyer and seller agents) are associated with lower selling prices.
- The gender of the agent is not associated with how long the real estate property is on the market.
- Attractive agents (both buyer and seller agents) are associated with higher selling prices.

The researchers believe that attractive agents get higher prices because they tend to obtain listings that sell better. These results suggest that physical attractiveness augments other personal characteristics, such as effort, intelligence, and organizational skills.

Customers in the real estate business are like the rest of us: visual creatures. People like to do business with an agent who is more polished and pleasing to the eye. One explanation for the results is that attractive agents use their physical attractiveness to attract higher-priced listings but are no better or worse at selling them than other agents.

What are the implications of the research findings on recruiting and selection? It would be too simple to say "hire attractive prospects." Physical attractiveness is like added-value. It may help, but it is certainly not a necessary criterion for selection.

[i] Adshade, M. (2012). Unattractive real estate agents achieve quicker sales. *Big Think*, April 11. Accessed from http://bigthink.com/ideas/unattrative-real-estate-agents-achieve-quicker-sales, on April 30, 2012. Gray, E. (2012). Real estate sale prices impacted by gender, attractiveness, and race. *Huffington Post*. Accessed from www.huffingtonpost.com/2012/04/13/real-estate-sale-prices-agents-attractiveness-race-gender-n-1424941.html, on April 30, 2012. Luido, G. (2012). Should you hire a hot real estate agent? *Chicago Now*. Accessed from www.chicagonow.com/getting-real/2012/04/should-you-hire-a-hot-real-estate-agent/, on April 30, 2012. Wheatley, M. (2012). Sexy real estate agents get higher prices. *Realty Biz News*. Accessed from http://realtybiznews.com/sexy-real-estate-agents-sell-homes-for-more/98711643/, on April 30, 2012.

Effective real estate salespeople are connected to family, friends, and associates. In other words, they are socially embedded in the community. They rely on their social network for referrals. They are socially stable, meaning that they are likely to be involved in steady relationships. Their social etiquette and manner should be impeccable, too. They are also financially stable: They are not in debt, they have not declared bankruptcy, and their credit ratings indicate good standing.

Legal Considerations

In the U.S., sales managers must consider the various laws governing recruitment. Among the important laws are the Civil Rights Act, the Age Discrimination in Employment Act, the Americans with Disability Act, the Equal Pay Act, the Pregnancy Discrimination Act, the Equal Employment Opportunity Commission (EEOC), and Affirmative Action.

The *Civil Rights Act* (of 1964) is a landmark piece of legislation in the U.S. that outlawed major forms of discrimination. Title VII of the Civil Rights Act is a specific law that prohibits discrimination by employers on the basis of race, color, religion, sex, or national origin. Employers also cannot discriminate against a person because of his interracial association with another (e.g., interracial marriage). However, the employer is permitted to discriminate if the employer can demonstrate a direct relationship between the protected trait and the ability to perform the duties of the job.[11]

The *Age Discrimination in Employment Act* (of 1967) is a law that forbids employment discrimination against anyone 40 years of age or older. The law prohibits discrimination in hiring, promotion, compensation, and termination of employment.[12]

The *Americans with Disabilities Act* (of 1990) is a law that prohibits discrimination based on disability. It affords similar protections against discrimination to Americans with disabilities as the Civil Rights Act did for other groups. The law defines *disability* as a physical or mental impairment that substantially limits a major life activity. Certain specific conditions are not regarded as "disabilities," such as substance abuse and visual impairment correctable by prescription lenses. Title I under this act prohibits employers from discriminating against a qualified individual with a disability. This applies to job application procedures, hiring, advancement and termination of employees, workers' compensation, and job training. Discrimination may include not making reasonable accommodations to the disabled employees. In recruitment, employers can use medical entrance examinations for applicants, after making the job offer, only if all applicants (regardless of disability) must take it.[13]

The *Equal Pay Act* (of 1963) is a law aimed at prohibiting sex discrimination in the work place. The law prohibits employers from paying wages to women at a rate less than the rate at which the employer pays wages to men for equal work on jobs, the performance of which requires equal skill, effort, and responsibility.[14]

The *Pregnancy Discrimination Act* (of 1978) prohibits discrimination on the basis of pregnancy, childbirth, or related medical conditions. In other words, women affected by pregnancy, childbirth, or related medical conditions shall be treated the same for all employment-related purposes, including receipt of benefits under fringe benefit programs, as other persons not so affected but similar in their ability to work.[15]

The *Equal Employment Opportunity Commission* (EEOC) is a law designed to enforce other laws against workplace discrimination. The EEOC takes action against employers who are found to have engaged in discrimination based on employee's race, color, national origin, religion, sex, age, disability, and genetic information, as well as retaliation for reporting, participating in, or opposing a discriminatory practice at work. The EEOC also treats acts of discrimination against lesbians, gays, and bisexual individuals as illegal. The Commission

171

not only files action in court against an employer who practices discrimination but also mediates and settles cases out of court.[16]

Affirmative action (also known as "positive discrimination" in the U.K. and "employment equity" in Canada) refers to policies that take factors (e.g., race, color, sex, religion, and national origin) into account to benefit an underrepresented group in employment and education. These policies aim at promoting employment opportunities of defined groups within a society—specifically groups that have a history of being disenfranchised. Affirmative action, thus, helps to compensate for past discrimination, persecution, or exploitation by the ruling class and address existing discrimination. Affirmative action in the U.S. dates back to 1961, when President John F. Kennedy required that government employers not discriminate against any employee or applicant for employment because of race, creed, color, or national origin; "take affirmative action" to ensure that applicants are employed; and treat those employed without regard to their race, creed, color, or national origin. Affirmative action was extended to women in 1967 by adding "sex" to the list of protected categories. The Canadian Employment Equity Act requires employers (in federally regulated sectors) to give preferential treatment to four designated groups: women, people with disabilities, aboriginal people, and visible minorities. In the U.K., any quotas or favoritism based on protected characteristics, such as sex, race, or ethnicity, is generally illegal in education, employment, and commerce.[17]

Prospecting Sources for Recruiting Candidates

Sales managers in real estate recruit candidates from several sources. These include job postings/job boards, newspaper advertising, employee/client referrals, employment agencies, educational institutions, career conferences, and professional societies.

Sales managers post job positions related to sales on many job boards. Examples of popular *job boards* in the U.S. include:

- *Indeed.com* (www.indeed.com/jobs?chnl=50&utm_source=publisher&utm_medium= organic_listings&utm_campaign=affiliate)
- *Beyond.com* (www.beyond.com/jobs/job-search.asp?aff=66DC799A-1214–4135–8D09– 9AEFC48C83C4&a=0&utm_source=Hasoffers%20&utm_medium=Affiliate%20 &utm_campaign=NatIntl14)
- *findtherightjob.com* (www.findtherightjob.com/?campaign_id=13565333mp6pzMCjTU)
- *the Ladders* (www.theladders.com/reg/signup_affiliate?cr=5994515&pl=CJ-00)
- *Monster.com* (http://login.monster.com/Login/SignIn?WT.mc_n=olmcjcon&ch=MONS &redirect=http%3a%2f%2fmy.monster.com%2flogin.aspx%3fWT.mc_n%3dolmcjcon)
- *Job.com* (www.job.com/my.job/btsup/us=164?AID=10281346&PID=5994515&SID= GozuI8SObi&refid=2828146)
- *Snagajob.com* (www.snagajob.com/job-seeker/registration?ref=affcj)
- *Employment Crossing* (www.employmentcrossing.com/lcsignin_cj.php?refid=1500&offer_ page=151)
- *Elance.com* (www.elance.com/employersignup?mpid=cj_10766276_5994515)
- *ExecutiveSearchOnline.com* (http://jobs.executivesearchonline.com/?PID=5994515)

Besides job boards, many real estate organizations use the local media, which means the local newspaper (especially the real estate section of the paper, in the form of display ads). There are, of course, employee/client referrals (i.e., referrals made by real estate agents within and outside the company, as well as referrals made by trusted clients), employment agencies (i.e., many sales organizations turn to employment agencies when recruiting for senior-level agents, sales managers, and other executive-level positions), educational institutions (i.e., many brokers and sales managers reach out to instructors at real estate training institutes and real estate faculty at colleges and universities to help them identify and make referrals from their student population), and real estate professional societies (i.e., many brokers use their professional association's newsletters, monthly/quarterly magazines, and communication and publication outlets to reach out to members of their societies to publicize available positions).

The various methods of recruitment described above can be viewed as active methods of recruitment. Of course, there are passive methods as well. For example, the company's reputation can be highly instrumental in attracting recruits.[18] Consider the fact that in many cities (e.g., New York) there are now websites established to help agents "shop around" for brokerage firms. The website allows brokerage firms to post company profiles with videos. Agents can browse those firms registered at the website and compare features (e.g., compensation plans). The website allows agents to send anonymous messages to certain brokers to ask more specific questions. The registered brokerage firms pay a monthly subscription fee. Brokers can contact agents who send messages through the system. They also keep track of the amount of traffic related to their company profile and what other firms the brokerage firm has been compared to. They can also identify the number of transactions particular agents have closed in certain time intervals.[19]

See Zooming-In Box 8.3 for more information about recruitment through good leadership and a caring culture. Another important method of passive recruitment is through the company's website. See Zooming-In Box 8.4 for more information of how a company's website can be designed to attract recruits. See Zooming-In 8.5 for information regarding the use of LinkedIn for recruiting purposes.

ZOOMING-IN BOX 8.3. RECRUITMENT THROUGH GOOD LEADERSHIP AND A CARING CULTURE[i]

Successful brokers can create a brokerage firm that has a reputation of nurturing its sales force. This reputation draws new recruits to the firm, and the broker does not have to work hard to recruit new sales associates. This reputation is not just built on enticements (e.g., a high commission split or cutting-edge technology); it is mostly based on exuding effective leadership and a culture that fosters productivity.

Exuding Leadership

Exuding leadership means that the broker is recognized as a change agent. Recruits are attracted to a firm with strong leadership; strong leadership means that the sales associates learn more and make more money. Strong leadership is like a drop of water in a calm lake that creates a ripple of excitement.

Brokers who exude leadership allocate their management time differently. Most brokers spend the majority of their time putting out fires. Brokers who exude leadership are those who spend much of their time inspiring their sales team. Good leaders set a vision of excellence, project values related to achievement and productivity, and build a team around excellence and professionalism. Doing so helps build a culture in the company that becomes like a magnet, attracting sales associates.

A Culture that Fosters Productivity and Caring

Sales associates are attracted to a brokerage firm that has a culture of productivity. A culture of productivity is a culture that says "success," a culture that says "we're number one," a culture that signals professionalism and perfection. Firms with a culture of productivity send a very clear message that the core beliefs and values reflect productivity, professionalism, achievement, and excellence. A culture that fosters productivity attracts sales associates who, to be successful, aspire to be the best they can be. In other words, a culture of productivity attracts associates who share the values of hard work, professionalism, and excellence.

Part of the culture that fosters this magnetism is caring. The sales associates know that the broker cares about them and their well-being. If the broker makes every attempt possible to show that he cares, to create esprit de corps, the sales associates are likely to feel good about the firm. And if the associates feel happy, they'll help with the recruitment of a new cadre of sales associates who share their values and the company's values.

To help in the recruiting effort, some brokers print the company's values on the back of the sales associates' business cards:

- a desire to be the best one can be
- working together as a family
- trusting your colleagues
- trusting your broker
- feeling passionate about real estate

Handing out business cards to other sales associates goes a long way to help with recruitment.

An important element of both leadership and a caring culture is revealed when brokers roll up their sleeves to consult with the sales associates, partner with them, and help build their business. Ultimately, doing so helps associates attain a better quality of life. Specifically, good brokers help associates establish goals and a strategic business plan, they act as mentor and coach, they monitor the associates' performance to ensure that the associates stay on track, and they suggest a course of action if the associates get off track. In other words, a good broker creates an environment in which associates feel that their broker cares about their personal growth, their growth within the company, and their growth within the profession.

[i] Filisko, G. M. (2007). Recruiting: Creating a buzz for your brokerage. *Realtor Magazine*, January 2007. Accessed from http://realtormag.realtor.org/for-brokers/feature/article/2007/01/recruiting, on January 30, 2012.

ZOOMING-IN BOX 8.4. HOW TO RECRUIT REAL ESTATE AGENTS ON THE WEB[i]

Here are some tactics a real estate sales manager or broker can use to find good recruits, get them to the firm's website, and make a convincing case about why they should join the firm. However, before describing the good tactics, let's describe the bad ones.

Bad Recruiting Tactics on the Web

Many sales managers and brokers use the Web as a recruiting tool but make significant mistakes. The following are the three biggest mistakes:

- **Using their main consumer website as their recruiting platform.** You can't appeal to qualified and perhaps seasoned real estate agents from other companies using the company's main consumer website for recruiting. Successful recruiting requires a separate website (or at least a special section of the main website). The main website customarily addresses buyer and sellers concerns (i.e., potential clients). Using the main portal to recruit real estate agents dilutes what is being communicated to both audiences (recruits and potential clients).
- **Endless bragging about their real estate business.** Information that focuses too much on the firm's success is not helpful to potential recruits. The information that should be presented in a portal dedicated to recruiting should spell out in specific terms what the firm can do for them. That is, it must explain the reasons the target sales organization is a great place to work and address the potential recruits' goals and concerns directly. Make the information about them or about what the company can do for them.
- **Generic copy.** Many websites use generic words, such as "call us for an appointment!" or "fill out an online application." Seasoned agents are not likely to consider switching companies using the generic material. Recruiting good people requires building trust, and building trust requires more sophisticated ways of engaging potential recruits.

Good Recruiting Tactics on the Web

Treat potential recruits like potential clients. In other words, consider the competition, and do better than they do. Potential clients "shop around" for a good brokerage firm company, and so do potential recruits—they "shop around" for a good sales organization. Here are some tactics.

- **Make it about them.** Make sure that much of the information presented on the web pages dedicated to recruiting addresses the needs and desires of the potential recruits. In other words, assume the perspective of the seasoned salesperson, try to understand his point of view, and present the information accordingly. And do it in a way to differentiate the firm from its competitors. Focus on how the potential recruit can make more money and enjoy the job more. For example, focus on the dividends a recruit may be able to accumulate for retirement.
- **Engage them, don't "tell" them.** It is best to write the text in second person ("you," "your," "yours"). Have many open-ended questions that may prompt the potential recruit to think about the job in new and creative ways. Help prospective recruits assess their careers and the progress they have made so far. Help them focus on future prospects. Avoid giving them answers (i.e., "telling" them), but phrase the copy in question format to allow them to think about their careers and develop their own answers. Encouraging them to develop their own answers is the most powerful form of persuasion.

- **Constantly reassure them.** When real estate agents consider jumping ship, they become apprehensive about other colleagues and especially about their sales managers finding out. Therefore, it is very important to reassure prospects that information about web visitors is treated with the highest levels of discretion and confidentiality. Assure them of this confidentiality by having a link to the privacy policy on every page (e.g., "We understand that discretion is of utmost importance to you as you explore your career options. Be assured that our staff respects your need for privacy.")
- **Have a follow-up system.** When potential recruits request additional information, the sales manager should respond directly and in a timely fashion, instead of relegating this responsibility to the office receptionist.

Eventually, potential recruits are likely to contact the sales manager for an appointment. When that happens, the recruiting website has done its job. It is time to meet face to face.

[i] Russer, M. (2007). The right way to recruit on the web. *Realtor Magazine*, November 2007. Accessed from http://realtormag.realtor.org/technology/mr-internet/article/2007/11/right-way-recruit-web, on January 30, 2012.

ZOOMING-IN BOX 8.5. HOW TO USE LINKEDIN TO RECRUIT REAL ESTATE AGENTS[i]

A few specific strategies can make your time on LinkedIn productive for recruitment:

Develop a Profile

Your first step is to develop a profile of your company in LinkedIn. The profile should contain lots of information—from company history to contact information.

Join One or More Groups

Get connected with other real estate agents by creating or engaging in discussion on a topic within your expertise. Suppose you have particular expertise in the psychology of negotiations. You could start a discussion on that topic and invite other real estate agents to join this discussion group. Alternatively, you can join an ongoing discussion.

Use the "Company Follow" Feature in LinkedIn

Use the "Company Follow" feature to find possible new recruits. This feature should allow you to understand your competitors and the real estate agents working for your competitors. You can follow moves made by certain agents and perhaps get a feel of the extent of turnover rate. Turnover may signal widespread dissatisfaction with the company. You may want to pursue agents who have a demonstrated record of achievement working for competitor firms with a high rate of turnover.

[i] Lewis, A. (2012). How to use LinkedIn for Recruiting and Talent Attraction. Clean Slate: Insights into the Real Estate Industry. *Better Homes and Gardens Real Estate*, February 9, 2012. Accessed from http://bhgrealestateblog.com/2012/02/09/how-to-use-linkedin-for-recruiting-and-talent-attaction, on May 2, 2012.

Selecting Prospects

The process of selecting prospects involves several steps. These are (1) initial screening, (2) reference checking, (3) in-depth interviewing, (4) employment testing, (5) follow-up interviewing, and (6) making the selection.[20] See Figure 8.2.

Initial Screening

In many cases when a sales position is formally advertised in some media, the sales manager becomes flooded with applications. Thus, the goal of initial screening is to develop a manageable pool of candidates. In other words, trim down the long list of applicants to a manageable handful. The sales manager uses an elimination process in the initial screening. That is, he develops a set of undesirable characteristics and starts eliminating applicants if they match the undesirable criteria. Criteria used to eliminate candidates from the onset may include:

- applicants who do not have their own means of personal transportation
- applicants who lack stability in their employment history (i.e., they jump from one job to another in no time)
- applicants who do not have a minimum level of education (i.e., high school)

Some real estate sales organizations use personality profiling (e.g., DISC) to help them with the initial screening. Applicants are directed to a website where they complete the personality test. The computer automatically scores each candidate based on their responses. Those who score below a specified level are not considered further.

With the growing popularity of social media, many managers are beginning to use Facebook and other social media portals to help them with initial screenings. The sales manager may check to see whether a candidate has made comments or posted pictures that reveal undesirable characteristics—physical (e.g., showing the body or body parts in sexually explicit ways), psychological (e.g., making lewd comments or denigrating others), and/or social characteristics (e.g., pictures showing the candidate in a bar intoxicated with friends or associates).

Figure 8.2
Steps in the Selection Process

177

Checking References

The second common step in selecting prospects in real estate marketing is checking references. Once a manageable pool of candidates is developed, the sales manager attempts to contact the references provided by the candidate. In some cases, reference letters are already attached to the applicant's résumé. Based on the quality of the letters of recommendation, the sales manager makes another attempt to further trim down that list by eliminating candidates that do not have very good letters of recommendation. In many cases the sales manager goes beyond those references that are provided by the candidate by contacting past employers to obtain specific information about the candidate's job performance, his personality, and reasons for leaving. Such information is likely to be more revealing about the exact qualifications of the candidates for the target position. However, many employers choose not to reveal information about their past employees for fear of legal liability (i.e., past employees may sue the company for character defamation). Hence, information from past employers may or may not be helpful, depending on how forthcoming the employer is in disclosing sensitive information about their past employees.

Furthermore, many sales managers check the accuracy of the information provided on the employment application or résumé. Of course, contacting past employers and verifying the dates of employment is a must. Another crucial verification check is on the details of the applicant's education: The sales manager may contact the educational institutions shown on the person's résumé to verify those claims.

In-Depth Interviewing

It is extremely uncommon to hire a real estate salesperson without a face-to-face interview (or what is traditionally called "in-depth interview"). In-depth interviews are used to reveal personal characteristics above and beyond what is on paper (i.e., the applicant's résumé and employment application). Common questions asked in the in-depth interviews involve:

- *Is the candidate qualified for the job?* The sales manager may conjure up a difficult personal selling situation and asks the candidate how he would act in that situation. The sales manager knows how top salespeople are likely to respond, and he judges the candidate accordingly.
- *Does the candidate really want the job?* The sales manager may probe the candidate's motives and determine that the candidate's only reason for seeking this job is to make money and that he does not have a passion for real estate. This is, of course, unfavorable for the candidate.
- *How does this job help the candidate meet his career goals?* The sales manager expects responses that reveal a desire for a career path in real estate sales, with greater specialization, additional education, and certifications, and aspirations to become a top performer in the recruiting company or in a firm of higher stature. Other favorable responses signal a career path that leads to becoming a sales manager or broker.
- *Will the candidate find the job challenging?* Good candidates tend to find the job they are applying for to be more challenging than their previous job. Applying for a more challenging job signals career aspirations and hope for career advancement.
- *What is the candidate's history of selling, as well as his past experiences, motives, and outcomes?* Sharing this information with the sales manager can reveal the candidate's depth of experience in selling, his motive to build a career in sales, and his passion for real estate.
- *How did the candidate operate in the past on tasks related to personal selling?* Did he do well in prospecting for new clients? How? Would his method of prospecting work in the context

of the new position? How did he go about assessing the needs of his clients? Would these methods work in the new position? How did he negotiate and handle objections from prospective clients? Would these selling techniques work in the context of the new position? And so on.

- *Will the candidate like particular aspects of the available position?* For example, real estate sales require a selling cycle that is longer than retail sales. Does the candidate have what it takes to do well in the longer selling cycle necessary in real estate (particularly in commercial real estate)?
- *What does the candidate expect in terms of a compensation plan?* Does the candidate have reasonable ideas about a compensation plan? Is there room for negotiation?

Employment Testing

Some real estate sales organizations use a variety of tests to further qualify candidates. We already discussed the use of personality tests. This is one form of employment testing. Some sales organizations use personality tests upfront in the selection process as an initial screening (in which applicants are directed to complete a questionnaire on a website); these include DISC (the personality test mentioned previously that is designed to measure the extent to which the applicant is high on four key personality traits: Dominance, Influence, Steadiness, and Compliance). Other organizations use personality tests after the in-depth interview.

There are, of course, other types of tests:

- real estate knowledge tests (a test to measure the extent of knowledge of real estate marketing, such as laws and ethics)
- sales knowledge tests (a test to measure the extent of knowledge of sales techniques, such as prospecting and negotiations)
- sales interest tests (a test to measure the level of interest and passion the person has for real estate marketing as a profession)
- attitude and lifestyle tests (a test to measure the extent to which the candidate lives a healthy lifestyle, such as eating healthy food, engaging in physical exercise, refraining from the use of drugs and other recreational substances)
- physical exam (real estate sales require a certain degree of physical activity and stamina; therefore, an applicant with poor health is likely not likely to succeed)
- drug tests (blood and/or urine analysis to check for evidence of drugs in the applicant's body)

Follow-Up Interviewing

Given that the candidate passes all the preceding tests and interviews, he is then re-interviewed before a formal hire. The follow-up interview is typically done with members of the sales team, especially the senior sales associates. The sales manager collects feedback from the sales team concerning the candidate. If the feedback is positive, a formal hire is initiated.

Making the Selection

This is the last step in the selection process—the formal hire. The sales manager reviews all the information up this point (screening, reference checks, interviews, tests, etc.) and makes a final decision. Once a selection is made, the sales manager invites the candidate in and makes a formal offer. Typically, the formal offer comprises (1) duties of the sales associate, (2) the compensation plan, (3) territory or market segment assigned, (4) length of employment relationship, and (5) termination of employment.[21]

With respect to *duties*, the sales manager spells out the company's expectations. These expectations can be grouped in terms of the various stakeholders the new hire has to deal with: internal and external. Expectations regarding internal stakeholders, such as other sales associates and management, may involve how to deal with complaints from other sales associates, how to handle problems with other members of the sales team, the company's policies related to sexual harassment, how to file a formal grievance against one's supervisor, expectations regarding attendance of weekly sales team meetings, etc. Expectations regarding external stakeholders, such as clients, the media, competitors, the community, and the local real estate board, are also spelled out.

With respect to *compensation*, the sales manager discusses with the new hire aspects of the compensation plan. This may include salary (whether there is one and when it is payable), draw (as applied to commissions, what the amount is and when it is payable, the terms related to situations in which the draw falls below commissions, the terms of draw related to resignation or termination of employment, etc.), bonus (whether there is any and the terms and expectations related to this aspect of the compensation plan), commission (the commission rate and when it is payable, how commissions will be determined given various selling scenarios involving more than one sales associate, any deductions from commissions based on specific services the company provides the sales associate, etc.), and expenses (the kinds and amounts of expenses that are reimbursable, the kind of documentation expected for reimbursement, etc.).

With respect to *territory*, the sales manager may assign the new hire to a specific geographic market. In doing so, the sales manager specifies the geographic boundaries and clarifies the expectations related to territorial infringements (calling on clients and prospects in territories that are assigned to other sales associates).

With respect to the *length of the employment relationship* and possible *termination*, the sales manager specifies the exact date when the employment of the new hire begins, the length of the employment contract (e.g., a 5-year contract), whether the new hire is obligated to remain under employment for a specified length of time, the nature and terms of contract renewal, and the relationship between the length of the employment contract and a minimum level of sales performance (e.g., meeting a minimum sales quota).

Validating the Process

The turnover rate in the real estate industry is horrendous. Real estate licenses in the industry are growing by 36% annually in the U.S. The retention rate in the business is very low. Data from 2004 indicate that the number of licensed agents increased by roughly 80% over the previous year; 14% of these were not renewed. Additionally, in the years following (2005–2008), 17.2%, 16.9%, 26.8%, and 31.8% of licenses granted were not renewed. This may be construed as a mass exodus of real estate agents—agents dropping out of the real estate marketing profession. Validating the recruitment process can help reduce this mass exodus.[22]

The validation process focuses on measuring subsequent success on the job and modifying (if necessary) any aspect of the previous processes (analyzing the job, determining job qualifications, selecting prospects, initial screening, checking references, interviewing, testing, making the final selection). For example, the sales manager hires a sales associate and several months down the road he observes that the associate is not performing as expected; he is floundering. Did he hire the wrong person for the job? What went wrong? What can be learned at this stage to avoid making the same mistake in the future? Perhaps the sales manager needs to revisit the job analysis part of recruitment? How about selection? In other words, the sales manager should make a concerted effort to search for reasonable explanations for this failure. Once a specific cause is identified, the sales manager takes corrective action to avoid making similar mistakes in future recruitment. The same goes for success. If a salesperson is deemed to be successful, then the steps, methods, and approaches the sales manager has used in relation to the successful candidate must be reinforced. This is the essence of validation. Measure the new hires' sales performance and develop explanations to account for success and failure in performance. Reinforce those recruitment approaches and methods applied to the successful hires, and make corrective changes in the recruitment system to reduce the likelihood of hiring salespeople who fail.

WHY IS TRAINING IMPORTANT?

Training is important because it enhances real estate agent *competency*. This, in turn, helps to improve customer relations, enhance job performance and sales, increase job satisfaction, increase organizational commitment, reduce absenteeism, reduce presenteeism, and reduce turnover.[23]

One particular study has clearly shown that real estate agents with much experience (partly due to training and continuing education) produce more than rookie agents. Specifically, the study compared the productivity of agents who acquired and maintained their real estate license for two years or less with those who have been licensed for 10 years or more. The rookie agents were found to sell their listings for roughly 10% less than their more experienced counterparts. The rookie listings stayed on the market for a longer duration before they are sold. The experienced agents sold approximately 2% more and 32% faster.[24]

THE TRAINING PROCESS

The training of the real estate sales force involves four distinct steps (see Figure 8.3). These are (1) assessing training needs, (2) budgeting for training, (3) developing training programs, and (4) evaluating the programs.

Figure 8.3
The Training Process in
Real Estate Marketing

Assessing Training Needs

The sales manager assesses the training needs of the sales force using a variety of methods, as shown in Table 8.3. These include: (1) sales force observation, (2) sales force survey, (3) management objectives, (4) customer information, (5) customer survey, and (6) company records.[25]

Sales force observation refers to an assessment of training needs based on the sales manager's personal observation of the sales associates' performance. For instance, a broker of a realty company has several conversations with a sales associate, and these conversations reveal that the sales associate does not know how to help his clients negotiate effectively. Therefore, based on these personal encounters with the sales associate, the broker concludes that this sales associate can use an educational course in negotiations.

Sales force survey is a more systematic way to identify specific needs in the sales force. The sales manager, in this case, administers a survey to all members of the sales team. The survey contains items that reveal the individual's needs for further instruction in specific areas of expertise (e.g., client prospecting, negotiations with real estate buyers and sellers, closing real estate deals, ethics and real estate law in real estate, or real estate appraisal). A typical survey describes the various areas of expertise and instructs the survey respondent to indicate the extent to which he might be interested in further instruction in these areas. Based on the survey results, the sales manager identifies areas of expertise that can be addressed by training programs.

Management objectives is an assessment of training needs based on a strategy that management has articulated and is trying to implement. For example, a realty company makes a strategic decision to position itself as a company that serves the needs of the retirement community. To do so, the sales force has to be better equipped to cater to the needs of the elderly and retired people in the community. Therefore, management decides to encourage or provide further training and instructions about the housing needs of the elderly and the retired. As such, management objectives to serve the retirement community dictate certain training needs.

Table 8.3 Assessing Training Needs of the Sales Force

Method	Description
Sales Force Observation	An assessment of training needs based on the sales manager's personal observation of the sales associates' performance
Sales Force Survey	An assessment based on a survey administered to the sales associates that reveals individuals' needs for further instruction in specific areas of expertise
Management Objectives	An assessment of training needs based on a strategy that management has articulated and is trying to implement
Customer Information	An assessment of training needs based on information provided by specific clients that may involve specific complaints about one or more sales associates
Customer Survey	An assessment of training needs based on a survey administered to past clients of specific sales associates asking the clients to rate their satisfaction with the various services provided
Company Records	An assessment of training needs based on sales associates' past sales performance, as documented in the company personnel records

Customer information refers to information provided by specific clients that may involve specific complaints about one or more sales associates. The nature of the complaint may reveal the training needs of the sales associates. For example, a broker of a realty firm runs into a seller client who was represented by a sales associate of the brokerage firm. The conversation reveals that the client was not particularly happy with the services provided by the sales associate—particularly the advertising of the property and the staging of the open house. This information signals that the sales associate in question may need further instruction in effective methods of advertising and the staging of an open house.

Customer survey is a more formal method in seeking information from past clients. The customer survey is essentially a customer "satisfaction" survey. Past clients are contacted by phone, mail, or e-mail and are requested to complete a short survey. The survey contains specific items related to the various tasks of the brokerage's sales agent. For example, if the sales associate was essentially a listing agent, then the various services provided may include listing the property in the MLS, developing and disseminating promotional sheets and brochures about the property, staging a series of open house events, advertising the property in the local newspaper, developing a virtual tour video of the property and posting it to the website, etc. These services are itemized in the survey and the survey respondent indicates the extent to which he was satisfied with each offering. Dissatisfaction ratings signal the need for training in relation to those offerings. For example, if a particular client indicates dissatisfaction with the quality and the extent of advertising that his property received, then perhaps the listing agent should obtain further instruction in advertising.

Finally, we have *company records* as another method to assess training needs. Here the sales manager simply inspects past sales performance of the sales associates. Those who have a poor sales performance record are then singled out. The sales manager would interview them to find out how training can help them enhance their performance.

Budgeting for Training

How do real estate sales managers develop a training budget? They usually do this using a variety of budgeting methods, shown in Table 8.4. These include (1) all you can afford, (2) percentage of sales, (3) percentage of profits, (4) competitive parity, (5) industry benchmark, and (6) objective and task.

Table 8.4 Budgeting Methods for Training

Method	Description
All You Can Afford	Allocate as much money for training as the sales organization can afford
Percentage of Sales	Allocate money for training based on a certain percentage of last year's sales commissions
Percentage of Profits	Allocate money for training based on a certain percentage of last year's profits
Competitor Parity	Allocate money for training comparable or better than what the main competitor has spent on its training programs last year
Industry Benchmark	Allocate money for training based on some industry average
Objective and Task	Allocate money for training by costing a set of training programs designed to achieve a specific stated sales objective for the sales organization

The *all-you-can-afford* method of budgeting is essentially a no-brainer. Allocate as much money as the sales organization can afford for training. Of course, this method is arbitrary and may not be effective in accomplishing the training goals of the organization. In other words, in many cases, the resources allocated for training may fall short of what is truly needed. See Zooming-In Box 8.6 for doing training on a very tight budget.

ZOOMING-IN BOX 8.6. HOW TO TRAIN YOUR REAL ESTATE SALES FORCE ON A VERY TIGHT BUDGET[i]

Here are tactics to maintain a semblance of training with few financial resources committed to training.

Sales Associates Training One Another

Seasoned sales associates who have acquired expertise in certain areas (e.g., foreclosure and short sales) can be used to train others. The sales manager can set up weekly or monthly training workshops. The topic can be selected by the sales associates, and the sales manager can identify a seasoned sales associate with particular expertise on that topic to conduct the workshop.

The Use of Local Professionals

Real estate agents should have much knowledge not only about selling, but about "the product." This means that to provide good advice to buyers and sellers, agents have to know a good amount about building and construction, home inspection, zoning issues, mortgage and financing, real estate law, and other relevant topics. Much of this knowledge can be acquired by inviting local professionals to conduct workshops or seminars for a firm's sales force. For example, a local attorney may be invited to conduct a seminar on real estate law, a local home inspector may be invited to do a seminar on home inspection, or a local banker can be invited to do a seminar on home mortgage.

Often, local professionals will lead such workshops and seminars without payment. They may do this as part of promoting their own business. Thus, the home inspector gains business referrals when a real estate agent recommends to his client a specific home inspection company, the attorney benefits when an agent recommends an attorney for closing, and the banker benefits when an agent recommends a mortgage company for financing.

The Use of Free Videos

There are many training videos related to real estate marketing available on YouTube and other social media. Training videos are also available through professional associations' websites. For example, the National Association of Realtors' (NAR) website has training videos on ethics, specifically the NAR Code of ethics. There are also cases and discussion questions posted to enhance training in real estate marketing ethics. Don't ask the sales associates to find and watch videos or read online material on their own. The sales manager should identify the best online material and present this material at a weekly or monthly meeting, and then use the videos and other online materials to generate a discussion among the sales associates. Much learning can occur through this process.

The Use of Professional Associations

Many professional associations related to real estate marketing (e.g., NAR) provide training resources that are either free or charge a minimal fee. Much of this type of training is in the form of webinars or workshops provided at the associations' regional or national conferences.

The Use of Online Tools

Many large real estate marketing organizations that have offices scattered in different geographic regions are beginning to use their online technology (e.g., SKYPE or Go-to-Webinar) to set up their own in-house training programs. For example, a training session can be conducted by a knowledgeable sales associate in one location using a webinar format that connects dozens, or perhaps hundreds, of other sales associates.

[i] Filisko, G. M. (2011). Training on a dime. *Realtor Magazine,* March 2011. Accessed from http://realtormag. realtor.org/for-brokers/solutions/article/2011/03/training-dime, on January 30, 2013.

The *percentage-of-sales* method of budgeting uses sales commissions of the previous year to allocate a certain percentage of sales to cover training expenses related to the current year. For example, the brokerage firm may have made $2 million in sales commissions in 2013. The broker may allocate 2% of the $2 million (i.e., $40,000) for training in 2014. The advantage of this method of budgeting is that it guarantees a tangible budget based on the stream of revenue that was generated the previous year: In years that the firm has done well, more money becomes available for training the following year. However, one can argue that if sales commissions have declined, this may signal that the sales force needs additional training. Additional training should help increase the performance of the sales team, which in turn should help generate higher levels of sales commissions. So this method of resource allocation works backward in the sense that, in good years, one ends up with more money available for training; in bad years, less. The opposite should be true. That is, in good years, one can afford to decrease the training budget because the sales force is doing well; in bad years, the sales force needs more training in order to perform better and close the revenue gap.

The *percentage-of-profits* method is similar to the percentage-of-sales method, with the exception that it is based on profits rather than sales commissions. In other words, the broker allocates a certain percentage of the company's last year profit (e.g., 10%) to cover training expenses for next year. For example, suppose that a brokerage firm made $800,000 in net profit (after all administrative expenses and taxes paid) in 2013. The broker allocates 10% of that profit (i.e., $80,000) toward training. This method of resource allocation is somewhat better than the percentage-of-sales method because it relies on profit rather than sales commissions. Profit is a better benchmark than sales commissions because it subtracts costs from the sales commissions. A brokerage firm can be doing well in terms of generating sales commissions; however, its operation costs may be too high. Allocating resources toward training based on sales commissions may be risky in the sense that too much may be allocated, causing the firm's profits to suffer.

The *competitive-parity* method relies on the competition to determine a training budget. In other words, the sales manager may feel that to survive and prosper, the company should spend as much or more on training than its major competitors in the local area. Consider a sales manager who believes that his firm is competing directly with XYZ Realty. His firm's market share is 30% and XYZ's share is 40%. He finds out that XYZ has spent $100,000 to cover training expenses in 2013. His firm spent only $60,000 in 2013. To gain a competitive advantage against XYZ, he decides to spend at least $100,000 or more in 2014. Doing so may help the firm attain competitive parity (i.e., increase its market share from 30% to 40% or more).

The *industry-benchmark* method relies on industry averages. Much information about training budgets in the real estate marketing industry is published in professional society magazines (e.g., *Realtor Magazine*, a professional magazine serving members of the National Realtors Association in the U.S.). Suppose that the published industry average for training for mid-size realty companies (those with approximately 30–40 sales agents) is $100,000 for 2013. This amount serves as a guide. If the sales manager is a broker in a mid-size realty company, then he may decide to allocate $100,000 to cover training expenses for 2014—or perhaps more, adjusting for inflation.

Finally we have the *objective-and-task* method. This method of resource allocation is based on the specific sales performance objectives that training can help achieve. In other words, training is a *task* that, when implemented effectively, should achieve a certain sales *objective*. Suppose the sales manager believes that the firm has the potential to achieve $5 million in sales commissions by serving the growing retirement community in 2014. However, to do this, the sales force should be better trained to serve this market. Training the sales force on the housing needs of the elderly and retired is estimated to cost $100,000 in 2014. Therefore, $100,000 should be budgeted for training in 2014. The $100,000 is what the sales manager needs to spend on the task (training the sales force on the housing needs of the elderly and the retired) that ultimately may achieve the objective ($5 million in sales commissions at the end of 2014).

Developing Training Programs

In developing training programs for the sales force, the sales manager asks and attempts to answer the following questions: (1) What topics of instruction? (2) Where to train? (3) Who should train? And (4) What training methods to use?

What Topics of Instruction?

In a typical real estate sales organization, the topics of instruction include company orientation, product knowledge, selling techniques, market-industry orientation, and professional designations. See Table 8.5.

A company orientation is typically the topic of instruction for incoming salespeople. A *company orientation* instructs incoming people about the company, its history, its chain of command, its mission, its markets and competitors, and other human resource management information, such as company policies regarding sales performance, ethics and law, and the compensation plan.

After the company orientation is a *product knowledge* seminar. Every business has its own product(s). Incoming people, especially the sales force, have to be educated about the product. Salespeople are expected to be intimately familiar with the product to be able to do a good job as sales agents representing buyers and sellers of that product. For example, if the sales organization is in the business of selling units in a comprehensive retirement community, ultimately the sales force has to be intimately familiar with all those retirement community offerings—housing and amenities related to independent living, assisted living, skilled nursing care, and long-term nursing care.

Table 8.5 Topics of Instruction for the Sales Force

Topic of Instruction	Description
Company Orientation	Information about the company, its history, its mission, organizational structure, company rules and procedures, and compensation plan
Product Knowledge	Information about specialty real estate, such as retirement homes, luxury homes, distressed properties, student housing, etc.
Selling Techniques	Information about techniques related to prospecting, client assessment, negotiations, customer service, closing, follow-up servicing
Market-Industry Orientation	Information about housing trends in relation to specific geographic and demographic segments; trends in commercial real estate in relation to land, office buildings, shopping centers, etc.
Professional Designations	Information that is relevant to particular specialty occupations in real estate marketing that have specific professional designations and certifications (e.g., CRS, GRI, SIOR, CCIM, IREM, RLI, and CRE)

Then we have *selling techniques*. Many real estate sales organizations encourage their salespeople to brush up on their selling techniques by taking seminars and courses related to a variety of topics such as prospecting new clients, assessing the needs of clients, qualifying clients, negotiations, closing techniques, and ethics and law, among others.

Market-industry orientation focuses on market aspects and trends. For example, if the real estate company is focused on serving the needs of the elderly and retired, a training course in the housing trends related to the retirement market may be in order. In this type of seminar, salespeople are exposed to international, national, and regional trends of housing for the retired and the elderly. The course may segment this market into various consumer segments and profile the housing needs of each, their financial situation, and what they can afford in terms of mortgage payments, the various communities they currently live in and want to retire to, and their media habits (i.e., the television programs they watch, the type of music they listen to, the local newspapers and magazines they read, the clubs and other associations they belong to and their publications, the social networks they use on the Internet, etc.).

The popular *professional designations* in real estate marketing include Certified Residential Specialist (CRS, https://crs.com) and Graduate Realtors Institute (GRI, www.varealtor. com/realtorinstitute). The CRS certification involves courses in marketing (e.g., Marketing Growth Strategies of the World's Top Agents), business planning (e.g., Business Planning and Marketing for the Residential Specialist), and customer service (e.g., Effective Buyer Sales Strategies). The GRI certification has the following curriculum: sales process (business development, sales and marketing, customer and client services, and cultural diversity), legal and regulatory issues (fair housing, brokerage relationships, contracts, and environmental issues), technology issues (communications, general technology, and real estate business technology), and professional standards (code of ethics, arbitration, and mediation).

Every state in the U.S. has its own set of real estate marketing associations (e.g., Virginia Associations of Realtors, or VAR). These professional societies offer a variety of designations or real estate specialties. For example, VAR offers several real estate specialties, such as Commercial Education, Property Management, Appraisal, Relocation, Broker/Manager, and the Realtor University.

Sales agents may choose to specialize in commercial real estate. In that vein, there are several professional designations related to this specialty: Society of Industrial and Office Realtors (SIOR, www.sior.com), Certified Commercial Investment Member (CCIM, www.ccim.com), Institute of Real Estate Management (IREM, www.irem.org), Realtors Land Institute (RLI, www.rliland.com), and The Counselors of Real Estate (CRE, www.cre.org). See Table 8.6.

Where to Train?

Many of the training seminars that are company-specific (e.g., company orientation) tend to be conducted *in-house*. Most of the other types of training seminars are conducted on location at conferences of professional associations. However, many real estate institutes provide their training courses in various formats: *classroom courses, webinars, online courses,* and *self-study*. For example, The Council of Residential Specialists (providing the CRS professional designation, https://crs.com/education) provides courses as webinars, classroom courses, online courses, and self-study.

Online courses are becoming increasingly popular, and many of the state-level associations of real estate marketing offer *online* courses. For example, the Virginia Association of Realtors (VAR) offers a 16-hour online course for salespeople (focusing on residential real estate sales). This course meets the 2-year license renewal requirement for salespeople in the State of Virginia. The course includes topics of instruction such as selling techniques, housing markets in Virginia rural and urban areas, and agency rules and procedures. VAR also offers a 24-hour online course for brokers (focusing on residential real estate). The course meets the 2-year license renewal requirement for brokers in Virginia.

Table 8.6 Institute, Societies, and Designations in Real Estate Marketing

Institute or Society	Designation
Counselors of Real Estate (CRE)	Counselor of Real Estate (CRE)
Institute of Real Estate Management (IREM)	Certified Property Manager (CPM) Accredited Management Organization (AMO) Accredited Residential Manager (AMO)
National Association of Realtors (NAR)	Graduate, Realtors Institute (GRI) Residential Accredited Appraiser (RAA) General Accredited Appraiser (GAA)
Real Estate Buyer's Agent Council (REBAC)	Accredited Buyer Representative (ABR) Accredited Buyer Representative Manager (ABRM)
Realtors Land Institute (RLI)	Accredited Land Consultant (ALC)
Realtors National Marketing Institute	Certified Real Estate Brokerage Manager (CRB) Certified Residential Specialist (CRS)
Society for Industrial and Office Realtors (SIOR)	Specialization in industrial and office real estate
Women's Council of Realtors (WCR)	Leadership Training Graduate (LTG) Referral and Relocation Certification (RRC)

Source: Adapted from Hamilton, D. (2006). *Real estate: Marketing and essentials.* Mason, OH: Thomson Higher Education (pp. 18–19).

In the U.S., the vast majority of community colleges (offering associate degrees in real estate) and universities teach courses in a classroom setting, and many are offering these courses online. For example, Virginia Polytechnic Institute and State University offers both an undergraduate major and a minor in real estate, and the courses taught in these degree programs are offered on campus in a classroom setting.

Who Should Train?

Who should lead the training depends on the topic. For example, a company orientation is typically taught by the *broker* or the *sales manager*.[26] However, other specialized topics are assigned to people who have developed considerable expertise in the designated area of specialty. Many specialized training seminars are conducted by *senior-level sales associates*. In other cases, professional *consultants* who have amassed a good reputation in conducting workshops on specific topics (e.g., ethics and law in real estate) are invited by a company or a local chapter of the state-level association of real estate agents location to conduct training workshops at a neighboring hotel or conference facility.

This, of course, does not mean to exclude degree programs offered at community colleges and universities in the U.S. The *professors* who teach these courses tend to hold at least a masters degree. Most faculty in universities' real estate programs have Ph.D. degrees.

Training Methods

There are many training methods commonly used in the real estate marketing industry. These include in-house group training, peer group training, field training, role playing, top-performer panels, seminars and workshops, and semester-like courses.[27] *In-house group training* is a training method that focuses on training salespeople in-house by bringing in a consultant and arranging a seminar or workshop on location. *Peer group training* refers to learning from peers. For example, the sales manager organizes a weekly meeting of the sales team and invites the sales associates at the meeting to share their experiences with others (the good, the bad, and the ugly). Much learning can take place by sharing experiences. The *field training* method is also hands-on learning. Here the junior-level sales associates learn by tagging along with the senior-level associates. The *role playing* method refers to mock sessions in which a sales associate is asked to rehearse and play a role he may not feel very comfortable with. For example, role playing is customarily used in teaching negotiations. One salesperson assumes the role of the buyer, the other the role of the buyer representative. The goal is to teach the salesperson playing the role of the buyer representative how to deal with buyer anxiety and how to counter objections in a controlled setting. *Top-performing panels* is assemble a panel of top performers that is moderated by the sales manager. The sales manager asks each panel member specific questions regarding best practices and takes questions from the audience (the lower-performing salespeople). The goal is to learn tidbits and best practice techniques from top-performing salespeople. Lastly, *seminars and workshops* involve a more formal method of instruction. In a classroom setting, the instructor uses traditional classroom teaching methods (lectures, class discussion, etc.) to impart knowledge.

Evaluating the Training Programs

Part of running an effective sales organization is making sure that the training programs are effective in achieving sales objectives. How does the sales manager evaluate the effectiveness of training programs? There are at least three evaluation methods that should be used in conjunction: (1) trainee satisfaction with the training, (2) trainee knowledge resulting from the training, and (3) trainee job performance.[28]

The *trainee satisfaction with training* is an evaluation method that is most commonly used in a variety of educational settings. The instructor administers evaluation forms at the end of the training session/seminar/workshop/course, and the trainee completes the form and turns it in to the administration anonymously. The evaluation form instructs the trainee to rate the instructor, the content of instruction, the materials used in the instruction, and the method of delivery using standardized satisfaction or quality rating scales (e.g., "Rate the instructor: very poor, poor, so/so, good, very good"). The training program can be evaluated as a direct function of the outcome of these ratings—good ratings signify that the training programs are effective, whereas poor ratings signify the opposite.

It is not sufficient to judge the effectiveness of a training program by asking the trainee to subjectively assess it. Satisfaction ratings of the training course may not be highly correlated with the ultimate bottom-line criterion: sales performance. There are many instances in which the training program is judged as inadequate not necessarily because of the lack of quality instruction but because of other characteristics: the instructor was overly demanding, the instructor did not interact with the trainees much, and so on. Thus, the effectiveness of a training program is not only judged based on trainees' quality ratings, but also based on some *knowledge* measure. Here the instructor administers an exam to the trainees to see whether the trainees gained the intended knowledge of the subject matter. For example, if the training program focused on sales negotiations, then the trainee is tested on the knowledge of negotiation principles covered by the training program. Passing such exams indicates that the training program is effective—effective in achieving the desired learning.

Again, the same argument leveled against the trainees' subjective evaluation of the program can be applied to the learning that trainees acquired. It may be that the salespeople have learned the material but failed to apply this knowledge in ways that boost their sales performance. Hence, the link between trainee aptitude and sales performance may not be strong. The ultimate measure has to be *sales performance*. As such, the sales manager has to monitor the level of sales performance of the trainee to ascertain improvements. If improvements in sales performance are observed, then the sales manager can infer that the training program is successful.

SUMMARY

This chapter covered many topics pertaining to recruitment and training of real estate agents. The chapter began by addressing the importance of recruiting and selection in real estate sales organizations. The primary goal of recruitment and selection is to fill vacant positions that ultimately ensure that the firm will gain more revenue through these positions. Other goals of effective recruitment and selection include high levels of job performance, job satisfaction, and organizational commitment, as well as lower levels of absenteeism, presenteeism, and turnover.

We then described the recruitment process in terms of five steps: analyzing the job, determining qualifications, prospecting sources for recruiting candidates, selecting prospects, and validating the process. Analyzing the job refers to the process which the sales manager develops a carefully crafted description for each position in the sales organization. A job description is essentially a description of activities and responsibilities inherent to a specified job. This may include position title, job objectives, who the person assuming this position reports to, the market territory or segment that the position is related to, the principal activities related to the job, and how the performance of the person in that position is evaluated and financially compensated.

Associated with each position in the sales organization are a set of qualifications. Job qualifications are customarily grouped in four major categories: knowledge, experience, personal characteristics, and legal considerations. Knowledge usually translates into a real estate sales license. Experience usually translates into previous employment either in real estate sales or a related sales profession. With respect to personal characteristics, we discussed these in terms of psychological characteristics, physical characteristics, and social characteristics. We also discussed legal considerations as part of qualifications. The sales manager has to abide by the many laws and policies governing recruitment and selection of real estate agents (e.g., in the U.S., these include the Civil Rights Act, the Age Discrimination in Employment Act, the Americans with Disability Act, the Equal Pay Act, the Pregnancy Discrimination Act, the Equal Employment Opportunity Commission, and Affirmative Action).

Then we shifted the discussion to prospecting sources. In that vein, we identified several sources, such as job postings/job board, newspaper advertising, employee/client referrals, employment agencies, educational institutions, career conferences, and professional societies. We also emphasized the importance of passive recruitment that emanates from the company's good reputation. In selecting prospects, we described a process that involves initial screening, reference checking, in-depth interviewing, employment testing, follow-up interviewing, and making the selection. The section on recruitment and selection was completed by a discussion on validating the process. The validation process focuses on measuring subsequent success on the job and modifying (if necessary) any aspect of the previous processes (job analysis, job qualifications, recruiting candidates, and selecting prospects).

The second section of this chapter covered training issues. In that vein, we started out this section by discussing the importance of training in real estate sales organizations. Training is important because it enhances real estate agent *competency*. This, in turn, helps to improve customer relations, enhance job performance and sales, increase job satisfaction, increase organizational commitment, reduce absenteeism, reduce presenteeism, and reduce turnover.

We described the training process as involving four distinct steps: assessing training needs, budgeting for training, developing training programs, and evaluating the programs. The sales manager assesses the training needs of the sales force using a variety of methods: sales force observation, sales force survey, management objectives, customer information, customer survey, and company records. With respect to budgeting for training, we described six methods: all you can afford, percentage of sales, percentage of profits, competitive parity, industry benchmark, and objective and task.

In developing training programs for the sales force, the sales manager asks and attempts to answer the following questions: What topics of instruction? Where to train? Who should train? And what training methods to use? In a typical real estate sales organization, the topics of instruction include company orientation, product knowledge, selling techniques, market-industry orientation, and professional designations. Many of the training seminars that are company-specific (e.g., company orientation) are conducted *in-house*. Most of the other types of training seminars are conducted on location at conferences. However, many real estate institutes provide their training courses in various formats: *classroom courses, webinars, online courses*, and *self-study*. Who should instruct the training depends on the topic. Training methods include in-house group training, peer group training, field training, role playing, top-performer panels, seminars and workshops, and semester-like courses. Finally, we discussed how the sales manager evaluates the effectiveness of training programs through trainee satisfaction with the training, trainee knowledge resulting from the training, and trainee job performance.

DISCUSSION QUESTIONS

1. What is the primary goal of effective recruitment and selection in real estate sales organizations? Secondary goals? Please explain.

2. What does the recruitment process entail? Describe the traditional steps that sales managers take in recruiting and selecting real estate agents.

3. What does it mean to analyze a job in the context of a real estate sales organization? Please explain.

4. Describe the personal characteristics of an effective salesperson in a real estate organization. How about physical characteristics? Social characteristics?

5. Describe some of the legal considerations related to qualifications in a traditional real estate sales organization.

6. How do sales managers or brokers go about publicizing the availability of sales positions in their firms?

7. What do we mean by "passive recruitment"?

8. How do sales managers go about selecting prospects? Describe the various steps involved in this process.

9. What do we mean by "validating the recruitment process"? Please explain.

10. Why is effective training of real estate agents important in a sales organization?

11. The training process involves four distinct steps. Please describe these.

12. How does the sales manager go about assessing training needs among the real estate agents?

13. How does the sales manager go about allocating resources for training?

14. How does the sales manager determine the topics of instruction?

15. Describe the various training methods deemed effective in training real estate agents.

16. How does the sales manager evaluate the effectiveness of training programs?

NOTES

1. Lehman, J. (2006). *The sales manager's mentor.* Seattle, WA: Mentor Press LLC. Cummings, B. (2005). Star search. *Sales & Marketing Management*, July, 23–27.
2. Baird, C. (2007). Recruiting: Tips for today's market and how it will affect your bottom line. *RISMedia.* Accessed from http://rismedia.com/2007–12–10/recruiting-tips-for-todays-market-and-how-it-will-affect-your-bottom-line, on April 30, 2012.
3. Greenberg, H., Weinstein, H., & Sweeney, P. (2000). *How to hire and develop your next top performer: The five qualities that make salespeople great.* New York: McGraw-Hill Professional Publishing.
4. Herd, R.L. (2003). *Real estate office management.* Mason, OH: South-Western (Chapter 5).
5. Introduction to the DISC Assessment. "Realtor Profiling System." Real Estate Champions website. Accessed from www.realestatechampions.com/DISC/, on July 29, 2013.

6. Introduction to the DISC Assessment. "Realtor Profiling System." Real Estate Champions website. Accessed from www.realestatechampions.com/DISC/, on July 29, 2013.
7. Introduction to the DISC Assessment. "Realtor Profiling System." Real Estate Champions website. Accessed from www.realestatechampions.com/DISC/, on July 29, 2013.
8. Introduction to the DISC Assessment. "Realtor Profiling System." Real Estate Champions website. Accessed from www.realestatechampions.com/DISC/, on July 29, 2013.
9. Introduction to the DISC Assessment. "Realtor Profiling System." Real Estate Champions website. Accessed from www.realestatechampions.com/DISC/, on July 29, 2013.
10. Zeller, D. (2012). Hiring—the ideal buyer's agent. *Realty Times*, April 27. Accessed from http://realtytimes.com/printpages/20120427_buyersagent.htm, on April 30, 2012.
11. Civil Rights Act of 1964. *Wikipedia*. Accessed from http://en.wikipedia.org/wiki/Civil_Rights_Act_of_1964, on July 30, 2013.
12. Age Discrimination in Employment Act. *Wikipedia*. Accessed from http://en.wikipedia.org/wiki/Age_Discrimination_in_Employment_Act, on July 30, 2013.
13. Americans with Disabilities Act of 1990. *Wikipedia*. Accessed from http://en.wikipedia.org/wiki/Americans_with_Disabilities_Act_of_1990, on July 30, 2013.
14. Equal Pay Act of 1963. *Wikipedia*. Accessed from http://en.wikipedia.org/wiki/Equal_Pay_Act_of_1963, on July 30, 2013.
15. Pregnancy Discrimination Act of 1978. *Wikisource*. Accessed from http://en.wikisource.org/wiki/Pregnancy_Discrimination_Act_of_1978, on July 30, 2013.
16. Equal Employment Opportunity Commission. *Wikipedia*. Accessed from http://en.wikipedia.org/wiki/Equal_Employment_Opportunity_Commission, on July 30, 2013.
17. Affirmative Action. *Wikipedia*. Accessed from http://en.wikipedia.org/wiki/Affirmative_action, on July 30, 2013.
18. Carter, M. (2011). Real estate recruiting: Don't let agents know they're in your sights. *inmanNEWS*. Accessed from www.inman.com/news/2011/07/28/real-estate-recruiting-don't-let-agents-know-theyre-in-your-sites, on April 30, 2012.
19. Brambila, A. V. (2013). Startup launches recruiting platform for brokers. *Inman News*, January 21, 2013, 21.
20. Hair, J. F., Anderson, R. E., Mehta, R., & Babin, B. J. (2009). *Sales management: Building customer relationships and partnerships*. Boston: Houghton Mifflin Company (p. 223).
21. Hair, J. F., Anderson, R. E., Mehta, R., & Babin, B. J. (2009). *Sales management: Building customer relationships and partnerships*. Boston: Houghton Mifflin Company (pp. 236–237).
22. Waller, B. D., & Jubran, A. M. (2012). The impact of agent experience on the real estate transaction. *Journal of Housing Research*, 21, 67–82.
23. Cron, W., Marshall, G., Singh, J., Spiro, R., & Sujan, H. (2005). Salesperson selection, training, and development. *Journal of Personal Selling & Sales Management*, 25, 123–136. Hair, J. F., Anderson, R. E., Mehta, R., & Babin, B. J. (2009). *Sales management: Building customer relationships and partnerships*. Boston: Houghton Mifflin Company (p. 250). Mendosa, R. (1995). Is there a payoff? *Sales & Marketing Management*, June, 64–71. Geber, B. (1995). Does your training make a difference? Prove it! *Training*, March, 27. Honeycutt, E. D., & Stevenson, T. H. (1989). Evaluating sales training programs. *Industrial Marketing Management*, 18, 215–222.
24. Waller, B. D., & Jubran, A. M. (2012). The impact of agent experience on the real estate transaction. *Journal of Housing Research*, 21, 67–82.
25. Hair, J. F., Anderson, R. E., Mehta, R., & Babin, B. J. (2009). *Sales management: Building customer relationships and partnerships*. Boston: Houghton Mifflin Company (pp. 255–257).
26. Kenner, W. F. (2006). Sales managers as trainers. *Selling Power*, September, 24–26.
27. Herd, R. L. (2003). *Real estate office management*. Mason, OH: South-Western (Chapter 6).
28. Leach, M., & Liu, A. H. (2003). Investigating interrelationships among sales training evaluation methods. *Journal of Personal Selling & Sales Management*, 23, 327–340.

Chapter 9

Motivation and Compensation Issues in Real Estate Marketing

LEARNING OBJECTIVES

This chapter is designed to help students of real estate marketing learn:

- how motivational policies and programs affect organizational outcomes
- the difference between intrinsic and extrinsic motivation
- extrinsic motivational factors and how they work
- intrinsic motivational factors and how they work
- how compensation plans affect organizational outcomes
- the compensation process
- common elements in a compensation plan
- how the sales manager packages a compensation plan for an individual salesperson
- how the sales manager goes about validating the effectiveness of the compensation plan

WHY ARE MOTIVATIONAL PROGRAMS IMPORTANT?

Motivational programs are aspects and conditions in the organization that help boost sales effort, which in turn raise sales performance. In this chapter we will discuss a variety of motivational programs, such as pay, company policies, working environment, interpersonal relationships, achievement programs, recognition programs, job enrichment, career advancement opportunities, and work-life balance programs.

Motivational programs play an extremely important role in maintaining and enhancing the financial viability of the real estate sales organization. Much evidence points to the fact that motivational programs are positively associated with the job performance of the sales force, job satisfaction, job morale, and organizational loyalty and commitment. The evidence also suggests that motivational programs are negatively associated with employee presenteeism (being present on the job without being productive), absenteeism (being absent from the job), and turnover (the rate of job quitting).[1]

INTRINSIC VERSUS EXTRINSIC MOTIVATION

Much of the research literature on motivation makes a distinction between intrinsic and extrinsic motivation. Human motivation can best be understood using the traditional distinction between low-order and high-order needs; see Figure 9.1. Low-order needs are essentially needs related to survival: biological needs (needs for food, water, air, sex, etc.), health and safety needs, social and family needs, and ultimately the minimum financial resources for the sustenance of oneself and one's family. Low-order needs are basic needs related to survival and the propagation of the human species. High-order needs are related to growth. They include a wide assortment of needs, such as social and relatedness needs, esteem and effectance needs, self-actualization needs, needs related to aesthetics and creativity, and intellectual and autonomy needs, among others. Many of these human needs are met through one's career. The workplace is recognized to be an important means to meet both low-order and high-order needs.[2]

Much research has demonstrated that a job that fails to meet basic needs causes much job dissatisfaction (negative feelings and emotions such as anger, fear, anxiety, despair, hopelessness, and depression) that spills over to cause life dissatisfaction as well. In addition, a job that meets only basic needs does not contribute much to job satisfaction or positive emotions (e.g., happiness and joy). A good job that meets basic needs can provide only relief, not joy or happiness. Conversely, a job that satisfies high-order needs can contribute significantly to positive emotions such as happiness that spill over to life satisfaction and psychological well-being. A job that fails to meet high-order needs is not likely to cause much job satisfaction.[3]

It should be noted that social needs are treated as belonging to both low-order and high-order needs. That is, social needs have the ability to generate strong positive and negative emotions. They act as both satisfiers and dissatisfiers. Satisfying social needs at work causes positive emotions (happiness, elation, joy, content, and laughter); at the same time, failing to meet social needs can be a source of much discontent, frustration, anxiety, anger, and unhappiness. Social needs are important sources of motivation in that they cut both ways— their satisfaction is related to strong positive emotions and their dissatisfaction is related to strong negative emotions.

Figure 9.1
High-Order
and Low-Order
Needs

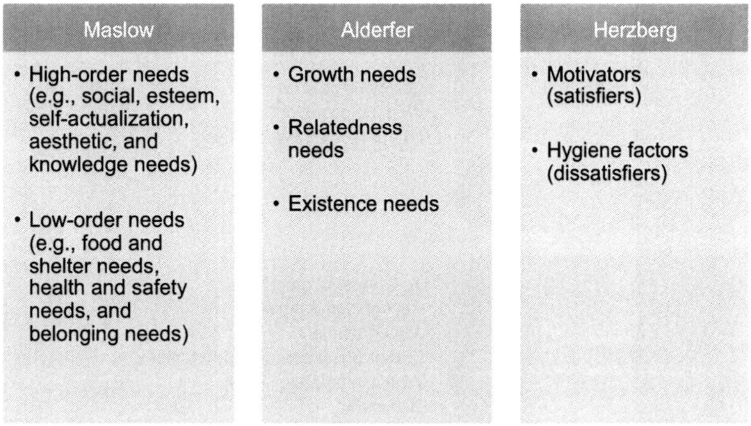

Maslow	Alderfer	Herzberg
• High-order needs (e.g., social, esteem, self-actualization, aesthetic, and knowledge needs)	• Growth needs	• Motivators (satisfiers)
	• Relatedness needs	• Hygiene factors (dissatisfiers)
• Low-order needs (e.g., food and shelter needs, health and safety needs, and belonging needs)	• Existence needs	

EXTRINSIC FACTORS

Extrinsic factors are essentially factors affecting the salesperson's level of performance motivation that are related to low-order needs (basic needs such as the need for economic resources to feed and shelter oneself and one's family and social needs). In the context of real estate marketing, several extrinsic factors will be discussed. These are pay, company policies, working environment, and interpersonal relationships. See Figure 9.2.

Pay

To ensure that salespeople do not experience job dissatisfaction, the sales organization has to provide adequate pay—adequate to meet the salesperson's basic needs (to provide for food, shelter, health, safety, and social needs).

How does pay motivate salespeople to do their best? Management psychologists use expectancy-value theory to explain this process. Expectancy-value theory asserts that pay is essentially a reward outcome. This reward outcome is used to achieve personal goals (i.e., a salesperson uses pay to satisfy a variety of needs, both low-order and high-order). The value of pay (what management psychologists refer to as *valence*) is established by its psychological association with achievement of personal goals. A salesperson perceives that a certain amount of pay can be attained through sales commission-based M dollar amount of sale. The psychological connection between pay and performance is thus established. This is what management psychologists refer to as *instrumentality*. That is, selling X number of units at Y dollars is instrumental in generating Z amount of sales commission. The question is, then: What effort would it take to sell X number of units at Y dollars? The effort-performance link is what management psychologists refer to as *expectancy*. For example, a salesperson may believe that if he contacts at least 10 new prospective clients a month, he may be able to meet or exceed the sales quota of selling X number of units at Y dollars. This is his expectancy that links his effort (contact at least 10 new prospective clients a month) with performance (sell X number of units at Y dollar), which in turn links with outcome (generate M dollars in sales commission or PAY), which links with personal goals (use money to maintain a desired standard of living)[4]. See Figure 9.3.

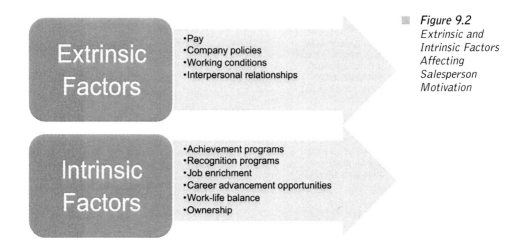

Figure 9.2
Extrinsic and Intrinsic Factors Affecting Salesperson Motivation

Extrinsic Factors
•Pay
•Company policies
•Working conditions
•Interpersonal relationships

Intrinsic Factors
•Achievement programs
•Recognition programs
•Job enrichment
•Career advancement opportunities
•Work-life balance
•Ownership

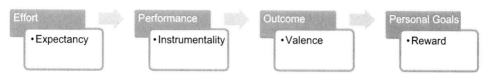

Figure 9.3
Expectancy Theory: The Motivational Effect of Pay

Most importantly, the sales manager has to ensure equity among members of the sales team. Members of the sales team who perceive that they are treated unfairly by the sales manager or other members of the sales team are likely to feel dissatisfied with their job, which, in turn, may become a significant factor in absenteeism and turnover.[5]

It is important to note that pay has intrinsic motivational features, in that higher pay can be used as a sign for achievement and recognition. In other words, salespeople typically use money as a gauge for achievement—the more they make, the more they feel proud of their performance. The high pay means "job well-done."[6] Also, in sales as a profession, the vast majority of salespeople (including real estate salespeople) enter this profession to "make lots of money." In other words, pay for salespeople is not simply about making a living. It is about getting rich—a life goal that has intrinsic motivational value. As such, salespeople respond particularly well to financial rewards; money motivates them in both intrinsic and extrinsic ways.[7]

Working Environment

Consider the following definition of the working environment: the physical workplace; the availability of relevant technological or communication resources; workplace security and safety; the cleanliness and ergonomics of the workplace; the level of noise and the lighting; the availability of canteen facilities; the proximity to public transport, shops, and facilities; the codes of behavior that define the workplace; the factors that make work physically possible, safe, pleasant, and convivial.[8] We can break down this definition further into the physical working environment, health and safety, ergonomics and space design, workplace design, privacy, lighting, and noise.

The physical working environment has an effect on motivation. An office that is dingy and dull is likely to adversely affect salesperson motivation. A work environment that is not healthy and safe is most likely to dampen salesperson motivation. Similarly, the ergonomics and space design (e.g., the color of the walls in the office, the furniture type and quality, the presence of an exterior view and natural light, the ability to see other salespeople working) are all factors that may affect motivation, too.[9]

How about the workplace design? Is an open-plan office conducive to motivation and performance in a real estate marketing firm? The level of noise in an open space may be detrimental to motivation and performance. Private cubicles may be the best workplace design. Much research has shown that a good physical environment supports a sense of well-being among workers.[10]

We also have the non-physical working environment, in addition to the physical environment. The non-physical environment may involve the company rules and regulations, the way things are done in the office, pressure from upper management to generate high dollar sales, and so on. This psychological environment plays an important role on salespersons' levels of motivation and performance.[11]

Interpersonal Relationships

Positive relationships between the sales manager and his sales team are essential to high motivation and productivity. So are relationships among members of the sales team. The sales manager has to ensure that relationships are collegial; otherwise, lack of collegiality and personality conflicts among members of the sales team can cause much dissatisfaction. Psychological theory posits that people engage in interpersonal relationships not only for goal attainment (i.e., accomplishing a task at work) but also because interpersonal relationships satisfy a very basic human need—relatedness needs. In a real estate marketing context, high quality relationships among sales associates as well as between the sales manager and sales associates (relationships that foster trust in others and confidence in oneself) serve to nourish the associates' psychological needs by providing social support and positive feedback about their selling competency. Ultimately, high-quality relationships in real estate sales organizations help maintain optimal job performance, whereas poor relationships usually lead to burnout.[12]

INTRINSIC FACTORS

Intrinsic factors are essentially factors affecting the salesperson's level of performance motivation that are related to high-order needs (growth needs such as social, esteem, self-actualization, aesthetic, and knowledge needs). For example, many of the top sales performers in real estate acknowledge that being an independent contractor—being one's own boss and managing one's own time—is very important.[13] We will discuss several intrinsic factors in this section: achievement programs, recognition programs, job enrichment, career advancement opportunities, work-life balance programs, and ownership. See Case/Anecdote 9.1.

CASE/ANECDOTE 9.1. REAL ESTATE AGENT RATED HAPPIEST JOB IN AMERICA[i]

A survey conducted by CareerBliss found that real estate agents have the happiest job in America. The survey was based on 65,000+ career reviews in which job holders were asked to rate their job happiness. Real estate agents rated their jobs higher on a "happiness scale" than other people rated their own jobs in positions such as senior quality assurance engineer, senior sales representative, and construction superintendent.

The CEO of CareerBliss attributed the result to the fact that the survey was conducted at a time when it was a seller's market (i.e., a boom in housing market demand). Real estate agent commissions were high, and compensation is an important factor in job happiness. Jobs rated very low on the happiness scale included clerk, associate attorney, and customer service associate.

[i] Rejoice: Real estate agent rated happiest job in America. *Commonwealth*, 2013, June/July, 9.

Achievement Programs

Achievement programs involve incentives or sales contests that reward sales performance. For example, the sales manager announces that sales associates who exceed their quarterly sales quota will be awarded a one-week vacation in Hawaii for two, paid in full. Such an incentive serves to motivate the sales force above and beyond bragging rights. When a salesperson excels in selling and does exceed the sales quota, he feels a sense of pride in his achievement. This sense of pride is further reinforced by the monetary reward related to the Hawaii vacation trip. In addition, he gains public recognition from the achievement, in the sense that his sales force partners witness this announcement, which again reinforces his feelings of pride.

There is much evidence in the personal selling and sales management literatures that suggest that sales contests do increase motivation and morale of salespeople. Increased motivation and morale leads to greater sales effort, which results in increased sales performance.[14] Furthermore, research has shown that a sales contest that has a limited number of winners (as opposed to too many or too few) is likely to have the most effect on salesperson motivation and performance. Also, cash awards seem to be more motivating than travel awards; awards with greater award value are more motivating than those of lesser value.[15] However, not all incentives and sales contests are effective. See Zooming-In Box 9.1 to understand why.

ZOOMING-IN BOX 9.1. CAVEAT! SALES CONTESTS CAN ALIENATE TEAM MEMBERS[i]

An example of a sales contest may be to divide the associates into two teams, and each week the teams are awarded points for making a sale. The winning team earns a paid vacation to some exotic place. Sales contests are an established tradition in the real estate marketing industry. Top producers shine through sales contests. Monthly, quarterly, and annual competitions are common in the realty business. The awards are in the form of inscribed wall plaques, ribbons, dollars, and even diamonds. A sales contest motivates sales associates to excel to greater heights. The sales associates feel energized to do their best in selling and obtain monetary reward for their achievement, as well as public recognition for their selling performance. Injecting some degree of competition is thought to motivate sales people to ramp up production. Sales awards are a good motivational tool. Sales agents love the recognition. Sales awards foster a sense of achievement and purpose, which is important in an industry fraught with much rejection. But contests can work as a motivational tool. Plaques, crystal, and other gifts are a cost-effective way of rewarding top producers and retaining their loyalty. Sales associates who win sales contests feel good about themselves; and they use their accolades as promotion tools, which do impress some clients.

However, it is not all rosy in the land of sales contests and achievement programs. Some evidence is pointing to the fact that sales contests have a downside. Sales contests can backfire by alienating the non-top performers—those who don't get the awards. Many sales people are downright emphatic about their disdain for sales contests.

Some real estate agents quit a brokerage firm because of sales contests and move to other firms that explicitly have no contest policies. They believe that competition

---

breaks down camaraderie among members of the sales force. Individual recognition is nice, but what is more important is working together as a team and focusing on meeting the needs of clients. Sales contests take away from focusing on serving clients and toward making more money, perhaps at the expense of clients' needs.

Another argument against sales contests is that competition does not increase sales production because some top producers boost sales figures by buying and selling their own properties. Also sales contests are typically based on dollar sales, not unit sales. Some sales associates may be huge producers in the low-end market but these associates are not recognized because the total amount of the sales dollars is not high, even though they may have high levels of transactions and happy customers.

Some junior-level sales associates feel discouraged when their sales performance figures are compared to the seasoned pros (the senior agents with much more experience). In other words, the sales contests ends up doing exactly the opposite of what it is supposed to do—motivate the sales force. And some sales associates who like privacy hate the limelight that comes with awards and recognition. Hence, sales contests may cause the second-string associates to feel less loyal to the brokerage firm because they may feel they are never going to be recognized. This translates into turnover that cuts into the firm's bread and butter.

The most effective contests are tailored to the personalities of the sales associates. This is due to the fact that not everyone is motivated by the same thing. For example, extroverts and introverts can both excel at sales, but each group is driven by different sources of motivation. Extroverts may enjoy applause and plaques on the wall. In contrast, introverts may take pleasure in being recognized by the sales manager, they don't enjoy the spotlight. Most associates are motivated by money, fun, relationships, or stability. The challenge is to match the reward with the personality. Such matches are likely to produce the best results.

Perhaps the sales manager can use DISC (Dominance, Influence, Steadiness, and Conscientiousness)—the oldest of the personality assessments. Specifically, the sales manager can tailor sales incentives based on personality. For example, don't offer a power lunch to associates who are high in conscientiousness, because they may prefer something low-key.

[i] McKuen, P.D. (2010). Do sales awards matter? Sales contests can alienate team members even when your goal is to boost morale. *Realtor Magazine,* accessed from http://realtormag.realtor.org/for-brokers/feature/article/2010/03/do-sales-awards-matter, on July 20, 2013. Motivate your sales force. *Realtor Magazine,* March 2006, accessed from http://realtormag.realtor.org/for-brokers/feature/article/2006/03/motivate-your-sales-force, on July 20, 2013.

Recognition Programs

The U.S. Department Labor Statistics has data showing that the "number one" reason employees leave their companies that they don't feel appreciated.[16] In real estate marketing, recognition programs take form in the public recognition of top performers. Typically, the sales manager places a display ad in the local newspaper announcing the winners of sales awards in different categories—the top category is the platinum award, followed by the diamond award, the gold award, the silver award, and the bronze award. The awardees are also given a plaque that is typically hung on the wall in the main office or in the salesperson's private office. Recognition programs, like achievement programs, contribute to feelings of satisfaction (e.g., pride, self-esteem, sense of achievement, social approval from significant others, a sense of belonging to top performers—an elite class). See Case/Anecdote 9.2 for an account of how important recognition is to incoming sales associates.

CASE/ANECDOTE 9.2. THE PERSONAL HONOR[i]

Tony Macaluso, CRB, broker-general manager of Schwartz Property Sales Inc., Better Homes and Gardens, Key Largo, Florida, swears that recognition is the most effective tool to motivate a sales force today. Macaluso gives plaques to the top producers on a monthly and quarterly basis. The recognition event inspires the newcomers. Just participating in these events at which they see accolades given to top-producing associates is a major source of motivation, especially for those who seek high regard from their peers.

Another source of motivation for newcomers is to make an actual sale. Making a sale can be inspiring and provides a major boost to help the newcomer continue on. The newcomers are likely to face plenty disappointments. Hence, helping them close a deal can be all-inspiring. This assistance can be in the form of training or personal coaching. Macaluso says, "If they make lots of calls but don't get appointments, they need more education in that area. If they get lots of appointments but don't bring in any listings, we know they need help there." Developing mentoring relationships with newcomers can boost morale in a major way.

[i] DeZube, D. (1999). Motivating your sales team: Has the rah-rah gone the way of the Gipper? *Realtor Magazine,* March 1999. Accessed from http://realtormag.realtor.org/for-brokers/feature/article/1999/03/motivating-your-sales-team, on July 20, 2013.

Job Enrichment

Job enrichment typically involves providing a salesperson with greater autonomy and responsibility. A senior salesperson, for example, might mentor a rookie salesperson. More examples: A broker asks a senior sales associate to head the recruiting committee, or suggests that she present a seminar involving best practices at a regional conference related to real estate marketing. Greater responsibility and autonomy on the job serve to satisfy growth-related needs, such as the need for autonomy and competence. Goal attainment related to these needs induces satisfaction and sense of well-being.[17]

A very popular model of job enrichment is called the *Job Characteristics Model*. The model argues that enriched or complex jobs are best composed of five core job characteristics: task significance, task variety, task identity, autonomy, and feedback. These five job characteristics are positively associated with job satisfaction. The extent to which these five factors of job enrichment influence job satisfaction depends on whether the employee has high or low growth needs. Employees with high growth needs have stronger high-order needs (social, esteem, actualization, knowledge, and creativity needs) and are likely to respond with enthusiasm to jobs enriched, whereas those with lower levels of growth needs are likely to respond with apathy to enriched jobs.[18] Going back to our discussion of real estate agents, we can say that autonomy and responsibility should enrich the job of sales associates, and this type of job enrichment is likely to motivate the associates to work harder. However, based on our understanding of the research related to the Job Characteristics Model, the boost of motivation related to job enrichment works best for those associates who have higher levels of growth needs (i.e., they have strong high-order needs).

Career Advancement Opportunities

Promotion within the sales organization can be a strong source of motivation to excel at the highest levels possible. In real estate sales organizations, the ranks may involve two tracks: one track related to personal selling and the other to sales management. The career track related to personal selling includes positions such as sales associate, climbing to senior sales associate, and ultimately to a partner position (in the case of brokerage firms). With respect to sales management, there is a sales manager position. However, in large real estate sales organizations, there is the associate sales manager position, which is used as a stepping stone to the full sales manager position. A sales manager can also achieve partner status. In large real estate organizations encompassing a large geographic territory (e.g., a national or international real estate development firm), the sales manager position can be further decomposed into a local sales manager (a sales manager overseeing a local sales force), a regional sales manager (a sales manager overseeing several sales managers over a large region), a vice president of sales (sitting at the corporate level), and possibly a senior vice-president of sales. These different positions allow sales associates and sales managers to climb up the corporate ladder in the real estate business. Doing so is a strong motivator for job performance and is a major source of positive emotions, such as sense of achievement, a sense of purpose in life, recognition from one's peers, and feelings of self-actualization. See Case/Anecdote 9.3 for one study on perceptions of real estate careers.

CASE/ANECDOTE 9.3. COLLEGE STUDENT PERCEPTIONS OF SALES CAREERS IN REAL ESTATE[i]

A survey was conducted at two traditionally African American universities in the U.S. and three predominantly Caucasian American universities. The survey focused on sales careers in six industries, including real estate sales. The other industries were consumer product sales, industrial product sales, insurance sales, international sales, and sales of securities/financial services. The survey showed that a sales career in real estate had great appeal among college students. It ranked second after international sales.

Most students believed that the starting salaries in real estate sales are second highest (international sales being ranked first). One striking finding is the perception of autonomy that a career in real estate sales affords. The majority of students perceived that real estate salespeople have a lot of autonomy; they ranked real estate sales first vis-à-vis the other sales careers.[i]

[i] DelVecchio, S., & Honeycutt, E. D., Jr. (2000). An investigation of African-American perceptions of sales careers. *Journal of Personal Selling & Sales Management, 20*, 43–52.

Work-Life Balance Programs

Work-life programs are designed to allow salespeople to attend to family needs and "have a life" outside of the office. Salespeople who do a good job at work plus have a full life outside the office are likely to be less stressed and more happy than those who fail to maintain a work-life balance. There are several work-life programs that sales organizations offer to their sales teams. These include flexible working hours, flexible leave arrangements, reduced working time, and flexible location. *Flexible working hours* refer to flexible schedules and management by results—the performance of the salespeople is judged not by how many prospective clients are contacted but by sales results. *Flexible leave arrangements* involve annual leave, maternity leave, paternity leave, emergency leave, employment or career break, sabbaticals, and educational leave. *Reduced working hours* refer to changing one's full-time job to part-time (with the right to return to full-time work in the specified future). *Flexible location* refers to telecommuting (i.e., working from home using e-mail, e-conferencing, and other telecommunication technologies).[19] See Zooming-In Box 9.2 for tips for real estate sales associates on how to manage work-life balance.

ZOOMING-IN BOX 9.2. ACHIEVING WORK-LIFE BALANCE IN REAL ESTATE SALES[i]

Achieving work-life balance is not easy for many real estate sales associates who have to juggle family and career. What makes it particularly difficult is the fact that sales associates are on call 24/7. The good thing is that the job offers some degree of flexibility in scheduling. The following tips are from real estate professional and time management experts.

One strategy to achieve work-life balance is the *team approach*. This approach is essentially having two or more associates team up to help one another. They swap referrals and assist each other's sellers and buyers. Partnering helps by allowing the associate to control his schedule by divvying up some responsibilities. Some associates use a *virtual assistant* to help out. Virtual assistants are personal assistants who work off-site to assist on a number of tasks—blogging, web design, writing listing ads, marketing coordination, etc.

Another major approach is to *prioritize*. This means time management. Time management means a process that involves (1) identifying priorities, (2) organizing the physical workspace, (3) creating more than one to-do list, (4) taking an "hour of power," and (5) planning ahead and maintenance. With respect to *identifying priorities*, one can do the following exercise. Draw a table with four cells (quadrants) and label these cells to reflect the most important life domains (e.g., job, family, health, and leisure). Then figure out the percentage of time you spend in these four areas of your life. Are you balanced enough in terms of spending time in these four areas? If for example, you notice that you spend 85 percent on your job; 5 percent in your family, health, and leisure life; then you must conclude that your life is imbalanced. To restore balance, make some strategic decisions to spend more time in your other important areas of your life and less in that life domain that seems to usurp much of your time and energy. The next step is to *organize your physical workspace*. This means organize, organize, and organize. An organized life is a well-managed life.

An organized life is time managed. For example, if your desk at brokerage firm is cluttered, try to organize it in a way that you can get to things easier—so it would not take you much time to do things. Lack of organizations is very wasteful, time-wise. The third step is to *create more than one to do-list*. One list focuses on today's activities, the second list is activities for the entire week, and the third list is essentially a master list that has a longer time span (perhaps 6 months or a year). The next step is *take an "hour of power."* This means that when you focus on a specific task, don't allow interruptions. One way to do this is to "unplug" from your digital telecommunication devices. Don't look at you e-mail. Don't answer your text messages immediately. The final step is *plan ahead and maintain*. This means that at the end of the day, you try to review what you have done and what needs to be done tomorrow and in the near future. Allotting some maintenance time every few days or a week to make sure that the things that keeps you functioning on a daily basis are working (e.g., computers, phones, printers, automobile). Maintenance activities may include taking care of your finances (e.g., paying bills, making deposits at the bank).

[i] Tracey, M. D. (2013). Yes, you can achieve work-life balance. *Realtor Magazine,* June 2013, accessed from http://realtormag.realtor.org/sales-and-marketing/relationship-management/article/2013/06/yes-you-can-achieve-work-life-balance, on July 21, 2013. Caputo, S. (2013). The productivity puzzle: What's your missing piece. An e-book that can be accessed from www.theproductivitypuzzle.com/.

Ownership

Ownership in the firm may take place in the form of providing the salespeople with stock options and/or partnership. We discussed partnership in the context of career advancement opportunities in the preceding section. Ownership is a great motivator and an important source of positive emotions (sense of pride and achievement, feeling that one's life has meaning, establishing a legacy, self-actualizing experience, etc.). Also, research has shown that ownership does motivate sales agents to higher levels of productivity.[20] See Case/Anecdote 9.4 for an example of a brokerage firm that used ownership to motivate its sales force.

CASE/ANECDOTE 9.4. STAKE IN OWNERSHIP GIVES BOOST TO SALES[i]

When Fidelity Real Estate Group Inc., in Colorado Springs, Colorado (www.fidelity-group.com), offered a stake in ownership to its sales associates, the firm noted a significant boost in sales. Monika Newman, president and managing broker, believes it is a winning strategy. Since August 1995, Fidelity has allowed each sales associate to become a shareholder in the company, with an equal vote in fees and budgeting.

This is how it works. Each sales associate buys into the company by paying $5,700. This is paid directly or through deductions from commissions over a 3-year period. Each shareholder then pays a monthly fee of $1,850. This fee covers all administrative expenses except advertising. (Different sales associates choose their own devices to advertise their listings; hence, the agents each pay their own advertising.) Each agent gets 100% of the sales commission.

Results show that this plan has increased sales performance: $67 million in 1995 with 64 salespeople, $50 million in 1996 with 18 associates, $46 million in 1997 with 15 associates, and $52.3 million in 1998 with 12 associates.

[i] Stake in ownership gives boom to sales. *Realtor Magazine,* June 1999, FM8.

WHY ARE COMPENSATION PLANS IMPORTANT?

Compensation plans play an extremely important role in maintaining and enhancing the financial viability of the real estate sales organization. Much evidence points to the fact that proper compensation is positively associated with the job performance of the sales force, job satisfaction, job morale, and organizational loyalty and commitment. The evidence also suggests that proper compensation is negatively associated with employee presenteeism, absenteeism, and turnover.[21]

THE COMPENSATION PROCESS

The compensation process involves three stages: (1) identification of the compensation plan elements, (2) packaging the elements for individual salespersons, and (3) validating the process. See Figure 9.4.

Identification of the Compensation Plan Elements

Elements of a compensation plan involve direct and indirect elements. Direct elements include salary, commission, wages, referral fees, expense account, incentives, bonuses, profit sharing, and stock options. Indirect elements include health and life insurance, social security benefits, retirement plans, paid time off, and other perks. See Figure 9.5.

Figure 9.4
The Compensation Process

Direct Elements of Compensation	Indirect Elements of Compensation
• Salary • Commission • Wages • Referral fees • Expense account • Incentives • Bonuses • Profit sharing • Stock options	• Health and life insurance • Social Security benefits • Retirement plans • Paid time off • Other perks

Figure 9.5
Direct versus Indirect Elements of a Compensation Plan

Salary is a form of financial compensation typically provided to the management and administrative staff of the real estate sales organization (e.g., sales manager, broker, accountant/bookkeeper, office manager, administrative assistant, receptionist, webmaster, and in-house legal counsel). Salary is a form of financial compensation involving an annual payment (usually disbursed in weekly, bi-weekly, or monthly payments). Salaries are typically conjoined with indirect elements of compensation (e.g., health and life insurance, social security benefits, retirement plan, paid time off, and other perks). About 5% of the real estate agents in the U.S. receive a salary (salary plus bonus, to be exact). The advantage of this method of compensation is control—the sales manager has more control over the sales team. It also works well when recruiting salespeople who are inexperienced in real estate.

Commission is a form of financial compensation in the form of a percentage of the sale of the property. For example, real estate agents tend to be compensated in terms of a commission from the total sale price of the real estate property. Typically, the brokerage firm representing the buyer claims 3% of the total sale price (paid by the seller), while the brokerage firm representing the seller claims another 3% of the total sale price—a total of 6% commission. The listing agent, in turn, gets an average of 70% of the money coming in to his brokerage firm; and similarly, the buyer agent usually gets 70% of the money claimed by his brokerage firm. In other words, the 3% of the total sale price that the seller pays to the listing broker is split 70–30 between the listing agent and the listing broker. The 30% that the listing broker keeps is considered overhead costs (cost to manage and run the sales organization). The same can be said in regards to the split of the commission between the buyer broker and the buyer agent. Today, there are many real estate sales organizations in the U.S. that are charging a fixed "administrative brokerage commission." This type of commission is not based on a percentage of the sale (e.g., 3% for listing the property), but a flat fee (e.g., $3,000). Furthermore, 17% of sales agents work with 100% commission. These sales agents are essentially "captains of their ships," in the sense that the sales agents pay a flat monthly fee to the broker for operating under his firm's umbrella, and they retain 100% of the commission. Re/Max real estate is one such sales organization. The 100% commission structure works best with highly experienced sales agents; it does not work with inexperienced agents.[22] Furthermore, many sales jobs in the real estate industry that are compensated through straight commission are accompanied by what is called "drawing accounts." A draw is a sum of money paid against future commissions. This advance against future commissions provides the service of giving salespeople the security of a fixed income while providing an incentive for productivity.[23]

Referral fees are fees paid to the agent for making a referral. Consider the following example. Associate X (belonging to Broker X) refers a buyer to Associate Y (belonging to Broker Y) in exchange for a 30% referral rate. Broker Y pays Broker X $3,000, which is 30% of the gross $10,000 commission. This referral fee ($3,000) is then split between the Broker X and Associate X based on some agreed-upon formula (e.g., 50–50 split). The remaining $7,000 from the gross commission ($10,000) is then split between Broker Y and Associate Y based on their agreed-upon formula (e.g., 50–50 split).

Wages are also a means of financial compensation, but in the form of short-term payments with little or no indirect elements. A wage is typically given to a staff member who may work part-time (e.g., webmaster, office receptionist, and accountant/bookkeeper).

An *expense account* is a form of payment designed to support professional expenses related to the selling task. Expense accounts are common in large sales organization involved in commercial real estate. The sales agent in this case usually incurs expenses related to travel, lodging, meals, and gifts. These expenses are paid from an expense account provided to the sales agent.

An *incentive* is a form of financial remuneration designed to reward sales agents for having achieved target sales objectives. It can be in the form of cash payment or a paid-for product or event (e.g., a one-week paid vacation in Hawaii covering travel and lodging expenses for two people).

A *bonus* is another form of an incentive that is typically given to the sales team at the end of the year (e.g., Christmas bonus). Traditionally, a bonus is contingent on the overall success of the sales team (i.e., if the entire sales team meets its management quota, then the entire sales team receives a year-end bonus).

Profit-sharing in real estate business works as follows: It is based on the number of salespeople an associate brings into the company. An associate sponsors a recruit, and the sponsoring associate earns a percentage of the commission that the recruit makes. Typically, the percentage the sponsoring associates makes from the recruit declines in a graduated rate and falls to zero in time (e.g., after a full year).

Stock options are another form of financial remuneration in which the employee is given the option to buy company shares. This is usually an attractive option in companies that are national or global in scope. In many companies, the purchase of company stock is paid for by both employee earnings and company contribution. In other words, the company helps employees buy company stock by perhaps matching their contributions (e.g., the employee pays 50% of the stock while the company pays the remaining 50%).

With respect to the indirect elements of the compensation plan, *health and life insurance* are common. Many real estate sales organizations, especially the large ones, offer health and life insurance to all full-time employees. The costs of a typical health and life insurance policy are typically shared between employer and employee (e.g., 50–50 split, where the employee pays 50% of the cost, with the remaining 50% covered by the employer). Many of the small-scale realty companies in the U.S. do not provide health and life insurance. Individual agents have to fend for themselves by securing their own policies. Even some large-scale realty operations (e.g., Re/Max) do not provide health and life insurance, but they also do not take money from the sales commission (the entire sales commission goes to the individual sales agent). Instead, the realty company charges each sales agent a monthly fee (a form of "rent").

Social Security benefits are provided to an employee when the employee is temporarily disabled (and cannot work) and as retirement benefits. Many real estate associates who are independent contractors (i.e., those who essentially work for a real estate brokerage firm by paying rent to the firm as independent contractors, receiving a 100% of the sales commission instead of receiving only part of the sales commission) feel the burden of having to pay from their own pocket Social Security and Medicare taxes. For those who are essentially employees of a real estate sales organization rather than independent contractors, the employer covers half of Social Security and Medicare taxes.

A *retirement plan* is another element of indirect compensation in which the employer contributes an agreed-upon amount of money toward the employee's retirement.

Paid time off refers to paid vacation days or sick leave. In other words, the employee is entitled to a certain number of days or weeks to be used for vacation and another number of days for illness.

Other perks may include educational expenses, expenses related to commuting, percentage of the company profits (in cases involving partnerships), etc.

Packaging the Elements of the Compensation Plan for Individual Salespersons and Groups

Customarily, the sales manager attempts to forecast the amount of unit and dollar sales anticipated from each additional salesperson. The sales manager then decides on a profit margin (e.g., 70%), after which the sales manager computes the cost to each salesperson (i.e., the amount of money compensated for each sales associate). For example, the expectation is that each salesperson is likely to generate $1 million in sales revenue per year. The cost for generating that amount is $500,000 (costs related to the product manufacturing, distribution, and promotion). The profit margin is set at 70%—that is, the sales organization is expected to make 70% of $500,000 ($1,000,000 – $500,000) from each salesperson, which amounts to $350,000. The remaining money ($500,000 – $350,000 = $150,000) should cover the salesperson's total annual compensation. Note that there is a direct tradeoff between the profit margin and the total compensation provided to each salesperson. The higher the profit margin, the lower the salesperson's compensation, and vice versa. Here, the sales manager has some latitude to decrease the profit margin to allow the salesperson to receive a higher level of compensation. How does the sales manager make these tradeoff decisions? Also, how does the sales manager divide the total compensation into its constituent direct and indirect compensation elements? To answer these questions, we should try to understand the various factors that influence the sales manager's decision-making: individual, organizational, market, and industry-related factors. Let's discuss these in some detail. See Figure 9.6.

Individual Factors

Individual factors are conditions unique to the individual salesperson such as the (1) exact *job position* that the individual assumes, (2) past sales performance, (3) learning periods, and (4) needs and preferences. In every sales organization, each job (junior sales associate, senior sales associate, sales manager, associate sales manager, etc.) is clearly identified and a range of pay is specified (e.g., pay scale of associate sales manager is $70,000–$90,000) based on qualifications. In other words, the person taking on the specified position would get paid within a range as a direct function of the level of experience and educational credentials—those with more experience and education getting paid at the higher end of the scale. This, of course, applies to positions that call for straight salary. Such pay scales cannot be set for jobs involving straight commission. In the latter case, the pay scale is highly individual—the pay is tied directly to sales performance. We also have to recognize that different sales

Individual factors	Organizational Factors	Market Factors	Industry Factors
•Exact job position •Past sales performance •Learning periods •Needs and preferences	•Financial resources •Marketing strategy and objectives •Organizational control •Fairness •Cooperation vs. competition •Length of time to close •Mixed promotional situations	•Market demand •Competitive landscape •Seasonal patterns •Economic fluctuations	•Industry norms for various sales positions •Changes within the industry at large

Figure 9.6
Packaging the Elements for an Individual Salesperson

jobs entail different methods of compensation. For example, a real estate company that sells time shares through a large-scale advertising campaign offers sales jobs that involve order-taking. People are hired to take orders over the phone. They follow a concrete script and procedure. Jobs involving order taking and closing deals are traditionally compensated using either straight salary or salary plus some form of performance incentive, as in a bonus or a small commission.

Job performance is another major individual factor affecting the compensation plan. A sales associate has been compensated based on commission with a 70–30 split (70% going to the sales associate and 30% going to the brokerage firm). The sales associate has been a top performer well exceeding his monthly sales quota. The sales manager decides to reward his exemplary performance by changing the 70–30 commission split to an 85–15 split.

Learning periods refer to the length of time for new sales associates to become acclimated to selling. Different professional tracks in real estate marketing require different learning periods. For example, selling commercial real estate is much more difficult than selling residential real estate. The time a sales associate takes to learn the ropes and become proficient in commercial real estate is longer than in residential real estate. Sales positions requiring a longer learning periods are typically compensated using straight salary or a mix of salary and commission. The converse is true for sales positions requiring shorter learning periods. That is, positions requiring shorter learning periods (e.g., selling of residential real estate) can best be compensated using straight commission.[24]

The sales manager may want to consider *individual needs and preferences*, too. Consider the results of a survey conducted in 2007 in the U.S.[25] that found women preferred benefits that maximize scheduling flexibility, such as paid leave, maternity leave, and child care, much more than men—up to four times as much as men did. In contrast, men are twice as likely as women to prefer perks like office space and profit sharing that maximize their income and professional status. The same survey also found that agents with long tenure place a great deal of importance on financial security. They value benefits such as life and disability insurance and 401(k) retirement saving plans much more so than those with shorter tenure (by margins of at least two-to-one). Looking closely at sales associates under 25 years of age, the survey uncovered the fact that financial security is also very important to this group—meaning a salary and a career plan. Although they are attracted by the prospect of earning a lot of money through commission, they place more value on financial security (51%) than on making much money (32%). Again, financial security for this age group translates to preference for salary over commission. Sales associates with five-plus years of experience are not bothered much by the uncertainty of commission; they are 15% more likely to choose a commission-based compensation plan. The survey also found that women prefer a salary (or salary-plus-bonus) more than men (about 22% more likely). They prefer the security of a salary than the uncertainty of commissions.

Cost of living for various geographic areas is another example of individual needs and preferences. Salespeople who are assigned to sell in certain locations (domestic or international) may have to pay more or less money to live comfortably. In other words, each city or geographic location has a different level of cost of living, and this information has to be taken into account in devising a compensation plan.

Organizational Factors

Organizational factors also play an important role in developing a compensation plan for each employee in the sales organization. Examples of organizational factors include (1) the organization's financial resources, (2) the organization's marketing strategy and objectives, (3) organizational control, (4) company policies and employment regulations concerning fairness, (5) cooperation versus competition among the sales associates, (6) length of time to negotiate and close a deal, and (7) mixed promotional situations.

The sales manager couched in a sales organization that has many *financial resources* and has bright prospects may decide to be a more generous in providing members of the sales team with direct compensation (e.g., a large end-of-year bonus) and/or indirect compensation (e.g., a greater percentage contribution the retirement plan). Real estate firms with few financial resources are less generous in the way they take risks. This often means a compensation plan mostly based on sales commissions.

The firm's *marketing strategy and objectives* is another major consideration in the formulation of a compensation plan. Consider the example of a real estate development firm that decides to position itself as a developer of retirement communities in certain neighborhoods of a large city. Based on a market analysis, the firm is optimistic that such positioning is likely to be highly profitable for the firm. To implement this marketing strategy, the sales manager is charged with training several members of his sales team to become experts on the housing needs of the elderly and retired. In doing so, he allocates more money for education and training for the selected sales associates. The additional money allocated is considered as an indirect element of compensation.

Organizational control refers to the need of the sales manager to control the extent to which the sales associates allocate their time. Sales associates in an outfit such as Re/Max are treated as independent agents who have total control over their time and the way they perform. They are only accountable to themselves, as long as they pay their monthly franchise fees. In contrast, agents in real estate development organizations with a large sales force (e.g., selling time share condominiums at a national or global level) would want to exert maximal control over how individual salespeople use their time. A common method of selling in such real estate development organizations is telemarketing—individual salespeople are either making telephone calls to a prescribed list of people (prescribed based on a demographic profile, such as middle-aged people with a middle-class household income) or receiving calls from potential customers who are responding to an advertisement. A compensation plan based strictly on a 100% commission may work best for real estate organizations that allow their sales associates to have total control of their time, whereas a compensation plan that is mostly salary based may work best in outfits that control the way their sales associates spend their time.

Company policies and employment regulations concerning *fairness* also play an important role in devising compensation plans. For example, the U.S. has laws that prohibit discrimination (discrimination based on gender, age, race, ethnicity, etc.) under the Equal Employment Opportunity Commission. For example, if male sales associates are compensated in a certain way, then it behooves the sales manager to compensate female associates similarly. Fairness is a major concern here, and it is commonly enforced internally by the human resources department within most real estate firms and by the government (e.g., the Labor Department in the U.S.).

The extent of *cooperation versus competition* required among the sales associates is yet another consideration. For example, many commercial real estate ventures require a sales team approach to selling. Professionals with varied expertise (e.g., marketing, architecture, building construction, finance, and facilities management) come together to form a sales team. This team works hard to develop a detailed proposal in response to a RFP (Request for Proposal) issued by a potential client. Such an arrangement requires a great deal of cooperation among the real estate professionals, which makes straight salary the most effective method of compensation. In contrast, a realty company specializing in residential real estate, with a large sales force and each salesperson assigned to a different neighborhood, may not require much cooperation among the sales associates. In this situation competition may be encouraged. As a result, straight commission is the best form of compensation.

There are many real estate deals that can be closed in a *short time frame* (e.g., in a matter of weeks) while others take a *longer time frame* (e.g., many months and perhaps years). For example, large-scale construction on a college campus or a military facility requires a very long planning horizon. It is very difficult to compensate marketing personnel working on real estate marketing projects requiring a long time span by straight commission. Such situations may call for straight salary.[26]

Finally, we have the *mixed promotional situation* as another organizational factor influencing the selection of the elements of the compensation plan. Consider the example of a real estate developer selling condominiums on some beach resort. The firm has just launched a large-scale but regional television advertising campaign. They are now recruiting a sales force to assist in the selling process. How should the salespeople be compensated? Straight commission? Straight salary? Or a mix between these two compensation methods? In situations in which there is minimal advertising, the sales manager can credit or blame the sales associates. If sales are low, this means that the sales associates failed to perform well; conversely, if sales are high, then the sales manager attributes this outcome to the hard work of the salespeople. In such situations, straight commission is most effective. In contrast, when the personal selling effort is confounded with advertising, then performance of the salespeople becomes hard to evaluate. Did the sales revenue increase as a result of increases in advertising or the personal selling efforts of the salespeople? In mixed promotional situations, it is best to compensate with either straight salary or salary-plus-commission.[27]

Market Factors
Market factors can influence the formulation of a compensation plan, too. Examples of market factors include (1) market demand, (2) the competitive landscape of the market, (3) seasonal patterns in the real estate market, and (4) economic fluctuations.

211

Market demand: let's face it, real estate properties sell like hotcakes in some markets but not in others. In other words, in some markets the demand is very high; in other markets, the demand is very low. Consider the low-end market (low cost housing) in most of the major cities around the world. The demand is very high. When such listings hit the market, they are snatched quickly. In "hot" markets (high demand markets) real estate salespeople do not have to work hard. Prospective buyers seek them out in a hurry; selling is effortless. Salespeople do not need to be motivated by financial means in high-demand markets. The opposite is true in low-demand markets. Consider the market for time share. Most real estate professionals characterize this market as "difficult," "hard sell," and "slow moving." Real estate salespeople need to be extra motivated in low-demand markets. Motivating a sales force in this situation takes money. That is, the sales manager has to devise a compensation plan that involves incentives, bonuses, and commissions to motivate the sales associates to work hard. Money can be a great motivator in low-demand markets. The sales manager has to appropriate a greater share of the sales revenue to compensate the sales force, and perhaps cut back on the profit margin.

The *competition* sometimes dictates what a sales organization should pay its sales associates. For example, if the vast majority of the brokerage firms in the area are doing a 70–30 commission split, the sales manager may decide to do an 80–20 split to a competitive advantage—to enhance organization loyalty and commitment and reduce turnover.

Seasonal patterns may be another consideration. In a real estate market marked by seasonal fluctuations (most of the sales occur in the spring and summer months), some salespeople are likely to be adversely affected. They may not be able to save during times of prosperity for rainy days. In such situations, the sales manager may decide to pay them a fixed salary, or perhaps a salary plus commission.

Furthermore, the housing market goes through booms and busts (*economic fluctuations*). Some sales managers decide to provide a financial cushion for sales associates to stabilize the cycles of booms and busts by placing members of the sales team on a fixed salary, rather than a commission.

Industry Factors

Finally, we have industry factors that influence the configuration of any individual compensation plan. Examples of industry factors include (1) industry norms for various sales positions and (2) anticipated changes within the industry at large. Industry norms refer to the average amount of pay for various positions. Information about the average pay related to the various positions (e.g., junior-level associate, senior-level associate, sales manager, and associate sales manager) is usually published in the professional magazines and newsletters of real estate marketing associations. The sales manager is, of course, influenced by this information. He tries to specify specific elements and levels of compensation guided by industry norms. The norms may be broken down by the type and size of the real estate firm. Such information is highly useful for the sales manager.

In addition to industry norms, *changes within the industry* may affect the selection of the compensation elements and levels too. Compensation plans in the real estate marketing industry are shifting. The traditional straight commission is longer the norm. Many real estate firms have now adopted more effective compensation plans that are implemented to enhance job satisfaction and organizational loyalty. Let's face it: A real estate firm that operates by straight commission stands to lose its people in the long run. Straight commission hacks away at loyalty. Salespeople who operate on straight commission are only loyal to themselves. Compensation plans that combine elements of salary, commission, bonus, and other incentives, in addition to indirect elements (e.g., health plans, retirement plans, other perks) are now commonplace. To remain competitive and to decrease turnover among the sales force, the sales manager has to keep up with the changing compensation plans in the industry.

Validating the Process

Validating the process means monitoring job performance and making changes to compensation components and levels as a direct function of this monitoring. See Case/Anecdote 9.5 for examples of effective compensation plans provided by five different brokerage firms in the U.S.

The validation process involves pretesting, administering, and evaluating the compensation plan at both individual and company levels. At the individual level, this may mean to try a compensation plan for an individual salesperson and see if it works for that individual. Does the plan make that person motivated? Does he generate a high level of sales? Is he committed to the organization? Is he happy with his job and his compensation plan? If the sales manager answers "yes" to these questions, then the sales plan for that particular salesperson seems well suited to him. Therefore, the plan should be maintained and re-evaluated periodically (every few months or so). If not, then the compensation plan should be further modified in an attempt to produce positive results in terms of the individual's level of performance, job satisfaction, and organizational commitment.

At the company level, a similar process can be instituted. The sales manager pretests a certain compensation plan devised for a small group of salespeople, perhaps in a certain geographic market. Again, the sales manager would ask questions such as: Does the compensation plan implemented in that region work? Does it produce high level of sales? Are the salespeople in that region happy with their job and their compensation plan? What is the rate of turnover among this group? If answers to these questions are positive, then the compensation plan should be implemented company-wide. If not, then the compensation plan should be revised and re-evaluated against the same performance metrics such as sales performance, job satisfaction, and turnover.

CASE/ANECDOTE 9.5. FIVE COMPANIES, FIVE DIFFERENT COMPENSATION PLANS[i]

In today's real estate sales, the traditional commission-based compensation is being supplanted by new forms of compensation plans. Five different plans are considered here.

@properties, Chicago, Illinois

@properties was able to subsidize health insurance for their sales team by pooling associates' premiums and obtaining insurance for less than they would as individuals. The company pays associates' insurance premiums, and associates reimburse the company for most of the premium. The insurance company that @properties deals with prefers this arrangement because it simplifies the process—the insurance company deals with the brokerage firm instead of many individual associates. A basic medical plan costs about $210 per month, and a very good plan costs about $300 per month. An associate with a family of four pays around $600. New associates keep 55% to 85% of commissions—the exact rate is determined by productivity.

The company also provides a 401(k) retirement account; it covers costs but doesn't match contributions.

Keller Williams Realty International Inc., Austin, Texas

Keller Williams' Realty is essentially a franchise in the United States. The company offers its franchisees (sales associates) compensation plans from which to choose. However, the company has a default plan for the company's 71,000+ associates. The plan includes four key elements: (1) Associates receive a baseline 70% commission, which can increase to 90% or more as a direct function of productivity. (2) Associates pay the firm an annual fee of $18,000-$30,000, after which they transition to a 100% split for the rest of the year. (3) There is also an annual franchise fee of $3,000. (4) There is a monthly profit-sharing program, based on the number of salespeople an associate brings into the company—a sponsoring associate earns profit from the sales commission generated by the sponsored associate (based on a graduated rate that declines to zero in time).

Lyon Real Estate, Sacramento, California

Lyon Real Estate is a brokerage firm that has 975 associates. The company's compensation plan focuses on incentives that motivate the associates to become high achievers. The commission splits range from 50% to 85%. Associates receive a bonus as a direct function of their productivity. The company also offers health and 401(k) plans. The company has managed to negotiate low group rates for the health plan; associates pay the premiums. The 401(k) plan is similar, in that the company does not contribute financially, but provides a financial adviser to the associates at no cost.

Prudential Carolinas Realty, Winston-Salem, North Carolina

Prudential Carolinas has 900 associates. The company offers a wide range of benefits, including health, dental, prescription, life, and disability insurance, as well as a retirement-savings plan. Of course, the associates pay for these benefits, but at a significantly lower cost than if they had to buy the same benefits on their own. In the retirement plan, associates invest in a pension program and 401(k) plans, but there is no company match. The company's commissions range from 50% to 100%. The 100% goal can be achieved by generating more than $200,000 in agent earnings within the past 12 months.

United Country Premier Brokers of Colorado, Salida, Colorado

All associates at United Country get 60% of their earned commissions. After working for the company for a year, an associate is eligible to buy an equity stake in the company. The cost per share of the buy-in is based on a formula involving the company profits of the preceding three years. At the end of the year, associates who have bought equity receive a bonus check based on the company's profit that year.

[i] Filisko, G. M. (2006). Compensation packages: Pull them in with benefits. *Realtor Magazine,* December 2006. Accessed from http://realtormag.realtor.org/for-brokers/feature/article/2006/12/compensation-packages, on July 15, 2013.

SUMMARY

This chapter covered many issues related to motivation programs and compensation plans commonly found in the real estate industry.

We started out by addressing the question of why motivational programs are important for a real estate marketing organization. The answer to this question lies with the evidence that shows that motivational programs are necessary to help boost sales effort, which in turn raises sales performance. Of course, sales performance is directly tied to the firm's bottom line (i.e., profitability).

We then made the distinction between intrinsic and extrinsic motivation. We explained this distinction in terms of low-order versus high-order needs. Low-order needs are essentially survival or basic needs, whereas high-order needs are growth related. Hence, extrinsic motives can be viewed as directly tied to basic needs (low-order needs) and intrinsic motives are associated with growth needs (high-order needs). Salespeople tend to express a great deal of dissatisfaction at work when their basic needs are not met, and not so much when growth needs are not met. Conversely, they express a great deal of satisfaction when their growth needs are met, not so much with basic needs.

In the context of real estate marketing, several extrinsic motives were discussed—pay, company policies, working environment, and interpersonal relationships. With respect to pay, we addressed the evidence that shows that inadequate pay is likely to lead to job dissatisfaction and turnover. Sales managers can understand how pay can be an important source of motivation for sales performance by applying expectancy-value theory to connect pay (valence) with performance (instrumentality) and effort (expectancy). Understanding these connections should help the sales manager use pay as an extrinsic motive to help salespeople achieve higher levels of performance. We also discussed the importance of the working environment as a source of extrinsic motivation. The working environment refers to the physical workplace; the availability of relevant technological or communication resources; its security and safety; its cleanliness and ergonomics; the level of noise and the lighting; the availability of canteen facilities; the proximity to public transport, shops and facilities; the codes of behavior that define the workplace; and the other factors that make work physically possible, safe, pleasant, and convivial. The non-physical environment involves the company rules and regulations, the way things are done in the office, pressure from upper management to generate high dollar sales, and so on. Both physical and psychological environments play an important role on salespersons' levels of motivation and performance. The same can be said with regards to interpersonal relationships. Positive relations among salespeople are essential to motivation. Similarly, positive relations between the sales manager and each salesperson are equally essential.

Then we shifted the discussion to cover intrinsic sources of motivation. Intrinsic factors are factors affecting the salesperson's level of performance motivation related to high-order needs (growth needs such as social, esteem, self-actualization, aesthetic, and knowledge needs). In that context we discussed achievement programs, recognition programs, job enrichment, career advancement opportunities, work-life balance, and ownership. Achievement programs involve incentives or sales contests that reward sales performance, and they increase motivation and morale of salespeople. Increased motivation and morale leads to greater sales effort, which results in increased sales performance. In real estate marketing, recognition programs take form in the public recognition of top performers. Job enrichment typically involves providing the salesperson with greater autonomy and responsibility. Promotion within the sales organization can also be a strong source of motivation to excel. In real estate sales organizations, the ranks may involve two tracks: one track related to personal selling and the other to sales management. Work-life programs are designed to allow salespeople to attend to family needs and have a social life outside of the office. There are several work-life programs that sales organizations offer to their sales team. These include flexible working hours, flexible leave arrangements, reduced working time, and flexible location. Finally, ownership in the firm may be assumed when salespeople earn stock options and/or partnership.

Then we turned our attention to compensation issues in real estate marketing organizations. We first discussed why compensation plans are important. The evidence points to the fact that proper compensation is positively associated with the job performance of the sales force, job satisfaction, job morale, and organizational loyalty and commitment. The evidence also suggests that proper compensation is negatively associated with employee presenteeism, absenteeism, and turnover.

Then we tried to understand the compensation process. This process involves three stages: (1) identification of the compensation plan elements, (2) packaging the elements for individual salespersons, and (3) validating the process. With respect to the first stage, we described the elements of a compensation plan in terms of two categories, direct and indirect elements. Direct elements include salary, commission, wages, referral fees, expense account, incentive, bonuses, profit sharing, and stock options. Indirect elements include health and life insurance, social security benefits, retirement plans, paid time off, and other perks. Then we tried to explain how sales managers go about packaging the direct and indirect elements to fit a particular group and individual salespeople.

We explained a host of factors that affect sales managers' decision-making: individual, organizational, market, and industry-related factors. Individual factors were described as conditions unique to the individual salesperson, such as the (1) exact *job position* that the individual assumes, (2) past sales performance, (3) learning periods, and (4) needs and preferences. Organizational factors also play an important role in developing a compensation plan for each employee in the sales organization. We described the following factors: (1) the organization's financial resources, (2) the organization's marketing strategy and objectives, (3) organizational control, (4) company policies and employment regulations concerning fairness, (5) cooperation versus competition among the sales associates, (6) length of time to negotiate and close a deal, and (7) mixed promotional situations. Market factors also influence the formulation of a compensation plan. We described the effects of (1) market demand, (2) the competitive landscape in the market, (3) seasonal patterns in the real estate market, and (4) economic fluctuations. Finally, we described industry factors, such as (1) industry norms for various sales positions and (2) anticipated changes within the industry at large.

The last stage of the compensation process is validation of the process. Validating the process means monitoring job performance and making changes to compensation components and levels as a direct function of this monitoring.

DISCUSSION QUESTIONS

1. Please explain why motivational programs are important for a real estate marketing organization.

2. Explain the distinction between intrinsic and extrinsic motivation.

3. How does extrinsic motivation affect job satisfaction and performance? In contrast, how does intrinsic motivation affect job satisfaction and performance?

4. How does pay affect job motivation? Please explain using expectancy-value theory.

5. How do factors such as interpersonal relations, the working environment, and company policies and procedures affect salespeople's job motivation? Please explain.

6. Describe the typical programs used in the real estate industry designed to boost intrinsic motivation.

7. Describe a common achievement program commonly used in real estate marketing.

8. What about recognition programs used in real estate marketing? Describe at least one.

9. What is job enrichment? How is job enrichment typically operationalized in real estate marketing?

10. What is work-life balance? Describe examples of work-life balance programs in real estate marketing firms.

11. What about ownership? Describe this concept in the context of real estate marketing firms.

12. The compensation process involves three major stages. What does this mean?

13. Describe the direct methods of compensation typically found in the real estate marketing industry.

14. Distinguish the direct elements from the indirect elements.

15. The sales manager attempts to customize a compensation plan for one or more sales associates by taking into account a host of individual factors. What are these individual factors and how do they affect the formulation of a compensation plan?

16. The same can be said for organizational factors. What are these factors and how do they affect the formulation of a compensation plan?

17. What about market factors? How do these influence the formulation of a compensation plan?

18. Industry factors! How do these affect the formulation of a compensation plan?

19. A good compensation process involves validation. What does this mean?

NOTES

1. Mathe, H., Pavie, X., & O'Keeffe, M.O. (2012). *Valuing people to create value: An innovative approach to leveraging motivation at work.* Hackensack, NJ: World Scientific. Durant, R.F., Kramer, R., Perry, J.L., Mesch, D., & Paarlberg, L. (2006). Motivating employees in a new governance era: The performance paradigm revisited. *Public Administration Review,* 66(4), 505–514. Latham, G.P., & Pinder, C.C. (2005). Work motivation theory and research at the dawn of the twenty-first century. *Annual Review of Psychology,* 56, 485–516.

2. Alderfer, C. (1969). An empirical test of a new theory of human needs. *Organizational Behavior and Human Performance*, 4, 142–175. Herzberg, F., Mausner, B., & Peterson, R.D. (1957). *Job attitudes: Review of research and options*. Pittsburgh, PA: Psychological Services of Pittsburgh. Maslow, A. (1962). *Toward a psychology of being*. New York: Nostrand. Amabile, T.M. (1993). Motivational synergy: Toward new conceptualizations of intrinsic and extrinsic motivation in the workplace. *Human Resource Management Review*, 3, 185–201.

3. Herzberg, F. (1979). The wise and old turk. *Harvard Business Review*, 52, 70–81.

4. Porter, L. W., & Lawler, E. E. (1968). *Managerial attitudes and performance*. Homewood, IL: Irwin.

5. Adams, J. S. (1965). Inequity in social exchange. *Advances in Experimental and Social Psychology*, 62, 335–343.

6. Herzberg, F. (1979). The wise and old turk. *Harvard Business Review*, 52, 70–81.

7. Darmon, R.Y. (1974). Salesmen's response to financial incentives: An empirical study. *Journal of Marketing Research*, November, 418–426. Corporate Meetings & Incentives. (2003). What motivates people? You might be surprised. *Corporate Meetings & Incentives*, March, 31–32.

8. Mathe, H., Pavie, X., & O'Keeffe, M.O. (2012). *Valuing people to create value: An innovative approach to leveraging motivation at work*. Hackensack, NJ: World Scientific (p. 96).

9. Pheasant, S. (1991). *Ergonomics, work and health*. London: Macmillan Publishing. Wiedenkeller, K. (2010). Some like it hot? Work environments impact productivity. *Film Journal International*, May 4.

10. Mathe, H., Pavie, X., & O'Keeffe, M.O. (2012). *Valuing people to create value: An innovative approach to leveraging motivation at work*. Hackensack, NJ: World Scientific (pp. 100–103).

11. James, K., & Arroba, T. (1999). *Energizing the workplace: A strategic response to stress*. Brookfield, VT: Gower.

12. Fernet, C., Gagne, M., & Austin, S. (2010). When does quality of relationships with coworkers predict burnout over time? The moderating role of work motivation. *Journal of Organizational Behavior*, 31, 1163–1180.

13. The survey was conducted in 2007 by *Realtor* magazine in partnership with CompensationMaster. The survey was based on 9,400 real estate sales people in the U.S. and focused on career goals. See Freedman, R. (2008). Survey: Sales associate pay. In real estate, the commission is sacrosanct. Or is it? *Realtor Magazine*, June. Accessed from http://realtormag.realtor.org/for-brokers/feature/article/2008/06/survey-sales-associate-pay, on July 16, 2013.

14. Murphy, W. H., & Dacin, P.A. (1998). Sales contests: A research agenda. *Journal of Personal Selling & Sales Management*, 18, 1–17.

15. Murphy, W. H., Dacin, P. A., & Ford, N. M. (2004). Sales contest effectiveness: An examination of sales contest design preferences of field sales forces. *Journal of the Academy of Marketing Science*, 32, 127–143.

16. Kouzes, J., & Posner, B. (1999). *Encouraging the heart*. San Francisco, CA: Jossey-Bass.

17. Hackman, J. R., & Lawler, E. E. (1971). Employee reactions to job characteristics. *Journal of Applied Psychology*, 55, 259–286. Vroom, V. H., & Deci, E. L. (1970). *Management and motivation*. Baltimore: Penguin Education. Crant, J. M. (1995). The proactive personality scale and objective job performance among real estate agents. *Journal of Applied Psychology*, 80, 532–537.

18. Hackman, J. R., & Oldham, G. R. (1980). *Work redesign*. Reading, MA: Addison-Wesley.

19. Maxwell, G. A. (2005). Checks and balances: The role of managers in work-life balance policies and practices. *Journal of Retailing and Consumer Services*, 12, 179–189. Taylor, R. (2002). *The future of work-life balance*. Swindon, U.K.: Economic and Social research Council.

20. Chinloy, P., & Winkler, D. T. (2012). Agent performance, incentives, and ownership. *Journal of Housing Research*, 21, 101–121.

21. Ryals, L. J., & Rogers, B. (2005). Sales compensation plans—one size does not fit all. *Journal of Targeting, Measurement & Analysis for Marketing*, 13, 354–362.

22. Common Commission Options. (2013). *Realtor Magazine*, April 10, 2013. Accessed from http://realtormag.realtor.org/tool-kit/retention/article/common-commission-options, on July 15, 2013.

23. Hair, J. F., Anderson, R. E., Mehta, R., & Babin, B. J. (2009). *Sales management: Building customer relationships and partnerships*. Boston: Houghton Mifflin Company (pp. 370–371).

24. Hair, J. F., Anderson, R. E., Mehta, R., & Babin, B. J. (2009). *Sales management: Building customer relationships and partnerships*. Boston: Houghton Mifflin Company (p. 368).

25. The survey was conducted in 2007 by *Realtor* magazine in partnership with CompensationMaster. The survey was based on 9,400 real estate sales people in the U.S. and focused on career goals. See Freedman, R. (2008). Survey: Sales associate pay. In real estate, the commission is sacrosanct. Or is it? *Realtor Magazine*, June 2008. Accessed from http://realtormag.realtor.org/for-brokers/ feature/article/2008/06/survey-sales-associate-pay, on July 16, 2013.
26. Hair, J. F., Anderson, R. E., Mehta, R., & Babin, B. J. (2009). *Sales management: Building customer relationships and partnerships.* Boston: Houghton Mifflin Company (p. 368).
27. Hair, J. F., Anderson, R. E., Mehta, R., & Babin, B. J. (2009). *Sales management: Building customer relationships and partnerships.* Boston: Houghton Mifflin Company (p. 368).

Chapter 10

Leadership Issues in Real Estate Firms

LEARNING OBJECTIVES

This chapter is designed to help students of real estate marketing learn:
- why it is important to have a manager with leadership qualities
- a good definition of leadership
- the distinction between managers and leaders
- leadership style
- leadership power
- leadership skills
- leadership action

WHY IS IT IMPORTANT TO HAVE A MANAGER WITH LEADERSHIP QUALITIES?

To manage a real estate sales force in a real estate development firm or through a real estate brokerage firm, certain human resource management goals have to be met. These goals include high job performance (i.e., exceeding the sales quota), high job satisfaction, high employee morale, high organizational loyalty and commitment, low employee absenteeism (i.e., not taking too much time off work), low employee presenteeism (i.e., minimal time being wasted on the job), and low employee turnover (low number of agents quitting their jobs). These human resource management goals ultimately affect the bottom line (i.e., the firm's financial health—its profitability).

DEFINITION OF LEADERSHIP

Good leadership in real estate marketing organizations (sales force management in real estate development firms and real estate brokerage firms) can be defined in terms of actions leading real estate marketing professionals toward achieving sales goals.

How? This is typically done by having the sales manager (in real estate development firms) or the broker (in real estate brokerage firms) set a clear vision for the sales force, motivate the sales force to achieve in selling, guide the sales agents through the selling process, and build morale among the sales agent.

MANAGERS VERSUS LEADERS

Management scholars typically make a distinction between management and leadership. Management is about *efficiency:* systems, processes, procedures, control, and structure. For example, sales managers in a real estate development firm direct the entire operation of selling real estate. They develop strategic plans, setting specific sales objectives on a monthly or quarterly basis; hire and train sales agents; provide agents with a compensation plan; evaluate their performance and adjust their compensation accordingly; implement sales incentive programs to motivate the agents; develop a set of policies that regulate the behavior of the agents; provide and manage organizational resources to assist the agents in accomplishing their goals; and so on. In other words, the sales manager is embroiled in systems, processes, procedures, control, and structure. Sales management is about efficiency; therefore, its goal is to maximize the performance of the sales agents—to use human capital to generate sales revenue—with minimum financial resources. Doing so ultimately enhances the bottom line—profitability (increase sales revenue while decreasing costs associated with the selling process).

Leadership, on the other hand, is about *effectiveness:* trust, inspiration, and people. A good leader is a sales manager who instills *trust*[1] among the sales agents. Trust is an important ingredient in sales performance. Sales agents have to trust the sales manager (who, in essence, represents the firm) to meet the social contract. That is, sales agents perform their selling function based on a set of guidelines, and as a result they are compensated based on their employment contract. Violating the social contract is a violation of trust. For example, when a sales agent believes that he has worked hard and met the sales quota for the month but has received little compensation and recognition for his efforts, his sense of trust in the firm is diminished. The firm does not stand by its promise. He can no longer count on the firm to do what is right. His sense of loyalty and commitment is compromised. His motivation to do his best is flushed down the drain.

Trust in the organization to do what is right is what good management is about, and that is leadership. A sales manager who is successful in instilling trust among the agents is a leader. A sales manager who is a good leader is therefore trustworthy. Trust is the degree to which the sales agents believe in the honesty, fairness, and benevolence of the sales manager. Thus, leadership goes above and beyond management. Good management is effectively guided by good leadership. Bad management is a result of bad leadership.

A good leader *inspires* his sales force. What is inspiration[2]? Inspiration refers to a burst of creativity. A sales manager who inspires his sales agents is one who makes them feel creative in the selling process. He motivates them to use their own mental faculties to approach selling situations differently. Each selling situation is unique in many ways. No standard solution can be applied to all selling situations. The sales agent has to be creative to find a unique solution to address the needs of each client. Some sales agents tend to be inherently creative in finding solutions for their clients. Others have to be inspired to become more creative. Good sales managers, as leaders, are good at inspiring their sales force.

Good leadership is about *people.* Sales managers who are good leaders tend to attend to the needs of the sales agents. They do not strictly focus on efficiency at the expense of effectiveness. Efficiency and effectiveness are equally important. Yes, what is important is to maximize the output of the sales force in terms of sales revenue, while minimizing the financial resources that are used to generate that output. And yes, what is important is the human capital that is used to generate the maximum output. The human capital factor is very important. A good sales manager who is a good leader is one who invests in human capital by treating the sales agent as an asset. The challenge is to build this asset to generate the maximum return on investment from the asset. Good sales managers build their human capital by effective recruitment, by effective training, by effective recognition programs, by effective compensation plans, by effective mechanisms of job performance, and by effective governance.

LEADERSHIP STYLE

There are many styles of leadership. Some of these styles are more suited for sales management in a real estate context than others. We will make distinctions between task-oriented and people-oriented leadership; transformational versus transactional leadership; and laissez-faire, democratic, and autocratic leadership.[3]

Task-Oriented versus People-Oriented Leadership Styles

Task-oriented leadership is an approach to sales management in which the manager focuses on the personal selling tasks to meet certain sales goals. In contrast, *people-oriented (or relationship-focused) leadership* is an approach in which the sales manager focuses on the satisfaction, motivation, and the general well-being of the sales agents. The latter approach is an indirect approach to achieving sales goals. In other words, task-oriented leadership focuses on the tasks that lead to sales achievement, whereas people-oriented leadership focuses on harnessing the power of people to perform the selling tasks and ultimately to achieve sales goals.

Specifically, task-oriented sales managers focus on the personal selling task at hand, or the process of the personal selling task, to achieve specified sales objectives. These sales managers are typically less concerned with catering to the needs of the sales agents, and more concerned with getting the job done to meet the monthly or quarterly sales objectives. Sales managers with this style of leadership will often actively define the nature of the personal selling task in precise detail; specify company rules and policies that regulate the personal selling task; assign sales agents to specific personal selling roles (e.g., certain geographic territories, demographic segments, and/or sales roles, such as being a listing agent versus a buyer representative); allocate resources to the various roles; and monitor and evaluate job performance in terms of measurable progress, such as sales revenue, units sold, number of closings, number of contacts with prospective buyers, and so on.

The task-oriented leadership style may be well suited for certain commercial real estate marketing organizations in which one sales project (e.g., development of a shopping mall) may involve breaking the project into many tasks, each task is guided by many policies and regulations, with specific hard deadlines. The sales manager in this situation may be well equipped to get the job done because he has an eye for details, he has the expertise, he knows the "nitty gritty" details related to company policies and the rules and regulations governing each task. The sales manager with this style may be effective in focusing on the necessary workplace procedures and delegating certain tasks to certain agents to ensure that every task is completed successfully and in a timely manner. Task-oriented leadership could also be suitable when the sales management is challenged with a sales force full of rookies who need much training and guidance. However, this style of management does not bode well for the well-being of the sales team. This may be due to the fact that this leadership style ignores the human side of management. The task-oriented sales manager may be autocratic in the way he interacts with the sales agents, resulting in motivation and retention problems.[4]

People, or relationship-oriented, sales managers focus more on the sales team and less on the task at hand (the business of selling). Much of their energy goes toward the support and personal development of the sales team and the interpersonal dynamics within the team. The people-oriented sales manager encourages good teamwork and collaboration. This is typically done through fostering positive relationships between the sales manager and individual members of his sales team, as well as positive relationships among the team members. People-oriented sales managers tend to be high on offering incentives like sales incentives and bonuses. They are eager to mediate conflict between members of the sales team. They favor casual or informal interactions. They devote much time and energy to learning the strengths and weaknesses of the sales agents. They favor strategies and methods that encourage sales agents to work cooperatively rather than compete against one another.[5]

This leadership style is highly suitable in situations in which the sales agents have to work together closely as a team. In a large brokerage firm involving a large sales force of listing agents and buyer representatives, establishing a climate of cooperation between buyer and seller agents within specific assigned geographic territories (i.e., certain neighborhoods) calls for this style of leadership. The results of this leadership style include minimizing personal conflicts, reducing dissatisfaction with jobs, and heightening loyalty and commitment to the firm. The downside of this leadership style is that, if taken too far, the development of team spirit may detract from carrying out the business of personal selling in a manner that is meticulous and "by the book."[6]

Table 10.1 further reinforces the narrative highlighting differences between task-oriented and people-oriented leadership styles.

Table 10.1 Task-Oriented versus People-Oriented Leadership Styles

	Task-Oriented Leadership Style	People-Oriented Leadership Style
Goals and Objectives	To achieve sales objectives in the short run	To achieve sales objectives in the long run
Means and Methods	Facilitating the personal selling function by guiding the sales agents through a maze of policies, procedures, and regulations Use of structures, roles, and tasks to assist sales agents in performing their personal selling functions	Facilitating interaction among members of the sales team Use of positive interpersonal dynamics to accomplish the goal of selling
Advantages	Gets the job done in the short run	Makes the sales team work as a cohesive unit to achieve long-term sales goals
Disadvantages	Getting the job done may come at the expense of stepping on the toes of the sales agents, causing job dissatisfaction in the long run	Members of the sales team may feel good about their job and the firm, but this may come at the expense of closing sales in a manner that adheres closely to a maze of policies, procedures, and regulations

Table 10.2 Combining Task-Oriented Leadership with People-Oriented Leadership

	Sales Manager's Concern for Sales (Low)	Sales Manager's Concern for Sales (High)
Sales Manager's Concern for the Agents (High)	Country Club Management	Team Management
Sales Manager's Concern for the Agents (Low)	Impoverished Management	Produce-or-Perish Management

Table 10.2 combines the task-oriented leadership style with that of the people-oriented style. The result produces four distinct types of leadership styles: (1) impoverished management, (2) country club management, (3) produce-or-perish management, and (4) team management. *Impoverished management* is characterized by a sales manager who is not highly concerned with either ensuring that his sales agents perform at their best (i.e., generate high levels of sales) or with the sales agents' well-being. This sales manager seems to be disengaged. *Country club management* is a style of leadership in which the sales manager focuses much on the well-being of the sales agents at the expense of sales performance. In other words, he has high concern for the sales agents but not much concern for sales results. *Produce-or-perish management* refers to a style in which the sales manager is concerned only with sales performance and does not give a hoot about the welfare of the agents. Finally, we have the team management approach to leadership. This style of leadership reflects a sales manager who is highly concerned with both the welfare of the agents and their performance.[7]

Of course, the team management approach seems to be highly effective in most circumstances. A sales manager who rallies the sales associates to excel in generating sales and is also highly concerned about the welfare of each individual salesperson is likely to be the most effective type of leader—effective in relation to the financial health of the organization.

Transactional versus Transformational Leadership Styles

Transactional leadership (also known as managerial leadership) is an approach of sales management in which the sales manager focuses on the role of supervision, organization, and performance. The essence of transactional leadership is best captured in the term *transaction*. Sales agents are hired to do a sales job and they get compensated accordingly. This, in essence, is a "transaction" between the employer and employee. "You do your job as expected by management, and you will get compensated according to how well your performance meets expectations." This is essentially a "social contract" that captures the nature of the transaction.

Transactional leadership can best be described in terms of two dimensions: (1) contingent rewards and (2) management-by-exception.[8] With respect to *contingent rewards*, transactional leaders attempt to influence their sales agents using rewards and punishments. Transactional leaders closely scrutinize sales agents' performance to ensure that performance meets management expectations. In doing so, the sales manager examines sales agents' performance along dimensions such as amount of dollar sales, number of units sold, number of closings, number of contacts with prospective clients, etc. He compares performance with expectations to measure compliance and deviations. Compliance (performance is as expected) and positive deviations (performance is better than expected) are rewarded, whereas negative deviations (performance is worse than expected) are reprimanded.

Management-by-exception refers to management designed to maintain the status quo. The status quo is the sales agents doing their job and meeting the sales quota. Deviations from the status quo require intervention by management. In other words, a sales manager who is transactional in nature provides enough latitude to his sales agents to do their job and intervenes only when the job is not done. In essence, transactional leadership is primarily passive. The focus is establishing criteria for rewarding sales performance (and punishing lack of performance) to maintain the status quo.

Transformational leadership, in contrast to transactional leadership, focuses on motivation and morale of the sales force to achieve sales goals. This is done by being a role model for the sales agents. The sales manager attempts to inspire them and challenges the agents to take greater ownership of their work. The sales manager attempts to understand the strengths and weaknesses of the agents in an attempt to align the agents with specific selling tasks that ultimately could enhance their selling performance.[9] Transformational leadership can best be described in terms of four dimensions: (1) individualized consideration, (2) intellectual stimulation, (3) inspirational motivation, and (4) idealized influence. *Individualized consideration* refers to the degree to which the sales manager attends to each agent's needs and listens to his concerns. The sales manager acts as a mentor or coach—helping each agent personally with empathy and support. He challenges the agent to excel in personal selling and celebrates his accomplishments.

Intellectual stimulation refers to the degree to which the sales manager challenges traditional assumptions and stimulates each agent to be creative to discovering solutions for their clients. Associates are encouraged to ask questions, think deeply about clients' concerns, and figure out better ways to deal with these concerns.

Inspirational motivation refers to the degree to which the sales manager articulates a vision that is inspiring to the sales associates. Sales managers challenge their sales associates to adopt high professional standards in everything they do. The sales managers communicate their sense of optimism about sales goals that can be achieved. They inject meaning and purpose into the various selling tasks at hand. Meaning and purpose provide the emotional energy that drives the salesperson to excel and do their best. The result is that sales associates are more willing to invest additional effort in their personal selling tasks; they become more optimistic about their future prospects and feel more confidence in their abilities, not only to achieve, but to exceed sales quotas.

Finally, *idealized influence* refers to the notion that a transformational leader becomes the role model. That is, the sales manager plays the role of the sales associate and "gets his hands dirty." By doing so, the sales manager demonstrates the art and science of personal selling performed with high ethical standards. This, in essence, is what psychologists refer to as *learning by modeling*. People learn to perform by emulating others they look up to and trust (i.e., role models). This is a very effective way to train and mentor sales associates to perform at the highest level of professional standards.[10]

Table 10.3 further highlights differences between transactional and transformational leadership styles.

Laissez-Faire versus Democratic versus Autocratic Leadership Styles

Laissez-faire leadership (also known as "delegative leadership") is a style in which sales managers take a very passive approach to the management of the sales force. It is essentially hands off, and the sales manager delegates much of the responsibility of overseeing certain sales tasks to the senior sales associates or assigns those tasks to committees. For example, the sales

Table 10.3 Transactional versus Transformational Leadership Styles

	Transactional Leadership Style	Transformational Leadership Style
Goals and Objectives	To fulfill the terms of the employment contract	To achieve the highest standards possible, in terms of both sales outcomes and processes
Means and Methods	Sales agents' performance is monitored; sales performance that meets management expectations is rewarded; conversely, performance below expectations prompts corrective action	Develop a vision for the sales organization that is meaningful and purposeful (e.g., be the best in real estate marketing). The vision is developed collectively (sales manager working closely with the sales associates)
	The sales manager maintains the status quo as long as the sales agents do what is expected of them; he intervenes only when the sales agents do not deliver on their promises	Connect the vision with a marketing strategy (be the best in serving the retirement community)
		Translate the marketing strategy into specific tasks and assign sales associates to different tasks
		Express optimism and confidence about the vision, strategy, and the plan of action
		Celebrate the small and large achievements collectively
Advantages	Works well for start-up real estate sales organizations in which the organization benefits significantly from setting and establishing performance standards and creating a monitoring and control system to ensure that the performance standards are met	Works well in sales organizations that have reached a certain level of maturity, in which sales performance requires creativity ("thinking outside the box") and in which the business environment is turbulent
Disadvantages	Does not work well in sales organizations that have reached a certain level of maturity, in which sales performance require creativity or in which the business environment is turbulent	Does not work well for start-up real estate sales organizations where the organization benefits from setting and establishing performance standards and creating a monitoring and control system to ensure that the performance standards are met

manager may form one committee charged with recruiting new agents, another committee that oversees training of the new recruits, another committee that handles personnel grievances, and another committee that is in charge of social events. In other organizations, many functional and administrative tasks are delegated to certain individuals within the sales force (typically senior sales associates) or to committees (made up of members of the sales force). Very little guidance is provided by the sales manager to the assigned committees or individuals. The sales manager simply provides the tools and resources needed to help the assigned individuals or groups get the job done. The assigned individuals or groups are expected to deal with the assigned tasks on their own.

Of course there are advantages and disadvantages of this style of leadership. Laissez-faire leadership works best when the sales organization is "top heavy" (i.e., when the sales force has a majority of seasoned sales associates who have amassed a great deal of knowledge and expertise in real estate marketing). These seasoned professionals are highly skilled and independent. They do not need much guidance from the sales manager. All they need are the tools and resources to get the job done. The obvious disadvantage is that such leadership style fails in sales organizations that are "bottom heavy" (the organization has a large number of new recruits). These rookies need more guidance and direction from the sales manager; hence, a laissez-faire style cannot be helpful in this situation.

In regards to *democratic leadership* (sometimes known as "participative leadership"), the sales manager leads the sales force by encouraging the sales associates to participate in many decisions involving the sales organization (recruiting new sales associates, deciding on who needs training and in which topics, who should be given a sales achievement award, etc.). By allowing the sales associates to have a say in most major organizational decisions, they identify with the organization, and identifying with the organization leads to a host of positive organizational outcomes, such as higher productivity, more creativity in selling, greater loyalty and commitment to the firm, etc.

Advantages of this style of leadership include identifying creative solutions to selling problems and making the sales associates more involved and engaged in the selling process and in handling other organizational tasks vital to the survival and prosperity of the sales organization. Therefore, this style of leadership works well for sales organizations in which creative problem-solving is instrumental to achieving high levels of sales and in selling situations that require a great deal of personal involvement and effort. Sales associates who feel that they are involved in many of the important decisions of the organization feel a sense of ownership, which in turn induces the sales associate to invest a great deal of time and energy in selling and other organizational tasks.

Democratic leadership does not work well in situations in which the sales associates are not skilled and experienced enough. Their input into important decisions is not likely to boost performance. Also, democratic leadership does not work when time is of the essence. For example, if there is a Request for Proposal to develop a shopping center that has a short deadline, it would be very difficult to meet the deadline with a quality proposal by encouraging the sales associates to contribute many ideas and to do this collaboratively. In this case, the sales manager may have to be somewhat autocratic in order to submit a quality proposal on time. Having said this, let us now turn to autocratic leadership.

Autocratic leadership (also known as "authoritarian leadership") refers to the situation in which the sales manager controls all important decisions of the sales organization, including all the selling-related tasks. Very little input is provided by the sales associates to these decisions. The sales manager, in this case, makes decisions based on his own set of beliefs and values, accepting little or no advice from the sales associates.

Advantages of this style leadership can be observed in selling situations dictated by short deadlines and in which input from the sales associates is not likely to be worth much (perhaps because they lack sufficient experience). There are other situations in which the sales organization may have been poorly managed in the past, and a sales manager with an autocratic style may be able to get the organization back on a healthy track. In doing so, he assigns tasks to different members of the sales force, establishes solid deadlines for completion, and follows through by ensuring that the assigned tasks have been completed satisfactorily and on time.

The major downside of autocratic leadership is that the sales associates may feel diminished. They may feel that the sales manager is bossy and overly controlling. Such feelings can translate into resentment and anger that reduce job satisfaction and organizational commitment. Job dissatisfaction and lack of organizational commitment may lead to turnover, which ultimately and invariably will affect the financial health of the sales organization.

Table 10.4 further highlights differences among laissez-faire, democratic, and autocratic leadership styles.

Table 10.4 *Laissez-Faire versus Democratic versus Autocratic Leadership Styles*

	Laissez-Faire Leadership Style	Democratic Leadership Style	Autocratic Leadership Style
Assumption about human nature and management effectiveness	If you leave the sales associates to do their own thing and get out of their way, they will organize and get things done on their own	Sales people do their best when they feel identified and engaged in the sales organization; encouraging them to participate in important decisions affecting the sales organization is a good way to get them to feel identified and engaged	Sales people do their best when they are assigned selling tasks that are clear; when the rules and procedures are made unambiguous; when explicit incentives and disincentives are announced; when they are provided with explicit feedback in the form of reward and punishment
Means and methods	Delegating many of the organizational tasks related to the sales organization to senior sales associates or forming committees; providing resources to the assigned individuals or committees	Setting up meetings to allow the sales associates to brainstorm and discuss issues that are important to the effective functioning of the sales organization; encouraging the sales associates to provide input on the issues, but ensuring that the final say is the manager's	Assigning selling tasks to the sales associates; telling them what rules and procedures to follow; using explicit incentives and disincentives to shape behavior; providing explicit feedback on sales performance in the form of rewards and punishments
Advantages	Works well for sales organizations involving many senior sales associates who are highly skilled and self-motivated	Works well for sales organizations involving both senior and junior sales associates with variable levels of skill and experience	Works well for sales organizations in which the vast majority of the sales force lack skill and experience
Disadvantages	Does not work well in sales organizations which have many sales associates who are inexperienced or unskilled	Does not work well in sales organizations which are either "top heavy" (the vast majority of the sales associates are highly skilled and independent) or "bottom-heavy" (the vast majority of the sales associates lack skill and experience)	Does not work well in sales organizations which have many sales associates who are highly experienced and skilled

LEADERSHIP POWER

Social scientists have long recognized that leadership power involves five sources: coercive, reward, legitimate, expert, and referent.[11] A good sales manager is one who is able to use these five sources effectively as a direct function of the situation. Let's discuss these five sources of power (see Figure 10.1).

Coercive Power

A sales manager can influence members of the sales force by using coercive power. Coercive power involves the use of punishing methods. For example, the sales manager can influence a sales associate by threatening to fire the associate if he does not meet his sales quota. The sales manager may threaten to demote him by shifting him to another geographic market that is less lucrative. The sales manager can use other adverse methods to influence selling performance by giving the sales associate a poor performance review. The use of coercive power is recommended only in dire situations when a sales associate can only learn from a direct reprimand or the threat of a reprimand. There are situations that call for the use of coercive power, such as a breach of ethics (violating one or more articles of the National Association of Realtors' code of ethics) or a violation of the law (breaking a real estate law, such as the fair housing law). These situations call for coercive action and the use of coercive power is justified.

Reward Power

A sales manager can influence his sales force by *rewarding* salespeople who meet management's expectations. This is customarily done through incentives (e.g., a one-week paid vacation in Hawaii), bonuses (e.g., a Christmas bonus check), raises (e.g., a change from 70% to 85% share of the sales commission), promotion (e.g., promotion to associate broker), etc. The power of reward can be used to motivate salespeople through the promise of a reward and, upon actual performance, the reward itself. Reward is customary in real estate marketing. Most of the rewards that are used are monetary rewards. They are very effective in influencing sales performance.

Figure 10.1
Five Sources
of Leadership
Power

Legitimate Power

Legitimate power emanates from the position of sales manager—the job's title and role within the organization. Consider the situation in which the current sales manager is retiring. He chooses his successor from the ranks of the senior sales associates. Once this senior sales associate assumes the role of the sales manager, his influence is extended by virtue of his new position. He now becomes the "boss." Through the authority of this position, he wields influence over the entire cadre of the sales force.

Expert Power

Expert power is based on the sales manager's professional experience, skills, and knowledge. As the sales manager gains experience in the art and science of personal selling and the management of a sales organization, he gains expert power. The sales manager uses expert power to help and guide the sales associates towards achieving higher levels of sales. A sales manager who has a long history of being a salesperson is likely to be more effective in leading his sales force than someone who does not have much experience in personal selling. Sales associates are more likely to seek guidance from a sales manager who has a lot of sales experience than one who has little experience.

Referent Power

A sales manager can wield influence over the sales force through the power of identification. That is, when a salesperson feels that the sales manager is "very much like him," a social bond is formed. This social bond allows the two people to communicate effectively with each other. Effective communication is the cornerstone to social influence. Consider the following example. The sales manager is a woman of who graduated from University of Virginia. One particular sales associate is also a woman who graduated from the University of Virginia. The sales associate identifies with the sales manager because the two belong to the same reference group (they both have the same alma mater, University of Virginia). A special bond is formed between the sales manager and the sales associate through this social identification. This bond allows the sales manager to influence the sales associate to take on certain selling-related tasks that ultimately result in greater sales revenue for the firm.

Using the Five Sources of Power

A sales manager with leadership ability is one who can leverage the five sources of power to wield influence over the sales force. Being able to use these sources of power as dictated by the situation is an art in itself. The sales manager has to understand the situation and identify the appropriate source of power to exert influence over one or more sales associates. Some situations call for coercive power, others reward power, and still others referent power. The challenge is to know what source of power to resort to as a direct function of the situation.

Consider the following situation: A sales associate has a personality conflict with another in the same realty firm. Both agents resent each other and make every attempt possible to avoid each other. However, a situation arises in which both have to cooperate to close a sale. One sales associate is the listing agent, the other is the buyer representative. They both have to cooperate to finalize the deal because they are both involved in the buying and selling of the same property—two sides of the same dyad. The sales manager may intervene in this situation using both *reward* and *legitimate* power. The sales manager, using his authority as "boss," summons both agents to his office and requests that both agents work together for the sake of closing this one deal. He impresses upon the two agents that this sale is large and therefore very important to the firm's survival and level of prosperity (legitimate power). He tells them that they are likely to receive "top performer" status given their own sales histories (reward power). Note that this situation calls for the use of legitimate and reward power. Using the power of his position as "boss" and offering the incentive of "top performer" status, the sales manager is able to nudge the two agents to work together. Coercive power, expert power, and referent power do not seem to be highly appropriate for this occasion.

LEADERSHIP SKILLS

Leadership scholars treat leadership skills in terms of four dimensions: (1) delegating, (2) participating, (3) persuading, and (4) telling. In other words, a skillful manager has to have skills related to these four behavioral dimensions. Let's describe these four leadership skills in some detail.

Skills Related to Delegation

A sales manager has to be able to delegate effectively.[12] *Delegation* involves assigning specific selling tasks to different members of the sales force. For example, a sales manager may assign different sales associates to different neighborhoods, and each sales associate is then expected to "farm" one community.

To be able to effectively delegate sales and/or administrative tasks to members of the sales team, the sales manager has to explicitly identify the tasks to be delegated (e.g., "we need a salesperson to help with recruiting new agents"). He needs to articulate the task requirements (e.g., the recruiter has to be highly skilled and knowledgeable about real estate marketing, has to have extra time away from his regular selling job to sink in a few hours per week for the recruiting task, has to be familiar with all the rules and regulations governing the recruitment process, etc.). Then the sales manager identifies potential candidates for the task to be delegated (Salespersons X, Y, and Z) and attempts to match the most qualified sales associate with the task.

Skills Related to Participation

A sales manager has to encourage members of the sales team to *participate* in important decisions affecting the sales organization. For example, there are many and varied tasks required to maintain the viability of a sales organization: recruiting new sales associates, training associates, developing a strategic plan for the sales organization, organizing professional events (e.g., the office caravan) and social events, recognizing top performers, acting on breaches of ethics, etc. Many of these tasks can be handled by the sales manager alone, but many managers opt to involve the sales team to participate in decision-making and provide input.

To facilitate participation in decision-making, the sales manager has to first select those decisions that can significantly benefit from group decision-making. For example, a firing decision because of an illegal breach (e.g., embezzlement) should not be a collective decision. Many of the sales team members who have befriended the person who have committed the breach may feel obligated to defend his action and feel sympathetic. Others may come down hard on the person in question. This type of situation may cause rifts among members of the sales team because people are likely to find themselves on opposite sides of the fence. Other types of decisions (e.g., recruiting decisions, organizing an office caravan, and organizing social events) may not only help the sales manager with administrative tasks but also may serve to build camaraderie among the team members.

The sales manager also has to choose which members of the sales team to have participate in each decision. The nature of the task may dictate the type of people most likely to significantly contribute to the discussion to help arrive at effective decisions. For example, suppose the sales manager wants to revise the overall strategic plan. He would like to assemble a committee that can examine the current strategic plan and make suggestions on how to improve it. Who should he select for this strategic plan committee? The nature of the task should provide the manager with sufficient cues: Associates who have much knowledge and skill in the industry and people who have long tenure in the company.

Skills Related to Persuasion

The sales manager uses effective communication skills to influence the behavior of the salespeople through *persuasion*. There are many instances in which the sales associates will not comply with management's expectations; they have to be convinced that management expectations are reasonable and sound. For example, a sales manager requests a member of the sales team to share his desk with a new recruit. The first reaction of the sales associate is likely to be: "Over my dead body." Here the sales manager has to make a case to make this request convincing. His reasoning may include the following arguments: By sharing the same desk, the rookie is likely to learn much from the senior sales associate. It is a good way for the senior agent to take the rookie under his wing. There are no desks available for the rookie; there are current plans to expand office space, and this is only a temporary measure. These arguments have to be effectively articulated and communicated to the sales associate to get him on board. Also, the sales manager has to motivate the sales associate to process the reasoning behind these arguments. In other words, he cannot simply make the arguments and expect the sales associate to accept them because they are well reasoned. The sales associate has to think about these arguments and weigh them accordingly.

Persuasion is not only an art but a science. That is, social scientists have long studied situational and personality factors that influence persuasion effectiveness. From this research, we find that there are two routes to persuasion: central and peripheral. The central route to persuasion tends to produce long-lasting and deep attitude change. Central route processing involves the message recipient becoming highly involved with the processing of the message, with adequate background knowledge to interpret the message as intended by the message sender. The result of central processing is enduring attitude change. Conversely, the peripheral route to persuasion involves a shortcut. The message recipient is not much involved with the processing of the message and does not really attend to the message content *per se*. He relies on cues such as the credibility and trustworthiness of the message sender to accept or reject the message. He does not allocate much cognitive capacity to process the pros and cons of the message claim, nor seek to scrutinize the evidence for the claim. The resulting attitude (acceptance or rejection of the message claim) is likely to be ephemeral and not enduring, compared to central route processing.[13]

What does this all mean in terms of persuasion skills that a good sales manager ought to muster? It means that for the sales manager to be persuasive, he has to first motivate the salesperson to listen, thus prompting the salesperson to allocate enough cognitive processing capacity to process the communication fully and without misinterpretations. Once the salesperson is motivated to listen, the sales manager communicates his position in ways supported by good and solid evidence. In other words, the message has to be of high quality. A high quality message involves a message claim supported by good arguments in favor of the message claim, and possibly good arguments against counterarguments to the message claim. Such communication is likely to be effective, in that the salesperson will accept the argument put forth by the sales manager and act on it. For example, suppose the sales manager has to convince a new sales recruit of the benefits of a 60–40 commission split (60% of the sales commission goes the salesperson and 40% goes to the brokerage firm), given the fact that the salesperson made a 70–30 split on his last job. The sales manager sets up an appointment with the salesperson and alerts him that he would like to discuss his pay. He asks the salesperson to do some homework and come in prepared to discuss this issue. Doing so should make the salesperson highly motivated to process the sales manager's argument through a central route. The sales manager makes a cogent argument about the 60–40 split, using good reasoning and data. He makes the argument (and supports it with data) that all incoming sales associates have started out at the firm with a 60–40 commission split; that the industry average for incoming sales associates is a 60–40 split; that giving him a 70–30 split is likely to cause feelings of inequity among other incoming sales associates; that a 60–40 split is likely to lead to a 70–30 split that should be earned through hard work, collegial behavior, and ethical conduct. The sales associate accepts the 60–40 split because he was motivated to process the message fully and because the argument was substantiated with good reasoning and data.

Skills Related to Telling

Finally, there is the *telling* skill. Good sales managers have to command certain members of the sales team to do (or not to do) certain things. For example, a sales associate may have committed an ethical breach (e.g., he failed to disclose to the buyer a structural defect in the property to protect the interest of the seller, his client). Such behavior has to be discouraged and an effective sales manager has to use his authority to quell any future ethics breach. Therefore, he may command the sales associate to disclose the defect, even though it is "after the fact," and formally apologize to the buyer. Effective leaders can "tell" their subordinates to do certain things with little resistance from the subordinates. This is, indeed, a leadership skill.

The skill of telling involves the use of three sources of power: legitimate, reward, and coercive power. The mix of these three sources of power allows the sales manager to "tell" a sales associate what to do and not to do, with little resistance or possible repercussions from the sales associate. Consider the following situation. A female sales associate approaches the sales manager to complain about a male associate. The male sales associate is a senior associate and is responsible for training the female associate, who is a rookie. The nature of the complaint is sexual harassment. The male senior associate has made unwelcome advances. She has repeatedly rebuffed him, yet he continues. Her complaint, she feels, is her last resort to deal with this problem. This situation calls for "telling" the male associate to cease and desist. He meets with the male associate and alerts him to the complaint filed against him. He takes action by reassigning the female associate to another senior associate—a female senior associate. He also "tells" the male associate to apologize to the female junior associate, and warns him that this complaint is now on file (i.e., similar future action may be cause for dismissal). The male associate, although defensive initially, succumbs to the order, apologizes to the female associate, and vows not to do this ever again. Note that in this case, the sales manager's "telling" effectiveness was based on the use of his authority and the threat of dismissal.

LEADERSHIP ACTION

What do sales managers do to guide the sales force towards achieving sales goals? They do the following: (1) plan, (2) organize, (3) motivate, and (4) control.[14] See Figure 10.2.

Planning

For the sales manager, planning involves determining quarterly or monthly sales objectives and articulating a general plan to achieve the stated objectives. Consider the following example. A realty firm has a history of providing a full range of services in residential and commercial real estate. Sales have been dwindling over the past five years. A new broker has been hired to breathe life into the brokerage firm. The new broker does some analysis on housing sales in the area and finds that the size of the elderly market has been on the rise and housing sales to the elderly have increased significantly. He examines the competition and comes to the realization that all of the other realty firms in the area provide the full range of services in both residential and commercial real estate. In other words, none of his immediate competitors specialize in serving the retirement community. He now believes that re-positioning the realty firm to serve the retirement community exclusively is likely to significantly enhance the financial well-being of the firm. He calculates that the market potential for the housing market in the local area serving the retirement community is approximately $250 million per year for the next five years. His firm can easily capture 30% of that market.

Organizing

Guided by an implicit or explicit plan, the sales manager spends much time and effort organizing the sales force to achieve the sales objectives and implement the plan. Using the example of the retirement community (see the preceding section on Planning), this may

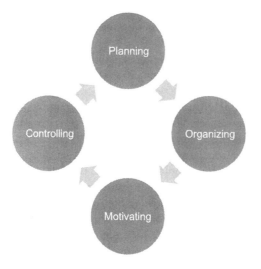

Figure 10.2
Five Dimensions of Leadership Action

mean that the new broker allocates resources to train (and/or retrain) his sales force to focus on serving the retirement community. He may allocate additional resources to recruit new sales associates and train them too. Perhaps he assigns different sales agents to farm specific neighborhoods. These activities are, in essence, "organizing activities"—activities designed to configure human and financial capital to help implement the firm's strategy and achieve the stated goals and objectives.

Motivating

The next set of tasks is related to sales force motivation. Here, the sales manager has to focus on ways to motivate the sales force to implement the overall plan and achieve the stated goals and objectives. The least common denominator in motivating a sales force is a good compensation plan. However, as we learned from the chapter on Motivation (Chapter 9), money is not the only source of motivation. Sales managers motivate their sales forces using a variety of monetary and non-monetary incentives. A monetary motivator may involve not only an increase in the share of sales commissions but perhaps also increased stock options or offering top performers partnership status in the firm. Non-monetary incentives may include developing and implementing a recognition program—top performers are publicly recognized in the local media with platinum, gold, silver, and bronze awards. Promoting junior sales associates to senior status with increased responsibility and pay is another motivator.

Turning back to our retirement community example, the new broker has to devise a motivation plan that includes monetary and non-monetary elements. He decides that to achieve the stated goals in capturing 30% of the retirement community market over the next five years, he will offer a compensation package involving a 60–40 split in sales commissions for junior sales associates (60% of the sales commission goes to the sales associate and 40% goes to the firm) and an 80–20 split for the senior associates. The senior associates who manage to maintain their sales quota (a minimum of 25 units sold in a year) will become partners, and partners will reap 50% of the annual profits among themselves (divided equally among the partners). Top performers among the junior associates will be promoted to senior associates after two years if they achieve top performer goals within the junior ranks (a minimum of 20 units sold in a year). Top performers will be publicly recognized in the media (with an advertisement in the local newspaper), through award banners in the firm's website, and at the annual award event organized by the firm around Christmas time. Also, top performers will be provided with their own private offices within the firm, comparable in size and style to the offices of the senior associates.

Controlling

Finally, we have a set of activities related to the control function. *Control* means that the sales manager monitors the performance of each sales associate and makes sure that those associates who meet or exceed their sales quota are properly rewarded. Of course, the rewards are administered based on the motivation and compensation plan, as discussed in the preceding section.

The sales manager may decide that a certain sales associate has not only failed to meet the sales quota but also has failed to cooperate with other sales associates within the firm. Perhaps the sales associate has a personality that is not congenial and seems to have become entangled in several quarrels with other associates. Perhaps one or two clients have contacted the sales manager to complain about the sales associate not returning calls. Such

235

situations may necessitate firing that sales agent. This means that the sales manager has to document evidence to legally justify the termination, and follow that with concrete action.

Remedial action is taken in relation to those who fail to meet their quota. What kind of remedial action, and for what end? Remediation may involve additional training or retraining. The sales manager attempts to uncover deficiencies in skills and experience that may account for the associate's low sales performance. Perhaps this agent lacks sufficient knowledge concerning how to go about finding new clients. Ah! Perhaps a training course on client prospecting may help boost his sales performance.

SUMMARY

This chapter covered a variety of concepts related to leadership. We started by discussing why is it important to have a manager with leadership qualities. Leadership is important because it assists the sales organization in attaining organizational goals, such as high job performance, high job satisfaction, high employee morale, high organizational loyalty and commitment, low employee absenteeism, low employee presenteeism, and low employee turnover.

What is a good definition of leadership? Good leadership in real estate marketing organizations is defined in terms of leading real estate marketing professionals toward achieving sales goals.

What is the distinction between management and leadership? Management is about efficiency: systems, processes, procedures, control, and structure. Leadership is about effectiveness: trust, inspiration, and people.

What is leadership style and which styles are suitable for which situations? We made distinctions between task-oriented and people-oriented leadership; transformational versus transactional leadership; and laissez-faire, democratic, and autocratic leadership. Task-oriented leadership was described as an approach to sales management in which the manager focuses on the personal selling tasks to meet certain sales goals. In contrast, people-oriented (or relationship-focused) leadership is an approach in which the sales manager focuses on the satisfaction, motivation, and the general well-being of the sales agents. The task-oriented leadership style is well-suited for certain commercial real estate marketing organizations in which one sales project may involve breaking the project into many tasks, each task is guided by many policies and regulations, with specific hard deadlines. This leadership style is highly suitable in situations in which the sales agents have to work together closely as a team.

Transactional leadership is an approach of sales management in which the sales manager focuses on the role of supervision, organization, and performance. Transactional leadership was described in terms of two dimensions: (1) contingent rewards and (2) management-by-exception. Transformational leadership, in contrast to transactional leadership, focuses on motivation and morale of the sales force to achieve sales goals. Transformational leadership was described in terms of four dimensions: individualized consideration, intellectual stimulation, inspirational motivation, and idealized influence.

Laissez-faire leadership is a style in which sales managers takes a passive approach to the management of the sales force. It is essentially hands off, and the sales manager delegates much of the responsibility of overseeing certain sales tasks to the senior sales associates or assigns those tasks to committees. Laissez-faire leadership works best when the sales organization is top heavy. With respect to democratic leadership, the sales manager leads the sales force by encouraging the sales associates to participate in many decisions involving the sales organization. Advantages of this style of leadership include identifying creative solutions to selling problems and making the sales associates more involved and engaged in the selling process, as well as in handling other organizational tasks vital to the survival and prosperity of the sales organization. Autocratic leadership refers to the situation in which the sales manager controls all important decisions of the sales organization, including all the selling-related tasks. Advantages of this style of leadership can be observed in selling situations dictated by short deadlines and in which input from the sales associates is not likely to be worth much.

What is leadership power? It involves five sources: coercive, reward, legitimate, expert, and referent. Coercive power involves the use of punishing methods. A sales manager can influence his sales force by rewarding salespeople who meet management's expectations (reward power). Legitimate power emanates from the sales manager based on the position and its role within the organization. Expert power is based on the sales manager's professional experience, skills, and knowledge. A sales manager can wield influence over the sales force through the power of identification (referent power).

What is leadership skill? There are four skills associated with good leadership: delegating, participating, persuading, and telling. Delegation involves assigning specific selling tasks to different members of the sales force. A sales manager has to encourage members of the sales team to participate in important decisions affecting the sales organization. The sales manager also has to choose which members of the sales team to participate in specific decisions. In persuasion, the sales manager uses effective communication skills to influence the behavior of the salespeople. Finally, there is the telling skill. Good sales managers have to command certain members of the sales team to do (or not to do) certain things.

What is leadership action? Leadership action involves four sets of activities: planning, organizing, motivating, and controlling. For the sales manager, planning involves determining quarterly or monthly sales objectives and articulating a general plan to achieve the stated objectives. Guided by an implicit or explicit plan, the sales manager spends much time and effort organizing the sales force to achieve the sales objectives and implement the plan. The sales manager has to focus on ways to motivate the sales force to implement the overall plan to achieve the stated goals and objectives. Control means that the sales manager monitors the performance of each sales associate, making sure that those associates who meet or exceed their sales quota are properly rewarded.

DISCUSSION QUESTIONS

1. Why is leadership important in the effective management of a real estate sales organization?

2. What is a good definition of leadership?

3. Distinguish between task-oriented and people-oriented leadership styles.

4. Under what conditions should a sales manager use a task-oriented style? Under what conditions should he use a people-oriented style?

5. Distinguish between transformational and transactional leadership styles.

6. Under what conditions should a sales manager use a transformational style? Under what conditions should he use a transactional style?

7. Distinguish among laissez-faire, democratic, and autocratic leadership styles.

8. Under what conditions should a sales manager use a laissez-faire style? Under what conditions should he use a democratic style? And under what conditions should he use an autocratic style?

9. What is leadership power? Describe the concept in general terms.

10. How should a sales manager use coercive power in the context of a real estate sales organization?

11. How should a sales manager use reward power in the context of a real estate sales organization?

12. How should a sales manager use legitimate power in the context of a real estate sales organization?

13. How should a sales manager use expert power in the context of a real estate sales organization?

14. How should a sales manager use referent power in the context of a real estate sales organization?

15. What is leadership skill? Describe the sets of leadership skills required to effectively manage a real estate sales organization.

16. What is leadership action? Describe leadership activities common in the management of a real estate sales organization.

NOTES

1. Hardin, R. (Eds.) (2002). *Trust and trustworthiness*. New York: Russell Sage Foundation.
2. Brogan, T. V. F. (1993). Inspiration. In A. Preminger and T. V. F. Brogan (Eds.), *The New Princeton Encyclopedia of Poetry and Poetics* (pp. 609–610). Princeton, NJ: Princeton University Press.
3. Bass, B. M. (1990). *Bass & Stogdill's handbook of leadership: Theory, research, and managerial applications* (3rd ed.). New York: Free Press.
4. Anzalone, C. (n.d.). Differences between task-oriented leaders and relational-oriented leaders, Houston Chronicle website: *Chron.com*. Accessed from http://smallbusiness.chron.com/differences-between-taskoriented-leaders-relationaloriented-leaders-35998.html, on July 5, 2013.
5. Anzalone, C. (n.d.). Differences between task-oriented leaders and relational-oriented leaders, Houston Chronicle website: *Chron.com*. Accessed from http://smallbusiness.chron.com/differences-between-taskoriented-leaders-relationaloriented-leaders-35998.html, on July 5, 2013.
6. Anzalone, C. (n.d.). Differences between task-oriented leaders and relational-oriented leaders, Houston Chronicle website: *Chron.com*. Accessed from http://smallbusiness.chron.com/differences-between-taskoriented-leaders-relationaloriented-leaders-35998.html, on July 5, 2013.
7. Blake, R., & Mouton, J. (1985). *The managerial grid III: The key to leadership excellence*. Houston: Gulf Publishing Co.
8. Bass, B. M. (2008). *Bass & Stogdill's handbook of leadership: Theory, research & managerial applications* (4th ed.). New York: The Free Press (pp. 50, 623). Hackman, J., & Michael, C. (2009). *Leadership: A communication perspective*. Long Grove, IL: Waveland Press (pp. 102–104).
9. Bass, B. M., & Avolio, B. J. (Eds.) (1994). *Improving organizational effectiveness through transformational leadership*. Thousand Oaks, CA: Sage Publications.

10. Bass, B. M., & Avolio, B. J. (Eds.) (1994). *Improving organizational effectiveness through transformational leadership*. Thousand Oaks, CA: Sage Publications.
11. French, J. R. P., & Raven, B. (1959). The bases of social power. In D. Cartwright (Ed.), *Studies in Social Power* (pp. 45–65). Ann Arbor, MI: University of Michigan Press.
12. Blanchard, K. H., Zigarmi, P., & Zigarmi, D. (1985). *Leadership and the one-minute manager: Increasing effectiveness through situational leadership*. New York: Morrow.
13. Petty, R. E., & Cacioppo, J. T. (1986). *Communication and persuasion: Central and peripheral routes to attitude change*. New York: Springer-Verlag.
14. Finch, L., & Maddus, R. B. (2006). *Delegation skills for leaders: An action plan for success as a manager* (3rd ed.). Boston: Thomson/Course Technology.

Part 5

Law and Ethics in Real Estate Marketing

Chapter 11

Real Estate Marketing Laws

⚡

LEARNING OBJECTIVES

This chapter is designed to sensitize the student to real estate marketing laws. Specifically, the student will learn the major laws in the U.S. (and other countries) related to real estate marketing. These laws include:

- Deceptive Trade Practices Act (DTPA)
- Interstate Land Sales Full Disclosure Act
- Real Estate Settlement Procedures Act (RESPA)
- Equal Credit Opportunity Act (ECOA)
- Truth in Lending Act
- Fair Housing Law
- National Environmental Policy Act (NEPA)
- Flood Disaster Protection Act
- Clean Air Act
- Federal Water Pollution Control Act
- Noise Pollution and Abatement Act
- Coastal Zone Management Act

THE DECEPTIVE TRADE PRACTICES ACT (DTPA)

The Deceptive Trade Practices Act (DTPA) is a U.S. law providing protection for consumers from unscrupulous business practices. Under this law, business firms, including real estate firms, can be prosecuted for false, misleading, or deceptive practices. Applied to real estate, the law addresses misrepresentations of real estate to claim that it possesses characteristics that it does not possess, and failure to disclose vital information concerning real estate known at the time of the transaction, such that the consumer would not have completed the transaction had the information been disclosed.

The Federal Trade Commission (FTC) in the U.S. is charged to regulate unfair or deceptive trade practices across a variety of industries, including real estate. Additionally, every state in the U.S. has its own consumer protection statutes, which allow state attorneys to sue business firms over false/deceptive advertisements and unfair business practices, as well as practices that jeopardize consumer safety. The state courts tend to use the federal act and interpretations of the FTC for guidance in the enactment of state laws. According to the FTC, a practice is deemed unfair if it results in substantial injury to consumers. A business practice is considered deceptive if it misleads consumers to the point where they take action that results in their own detriment. Case in point is false advertising. For example, a buyer of real estate has the right to sue a false advertiser for fraud, if the advertiser made false representations about the advertised real estate that the buyer relied on, and the buyer was harmed as a result. Typically, the FTC regulates false advertising at the national level. For example, the FTC can issue a cease and desist order, forcing the real estate firm that is running a national advertising campaign to stop advertising, or compelling the firm to make corrections or disclosures informing the public of the misrepresentations.[1]

Let's be specific about the concept of misrepresentation and how it is applied in real estate marketing. Misrepresentation is a false statement that can occur by commission or omission. Commission refers to the stating something that is not true (e.g., a real estate agent representing a buyer of a residential property says that "this house has been examined by a home inspector and was deemed in good condition," when this did not actually occur). Omission refers to leaving out information (e.g., an agent representing a buyer knows that there are significant monthly dues that the home owner has to pay to the homeowners association, but has failed to inform his client). See Cases/Anecdotes 11.1 and 11.2, which are related to failure to disclose, and learn more about laws about failure to disclose across countries in Case/Anecdote 11.3. Ethics Box 11.1 further illustrates the issue of failure to disclose.

CASE/ANECDOTE 11.1. REALTORS ARE SHIELDED FROM LIABILITY IF A PUBLIC RECORD IS INACCURATE[i]

The current state of the law allows real estate brokers and their agents to be sued if they use information from the Multiple Listing Service (MLS) in their advertising and it turns out that this public record is inaccurate. In Virginia, there is now a new law that protects real estate brokers and agents from being sued for any misrepresentation in their advertising and other promotion efforts if this information is obtained from other professional sources, such as surveying firms, client disclosures, and the MLS. When citing such sources, brokers and agents cannot be held liable in a civil suit or regulatory proceeding at the Real Estate Board.

[i] Dicks, J. G. (2013). New laws for 2013: Realtors are shielded from liability if a public record is inaccurate. *Commonwealth*, June/July, 30–32.

CASE/ANECDOTE 11.2. YOU MUST DISCLOSE IF A HOME WAS ONCE A METH LAB[i]

Much research has demonstrated that once a home is used as a meth lab, it is difficult to get the meth residue out of the home to make it safe for the new occupants. The Virginia legislature has passed a new law stating that real estate brokers and agents have to disclose to prospective buyers the fact that a home was used as a meth lab. However, there is a loophole: Brokers and agents do not have to disclose this fact if the home was restored and declared clean and safe, as certified by the Virginia Department of Health.

[i] Dicks, J. G. (2013). New laws for 2013: You must disclose if a home was once a meth lab. *Commonwealth*, June/July, 30–32.

CASE/ANECDOTE 11.3. CONSUMER PROTECTION LAWS IN DIFFERENT COUNTRIES

As a member state of the European Union, the U.K. is bound by the consumer protection directives of the E.U. Consumer protection complaints are made to the Director-General of Fair Trade. If a complaint is accepted, the Office of Fair Trading (OFT) then investigates and, if the case warrants, litigates against the real estate business. The OFT can also be engaged by consumer advocacy groups. The OFT also acts as the U.K.'s official consumer and competition watchdog at the national level and the Trading Standards departments at the municipal level.[i]

In Australia, the Australian Competition and Consumer Commission or the individual State Consumer Affairs agencies focus on consumer protection related to real estate transactions.[ii] In Germany, there is a federal ministry responsible for consumer rights and protection called "Verbraucherschutzminister."[iii] In India, there is the Consumer Protection Act of 1986, which encourages the formation of consumer tribunals in each and every district, in which a consumer can file a complaint against any company, including real estate. The complaint is reviewed by a presiding official at the district level. This setup is less formal, but speedier, than in other developed countries.[iv]

[i] Consumer Protection, *Wikipedia*. Accessed from http://en.wikipedia.org/wiki/Consumer_protection, on June 9, 2013.

[ii] Consumer Protection, *Wikipedia*. Accessed from http://en.wikipedia.org/wiki/Consumer_protection, on June 9, 2013.

[iii] Consumer Protection, *Wikipedia*. Accessed from http://en.wikipedia.org/wiki/Consumer_protection, on June 9, 2013.

[iv] Consumer Protection, *Wikipedia*. Accessed from http://en.wikipedia.org/wiki/Consumer_protection, on June 9, 2013.

ETHICS BOX 11.1. FAILURE TO DISCLOSE[i]

Consider the following scenario. A home buyer is threatening to sue the seller for discovering a leak hidden behind the wall under the kitchen sink. The seller didn't know about the leak. The seller does not want to go to court over this issue; he would rather settle with the buyer.

In this situation, the real estate agent representing the buyer could have done a few things to avoid this very "common situation." First, the buyer agent should have made sure that a home inspection was done by a professional. (Make sure not to recommend only one home inspection firm because this may be misconstrued as a "conflict of interest" — that is, a business arrangement with that particular firm that provides you with referral fees. Instead, recommend several home inspection companies, and have the home buyer select one on his own.) In other words, this situation could have been avoided if the home had been inspected by a professional before the transaction was completed.

Second, the listing agent may have done a better job by encouraging the seller to complete a disclosure statement. This is a document that describes the property in some detail. The seller makes an honest attempt to identify all problems related to the property and documents these problems in this "disclosure statement." The listing agent should not get involved in completing this disclosure statement because the agent may be held liable if any of the information disclosed is found to be inaccurate. Nevertheless, the listing agent should simply encourage the seller to complete this statement to the best that the seller can, "honestly."

Third, the buyer representative should have gathered information about the property from independent sources, such as government websites and offices. The government usually publishes information about properties located in specific areas, such as flood maps, zoning restrictions, and tax records. Part of the tax record of the property in question should reflect information about the age of the property and its geometric area (the property itself and the lot at large). In this instance, the tax record could have alerted the buyer representative of potential problems, such as plumbing leaks, because the house may have been "too old."

Some real estate agents engage in practices that do not hold up in a court of law. For example, some listing agents encourage their seller clients to extract a waiver from the buyer. The waiver is essentially a written document signed by the buyer that states that the buyer is waiving his right to sue. These waivers do not work in a court of law—and, in addition, they may be construed as "unethical."

Another tactic commonly used by real estate agents is the "as-is" clause. The buyer is asked to sign an "as-is" document, which means that the buyer will not sue the seller even though the buyer is not fully aware of any problems with the property. It is more ethical to disclose all problems that the seller may be aware of than to convince the buyer to sign a document that says that the buyer is accepting to purchase the property "as-is."

Remember that in real estate "caveat emptor" (let the buyer beware) does not usually work. The buyer does not have to show due diligence; the seller does. Therefore, the advice here is for the seller to disclose any and every defect that he is aware of. This is the right thing to do.

[i] Hamilton, D. (2006). *Real estate: Marketing and sales essentials*. Mason, OH: Thomson Higher Education (pp. 315–319).

INTERSTATE LAND SALES FULL DISCLOSURE ACT (ILSFDA)

The Interstate Land Sales Full Disclosure Act (ILSFDA) is a U.S. law passed in 1968 to regulate interstate land sales—specifically to protect consumers from fraud and abuse in the sale or lease of land. The target firms of this law are real estate developers involved in developing subdivision lots and selling these lots to home buyers. The federal agencies in charge in enacting this law in the U.S. include the Department of Housing and Urban Development (HUD) and the Bureau of Consumer Financial Protection (CFPB). A real estate developer is required to provide buyers with a disclosure document that contains relevant information about the subdivision before the sale transaction is completed. ILSFDA gives the buyer a minimum of seven days to cancel the purchase agreement.[2]

Specifically, ILSFDA states that it is unlawful to sell subdivision lots over interstate boundaries without having this real estate registered with HUD. The registration includes the name and address of the seller, a legal description of the lot in terms of topography, the exact title to the land, access to the subdivision, article of incorporation of the seller, the deed establishing the title to the subdivision, means used to sell the lots, possible easements and other restrictions to the lots, financial statements, and other information, such as terms, conditions, prices, and rents. Thus, when a home buyer purchases a lot in the specified subdivision, the buyer is provided with all that information to guide his decision-making.[3]

REAL ESTATE SETTLEMENT PROCEDURES ACT (RESPA)

The Real Estate Settlement Procedures Act (RESPA) is a U.S. law created in 1974. The purpose of the law is to crack down on real estate companies (e.g., lenders, real estate agents, real estate developers, and title insurance companies) that engage in providing undisclosed kickbacks to each other. These kickbacks often result in inflating the costs of real estate transactions. For example, a home lender advertises a 5% interest rate. When buyers apply for the loan, they are then told that they have to use the lender's affiliated title insurance company at a rate of $1,000. The lender gets a certain amount of money from the title insurance company as kickback.

RESPA is administered by the Department of Housing and Urban Development (HUD) and enforced by the Consumer Financial Protection Bureau (CFPB). RESPA requires lenders to provide a good faith estimate (GFE) for all the approximate costs of a particular loan.[4]

Under RESPA, lenders have certain duties and obligations. Specifically, lenders have a duty to provide borrowers with three types of information: (1) settlement cost booklet, (2) advance disclosure of settlement costs, and (3) uniform settlement statement. The settlement cost booklet is a document from the U.S. Department of Housing and Urban Development (HUD) that provides information about settlement costs, the standard form that is used by lenders, escrow accounts, choices that borrowers have in selecting a lender, and lenders' possible unfair practices. The advance disclosure of settlement costs is usually referred to as *good faith estimate*, a statement provided by lender estimating settlement charges. Good faith estimates have to be in sync with the final settlement costs. The uniform settlement statement is a statement that lists settlement charges such as real estate commissions, lender's fees, escrow reserve amounts, and title charges.[5]

EQUAL CREDIT OPPORTUNITY ACT (ECOA)

The Equal Credit Opportunity Act (ECOA) is a United States law enacted in 1974. In the context of real estate, this law applies mostly to mortgage lenders. The law states that it is unlawful for any mortgage lender to discriminate against any applicant who has the capacity to contract on the basis of race, color, religion, national origin, sex, marital status, or age. Lenders who fail to comply with the ECOA are subject to civil liability for actual and punitive damages. Liability for punitive damages can be as much as $10,000 in individual suits and the lesser of $500,000 or 1% of the lender's net worth in class action suits.[6]

Specifically, the ECOA makes it unlawful to discriminate based on any of the aforementioned social categories (race, color, religion, etc.). Lenders have to abide by the following guidelines: (1) lenders cannot use sex or marital status in credit scoring; (2) lenders should provide explicit reasons to borrowers for denying credit; (3) lenders cannot inquire about borrowers' use of birth control practices, childbearing capabilities or intentions, and actual and anticipated labor force participation as a direct result of childbearing; (4) lenders cannot discount part-time employment, and (5) lenders should maintain applicant records (including complaints and charges of discrimination) for at least 15 months following the date of notification of denial of credit.[7]

TRUTH IN LENDING ACT (TILA)

Home mortgage is fraught with complicated finance terms preventing home buyers from fully knowing amounts they are paying and where that money goes. The Truth in Lending Act (TILA) is a law enacted in the U.S. in 1968 requiring lenders to fully disclose and explain the loan terms in understandable language. TILA also gives consumers the right to cancel a transaction involving a lien on a consumer's principal dwelling. TILA regulates the charges that may be imposed for consumer credit of high-cost mortgage loans. It requires standardized disclosure of costs to encourage consumers to shop around.[8]

The Federal Trade Commission in the U.S. enforces advertising by real estate agents and brokers. For example, real estate brokers may not advertise that a buyer "may assume an 8% mortgage." The ad should correctly state that a buyer can "assume an 8% annual percentage rate mortgage." Annual percentage rate (APR) is based on the total finance charge for obtaining the loan, in that it includes the interest rate and other charges, such as loan application fees, loan finder's fee, inspection fees, discount points, prepayment penalties, cost of credit insurance, private mortgage insurance fees, and prepaid mortgage insurance fees. The law requires lenders to include these charges in their calculation of the advertised interest rate, captured by the APR acronym. The goal here is to ensure that the *cost of credit* is transparent to the borrower.[9]

FAIR HOUSING LAW

In the United States, the fair housing law dates back to the 1960s. The goal of this legislation was to outlaw discrimination in all types of housing-related transactions, such as advertising, mortgage lending, homeowner's insurance, and zoning. Specifically, the Fair Housing Law of 1968 is designed to protect buyers (or renters) of a residential property from discrimination by property owners (or managers). The law makes it illegal to:

- refuse to sell or rent a residential property to any person because of race, color, religion, sex, familial status, or national origin
- discriminate based on race, color, religion, sex, familial status, or national origin in the terms and conditions of the sale or rental of the residential property
- advertise the sale or rental of a residential property indicating preference based on race, color, religion, sex, familial status, or national origin
- use coercive methods (i.e., coercion, threats, intimidation, interference, or retaliation) against a person or organization that supports fair housing[10]

The law requires real estate brokers to display fair housing posters at their place of business and wherever a property is offered for sale or rent. The U.S. Department of Housing and Urban Development (HUD) is the federal agency in charge of enforcing the fair housing law. Furthermore, HUD requires an affirmative fair housing marketing plan for real estate developers of federally assisted or insured housing. The goal of the plan is to encourage minority applicants. Specifically, the developer must maintain an equal opportunity hiring policy for sales agents and display the HUD fair housing sign in all forms of promotional material (including project sites).[11] See Case/Anecdote 11.4 for what fair housing laws say about *overcrowded* housing. Case/Anecdote 11.5 deals with fair housing laws in relation to ethics violations by realtors. Case/Anecdote 11.6 describes a study that documented continuing housing discrimination in the state of Virginia.

CASE/ANECDOTE 11.4. "OVERCROWDED" IS NOW CLEARLY DEFINED[i]

Based on the Fair Housing Laws in the U.S. that are enforced by the Department of Housing and Urban Development (HUD), current law prohibits overcrowding in apartment rentals. The current law specifies "two persons per bedroom" as maximum occupancy standard. Violating this standard breaks the fair housing law. The 2013 Virginia Uniform Statewide Building Code has now further clarified the HUD guideline as follows: "50 square feet per person per bedroom."

[i] Dicks, J. G. (2013). New laws for 2013: "Overcrowded" is now clearly defined. *Commonwealth*, June/July, 30–32.

CASE/ANECDOTE 11.5. ALL FAIR-HOUSING CASES INVOLVING REALTORS WILL BE HEARD BY THE REAL ESTATE BOARD[i]

In Virginia, it used to be that fair-housing cases filed against people without real estate licenses were processed by the Fair Housing Board and cases filed against real-estate licensees were heard by the Real Estate Board. Virginia law now states that the Real Estate Board will process all fair-housing cases involving real estate licensees and non-licensee property owners. In other words, the new law gives the Real Estate Board jurisdiction over all fair-housing cases involving real estate licensees.

[i] Dicks, J. G. (2013). New laws for 2013: All fair-housing cases involving realtors will be heard by the real estate board. *Commonwealth*, June/July, 30–32.

CASE/ANECDOTE 11.6. STUDY FINDS DISCRIMINATION ACROSS VIRGINIA[i]

A new study conducted by the Equal Rights Center and the law firm of Drinker Biddle & Reath has uncovered disturbing information about housing discrimination against Latinos in the state of Virginia. The study involved an experiment involving a matched-pair test. Two groups of people pretending to seek housing rentals were assembled: one group involved Caucasians and the other Latinos. Both groups were matched in age, marital status, income, and other major characteristics.

Both groups were dispatched to seek housing for rent in different parts of Virginia. The study results showed that 55% of the Latino apartment seekers received "adverse, differential treatment" compared to the Caucasian seekers. Specifically, the study found that (1) Latino seekers were quoted higher rents or higher fees for the same rental units than Caucasian, (2) Caucasian seekers were offered more "specials" or other incentives than the Latino seekers; (3) Latino seekers were offered fewer available units or later availability dates than their counterparts; and (4) landlords required additional application requirements (e.g., credit check, Social Security card) from the Latino compared to Caucasian apartment hunters.

i Study Finds Discrimination across Virginia. (2013). *Commonwealth,* June/July, 6.

NATIONAL ENVIRONMENTAL POLICY ACT (NEPA)

NEPA is a U.S. law promoting the protection and enhancement of the environment. That is, the law is designed to prevent or eliminate damage to the environment as a direct result of real estate development performed on behalf of the federal government. Thus, NEPA ensures that environmental factors are weighted in the decisions made by federal agencies. For instance, NEPA requires federal agencies prepare an environmental impact statement to accompany requests for funding from Congress. The purpose of an environmental impact statement is to help decision makers better understand the environmental consequences of a particular real estate development and identify alternatives that may be less damaging to the environment.[12]

The environmental impact statement must cover any adverse environment effects that cannot be avoided if the proposed development plan were to be implemented, and possible alternatives that are likely to cause less adverse impact. For example, environmental impact statements are required for the construction of power plants, high voltage transmission lines, the routing of highways, public parks, subsidized housing, ski resorts, among others. HUD is the agency that oversees and enforces this law.[13]

National Flood Insurance Act

This piece of legislation was enacted in 1968 in the United States. It led to the creation of the National Flood Insurance Program (NFIP). NFIP was designed to accomplish two goals: (1) to provide flood insurance for real estate property in floodplain communities, and (2) to identify areas of high and low flood hazard and, based on this assessment, establish flood insurance rates for real estate properties inside each flood hazard area. Thus, NFIP helps property owners in floodplain communities buy insurance from the government to protect losses from flooding. As of April 2010, NFIP insured about 5.5 million homes—most of them were in floodplain communities in Texas and Florida.

In 1973, NFIP was amended by the Flood Disaster Protection Act, which made the purchase of flood insurance mandatory for the protection of real estate property within flood hazard areas. In 1982, the law was amended by the Coastal Barrier Resources Act (CBRA) to cover flood insurance for new or significantly improved structures. The National Flood Insurance Reform Act of 1994 established an incentive program encouraging communities to purchase flood insurance to exceed the minimal federal requirements for development within floodplains within the NFIP. NIFP was further amended by the Flood Insurance Reform Act of 2004. The goal of the latter law was to reduce losses to real estate properties for which repetitive flood insurance claim payments are made.[14]

The federal agencies involved in overseeing and enforcing these laws include the Federal Insurance Administration (FIA) and the Department of Housing and Urban Development (HUD). Typically, FIA (under the auspices of HUD) notifies a community that it is in a flood-hazard area and requires this community to take action by taking zoning measures to prevent new construction and enacting regulations that require buildings to be placed on stilts, piling, or additional land fill. Once these measures are adopted by the community, residents may purchase flood insurance at highly discounted rates. It should be noted that flood insurance under these laws is required for FHA and VA mortgages in these communities.[15]

CLEAN AIR ACT

This is legislation related to clean air governing the emission of air pollutants from human sources into the atmosphere. It also includes regulation of indoor air quality. In the U.S., the Clean Air Act Amendments of 1990 addressed air quality standards in terms of CO_2 emissions, acid rain, ozone depletion, and toxic air pollution. In doing so, the law established a national permits program.[16]

Guided by this law, the Environmental Protection Agency (EPA) regulates transportation in major cities in the U.S. The goal is to reduce air pollution by promoting mass transportation, car pools, and staggered working hours. Other measures include reducing the number of parking spaces to discourage people from driving their own vehicles and to encourage the use of mass transportation.[17]

This legislation has important ramifications for real estate developers in highly dense urban areas. For example, real estate developers in dense cities are discouraged from building structures with parking spaces. Parking space encourages people to use vehicles that further exacerbate air pollution.[18] Case/Anecdote 11.7 describes clean air laws in different countries, such as the U.K. and Canada.

CASE/ANECDOTE 11.7. AIR POLLUTION LAWS IN DIFFERENT COUNTRIES[i]

In the U.K., the British Parliament passed the Clean Air Act in 1956. This law identified zones where smokeless fuels had to be burnt; it also identified certain rural areas for power stations. Of course, such area designations have important implications for the way real estate developers select development sites. The law has been amended several times since.[i]

In Canada, there are two laws addressing air quality. The first Clean Air Act enacted in 1970 is designed to regulate the release of four specific air pollutants: asbestos, lead, mercury, and vinyl chloride. This law was replaced by the Canadian Environmental Protection Act of 2000. In 2006, the second Clean Air Act was passed. This law focuses on fighting smog pollution and greenhouse gases.

[i] Clean Air Act (United States), *Wikipedia*. Accessed from http://en.wikipedia.org/wiki/Clean_Air_Act_ (United_States), on June 9, 2013.

The Clean Water Act

The Clean Water Act (CWA) of 1972 is the national law in the United States related to water pollution. The goal is to eliminate releases of high amounts of toxic substances into water systems and to ensure that surface water (e.g., rivers and lakes) meets standards necessary for human sports and recreation. The law was based on the Federal Water Pollution Control Amendments of 1948. Amendments were further made in 1977 and 1987.

The Environmental Protection Agency (EPA) is charged with implementing water pollution laws. For example, EPA oversees area-wide waste water management plans to control water pollution from point (direct discharges of waste into navigable waters) and nonpoint sources (water runoff from agriculture, mining, construction, etc.). These plans may restrict construction permits (building, roads, parking places, etc.). EPA has to approve the building and operation of public sewage waste water facilities. EPA also regulates rapid-growth communities in urban areas by requiring the expansion of waste treatment facilities, restricting land use and building permits, and creating zoning ordinances, among other measures.[19]

NOISE POLLUTION AND ABATEMENT ACT

The Noise Pollution and Abatement Act of 1972 is a U.S. law designed to regulate noise pollution to protect human health and minimize noise annoyance to the general public. The law established standards related to virtually every source of noise (e.g., motor vehicles, aircraft, and equipment related to heating, ventilation, and air-conditioning).

The Environmental Protection Agency (EPA) is commissioned to conduct research and publish information on noise and its effects on the public. EPA also regulates noise pollution. Noise regulation is designed to restrict the amount of noise, the duration of noise, and the source of noise. For example, many city ordinances prohibit sound above a threshold intensity from trespassing over property lines at night; during the day, they restrict the noise at a higher sound level.[20]

Many U.S. states and cities have stringent building codes in relation to construction of new apartments, condominiums, hospitals, and hotels. The building codes require developers to perform acoustical analysis to protect building occupants from exterior and interior noise sources. For example, the codes require measurement of the exterior acoustic environment to determine the performance standard required for exterior building skin design. Building architects work with the acoustical analyst to identify the best, most cost-effective means of creating a quiet interior. There is noise from within the building to consider. Noise travels through walls or floors/ceilings and originates from either human activities in adjacent living spaces (e.g., amplified sound systems) or from mechanical noise within building systems (e.g., elevators, boilers, refrigeration or air conditioning systems).[21] Furthermore, HUD prohibits federal assistance to residential buildings that are besieged by high noise levels.[22] See Case/Anecdote 11.8 for information about noise pollution laws in different countries.

> ## CASE/ANECDOTE 11.8. NOISE POLLUTION
> ## LAWS IN DIFFERENT COUNTRIES
>
> U.S. noise control legislation has been the basis of several European countries' noise control laws, such as those in Netherlands, France, Spain, and Denmark. In some

cases, product innovations have led to quieter products, such as hybrid vehicles and washing machines.

Japan has a national noise control law, but its scope is much more limited than the U.S. law. The law in Japan is focused mainly on workplace and construction noise.

Britain's National Environmental Protection Act of 1990 is stimulating research that is aimed at setting certain definitive noise standards. However, this effort is lagging behind U.S. laws. Russia, China, and developing countries lag even further behind.[i]

[i] Noise Regulation, *Wikipedia*. Accessed from http://en.wikipedia.org/wiki/Noise_regulation, on June 13, 2013.

COASTAL ZONE MANAGEMENT ACT

This law was passed in 1972. It is relevant to land use and water development in 30 states that have coastal waters. The law is designed to preserve, protect, and develop coastal resources. The law is subject to state administration. That is, federal grants are provided to states with coastal waters to set up local authorities that can control land and water uses that have a significant impact on coastal waters. In this context, real estate development is significantly affected in coastal regions around the country.[23]

SUMMARY

This chapter was designed to sensitize the student to real estate marketing laws. The Deceptive Trade Practices Act is a law by which real estate firms can be prosecuted for false, misleading, or deceptive practices. The Interstate Land Sales Full Disclosure Act regulates interstate land sales to protect consumers from fraud and abuse in the sale or lease of land. The Real Estate Settlement Procedures Act is a law whose purpose is to crack down on real estate companies (e.g., lenders, real estate agents, real estate developers, and title insurance companies) that engage in providing undisclosed kickbacks to each other. The Equal Credit Opportunity Act is a law making it illegal for any mortgage lender to discriminate against any applicant who has the capacity to contract on the basis of race, color, religion, national origin, sex, marital status, or age. The Truth in Lending law requires lenders to fully disclose and explain loan terms in understandable language. The goal of the Fair Housing law is to make illegal to discriminate in all types of housing-related transactions, such as advertising, mortgage lending, homeowner's insurance, and zoning. The National Environmental Policy Act is designed to protect and enhance the environment. The National Flood Insurance Act helped accomplish two goals: (1) to provide flood insurance for real estate property in floodplain communities, and (2) to identify areas of high and low flood hazard and, based on this assessment, establish flood insurance rates for real estate properties inside each flood hazard area. The Clean Air Act was designed to regulate the emission of air pollutants from human sources into the atmosphere. The Clean Water Act is designed to minimize water pollution. The Noise Pollution and Abatement Act was designed to regulate noise pollution to protect human health and minimize noise annoyance to the general public. The Coastal Zone Management Act was designed to preserve, protect, and develop coastal resources.

DISCUSSION QUESTIONS

1. Describe the Deceptive Trade Practices Act. Think of a real estate marketing example that can best illustrate violation and enforcement of this law.

2. Do the same with the Interstate Land Sales Full Disclosure Act.

3. Do the same with the Real Estate Settlement Procedures Act.

4. Do the same with the Equal Credit Opportunity Act.

5. Do the same with the Truth in Lending Act.

6. Do the same with the National Environmental Policy Act.

7. Do the same with the National Flood Insurance Act.

8. Do the same with the Clean Air Act.

9. Do the same with the Clean Water Act.

10. Do the same with the Noise Pollution and Abatement Act.

11. Do the same with the Coastal Zone Management Act.

NOTES

1. Pridgen, D., & Alderman, R. M. (2012). *Consumer protection and the law: 2012–2013 edition*. Danvers, MA: Thomson Reuters.
2. Interstate Land Sales Full Disclosure Act of 1968. *Wikipedia*. Accessed from http://en.wikipedia.org/wiki/Interstate_Land_Sales_Full_Disclosure_Act_of_1968, on May 22, 2013.
3. Shenkel, W. (2004). *Marketing real estate*. Mason, OH: Southern Educational Publishing (pp. 262–264).
4. Based on Real Estate Settlement Procedures Act. *Wikipedia*. Accessed from http://en.wikipedia.org/wiki/Real_Estate_Settlement_Procedures_Act, on May 22, 2012.
5. Shenkel, W. (2004). *Marketing real estate*. Mason, OH: Southern Educational Publishing (pp. 265–268).
6. Equal Credit Opportunity Act. *Wikipedia*. Accessed from http://en.wikipedia.org/wiki/Equal_Credit_Opportunity_Act, on May 22, 2012.
7. Shenkel, W. (2004). *Marketing real estate*. Mason, OH: Southern Educational Publishing (pp. 268–269).
8. Truth in Lending Act. *Wikipedia*. Accessed from http://en.wikipedia.org/wiki/Truth_in_lending_act, on May 22, 2013.
9. Shenkel, W. (2004). *Marketing real estate*. Mason, OH: Southern Educational Publishing (pp. 269–271).
10. Fair Housing (United States). *Wikipedia*. Accessed from http://en.wikipedia.org/wiki/Fair_housing_(United_States), on June 9, 2013.
11. Shenkel, W. (2004). *Marketing real estate*. Mason, OH: Southern Educational Publishing (pp. 271–275).
12. National Environmental Policy Act (United States). *Wikipedia*. Accessed from http://en.wikipedia.org/wiki/National_Environmental_Policy_Act, on June 9, 2013.
13. Shenkel, W. (2004). *Marketing real estate*. Mason, OH: Southern Educational Publishing (pp. 275–276).
14. National Flood Insurance Program. *Wikipedia*. Accessed from http://en.wikipedia.org/wiki/National_Flood_Insurance_Program, on June 9, 2013.
15. Shenkel, W. (2004). *Marketing real estate*. Mason, OH: Southern Educational Publishing (p. 276).

16. Clean Air Act (United States). *Wikipedia.* Accessed from http://en.wikipedia.org/wiki/Clean_Air_Act_(United_States), on June 9, 2013.
17. Shenkel, W. (2004). *Marketing real estate.* Mason, OH: Southern Educational Publishing (pp. 276–277).
18. Clean Air Act (United States). *Wikipedia.* Accessed from http://en.wikipedia.org/wiki/Clean_Air_Act_(United_States), on June 9, 2013.
19. Shenkel, W. (2004). *Marketing real estate.* Mason, OH: Southern Educational Publishing (p. 277).
20. Noise Control Act. *Wikipedia.* Accessed from http://en.wikipedia.org/wiki/Noise_Control_Act, on June 13, 2013.
21. Noise Regulation. *Wikipedia.* Accessed from http://en.wikipedia.org/wiki/Noise_regulation, on June 13, 2013.
22. Shenkel, W. (2004). *Marketing real estate.* Mason, OH: Southern Educational Publishing (p. 278).
23. Coastal Zone Management Act. *Wikipedia.* Accessed from http://en.wikipedia.org/wiki/Coastal_Zone_Management_Act, on June 13, 2013.

Chapter 12

Ethics in Real Estate Marketing

LEARNING OBJECTIVES

This chapter is designed to sensitize the student to ethical issues in real estate marketing. Specifically, the student will learn how to apply a series of ethics tests to address ethical dilemmas in real estate marketing. These tests include:

- the legal test
- the consequence and utilitarian test
- the social contract test
- the moral reasoning test
- the consumer rights test
- the social justice test
- the stakeholders test
- the duties test

THE LEGAL TEST

Although real estate laws in the U.S. tend to vary slightly from state to state, there are some laws that apply across the board. These include:

- Deceptive Trade Practices Act
- Interstate Land Sales Full Disclosure Act
- Real Estate Settlement Procedures Act (RESPA)
- Equal Credit Opportunity Act
- Truth in Lending Act
- Fair Housing Law
- National Environmental Policy Act (NEPA)
- Flood Disaster Protection Act
- Clean Air Amendment
- Federal Water Pollution Control Act
- Noise Pollution and Abatement Act
- Coastal Zone Management Act

The legal test, in essence, asserts that an ethical decision is the decision to engage in action that is consistent with real estate laws. Chapter 11 covered these laws in some detail. What we are saying here is that real estate agents and brokers should ask themselves, "Is my contemplated action consistent with spirit of the law?" Many business ethicists consider the legal test to be a "minimal test" (i.e., the least one can do to ensure ethical action).

THE CONSEQUENCE AND UTILITARIAN TEST

Consequentialism is a group of ethical theories asserting that the consequences of one's action should be the basis for any judgment about the rightness or wrongness of that action. Thus, from a consequentialist perspective, a moral action is one that produces a good outcome, or consequence.[1] The history of consequentialism can be traced as far back as Jeremy Bentham; many philosophers and ethicists recognize his writings as providing the foundation for this approach to ethics.[2] Jeremy Bentham asserted that people make ethical decisions based on their interests and their fears; they make decisions by anticipating the consequences of their actions. Thus, an action considered ethical is one that maximizes pleasure and minimizes pain for all who are affected by that action.

Another term for "consequentialism" is "teleological ethics," which is usually distinguished from "deontological ethics," in that deontological ethics assert that the rightness or wrongness of one's action is judged based on rules or conventions related to that action. Later on, we will discuss the duties test, which involve a set of professional rules of conduct. We will devote some time to describing the National Association of Realtors code of ethics, which contains a set of duties to clients, the general public, and other real estate professionals. Teleological ethics is also distinguished from virtue ethics, which focuses on the character of the broker or agent. An action is considered ethical if the action is considered "virtuous," reflecting the character of the broker or agent.

Going back to teleological ethics (or the consequences test), an action is judged as unethical if it generates more harm than good in relation to all the stakeholders. For example, an agent has to decide whether to conceal from a buyer the fact that the property is located next to an environmentally hazardous land fill. Who are the stakeholders? How is each stakeholder affected by this concealment? How bad are the consequences arising from this action?

In this example, the real estate agent in this case is focusing on the consequences of his action (possibly concealing the information). In other words, he examines what the consequences of his actions would be in order to determine what is right or wrong. Ethicists refer to this brand of ethics as teleological ethics or "end results ethics." Typically, when faced with an ethical dilemma, the ethical decision-maker attempts to identify all the pros and cons of a possible action (e.g., concealment). The pros and cons are not restricted to him but also to others (e.g., client, his firm, the community, even competitors). Thus, the action is deemed ethical if it generates a whole lot more good than bad in relation to oneself and others. Let's again revisit the example of concealing the fact that the property is located close to a hazard environmental landfill. What may be the consequences of concealment? The "pros" are that concealment may lead to:

- a sale
- an immediate sale
- a higher-priced sale
- an expedited closing
- a higher commission on the sale

The "cons" are that concealment may lead to:

- members of the home buyer's family may eventually get sick
- an eventual law suit against himself and the firm
- the agent's professional reputation (as a person with high integrity) may suffer if people in the community find out about this

- his success in making sales in the same community may suffer as a result
- his family and friends may treat him as a pariah if they find out
- his colleagues and associates at his brokerage firm may think of him as "shady"

Note that in this situation, all the pros and cons are stated in terms of possible consequences. An ethical decision is the decision that considers all the consequences, not only related to self but also to others. In this case, the "right" decision is avoid concealment.

The *utilitarian test* is, in essence, a variation of the consequence test in that the decision maker does not only consider the consequences of a given action but evaluates the consequences of alternative actions. An action is judged as unethical if alternative actions could have been taken that were more likely to produce more good than harm in relation to all stakeholders. This utilitarian test is referred to as "preference utilitarianism" among philosophers and ethicists.[3] Going back to the example of concealment discussed above, the utilitarian test guides the decision maker to consider alternative courses of action and to evaluate the possible positive and negative consequences of each action. In this case, the real estate agent should not only evaluate the consequences of concealment but also disclosure and additional possible courses of action, such as providing the information to a third party and encouraging the third party to disclose the facts to the prospective buyer.

THE SOCIAL CONTRACT TEST

An action is judged as unethical if the action violates community norms and standards. This is the essence of the social contract test. Real estate agents and brokers who consider the interests of the community they serve may feel most comfortable applying this ethics test. In doing so, they consider the customs and values of the people making up the community. The "community" in the context of real estate marketing can be defined in terms of a geographic locale (as in a neighborhood, town, or larger area such as a metropolis) or a demographic market segment, such as the elderly, people belonging to a certain religious faith, etc. The idea is to understand the customs and values of the target market and make moral decisions consistent with these customs and values.

The social contract test is based on the philosophy of Jean-Jacques Rousseau in his most popular book, *The Social Contract*.[4] Rousseau believes that people are born free but are ultimately bound by social convention. To accept living in civil society, people have to give up a certain degree of personal liberty and accept certain rules of society that bind people together and allow smooth social functioning. Rousseau argued that legitimate authority (e.g., that wielded by governments) has to be based on a "social contract," which is essentially a psychological agreement by which the people band together for their own good. By banding together, they form a human collective called the "sovereign." Thus, the people belonging to that collective or "sovereign" are the supreme authority. In other words, all state institutions serve the people (a government by the people and for the people). Rules and laws are created to reflect the sovereign's interest, or the general will of the people, promoting the common good. Thus, the private will of each citizen is subject to the general will of the people. Rousseau asserts that the sovereign establishes and employs government as a representative of the people to carry out the general will of the people. Based on this definition, the sovereign has the right to change a government and replace its leaders.

Real estate marketing professionals accept company policies regarding rules of ethical conduct. These rules are essentially a "social contract." For example, many brokerage firms develop policy and procedure guides to promote ethical conduct among their employees (the sales associates). The sales associates may not have anything to do with the development of the policy and procedure guide; they accept them by virtue of their employment contract. In other words, by accepting employment in that brokerage firm, they implicitly accept and agree to follow the rules of conduct spelled out in the policy and procedure guide. Similarly, real estate agents and brokers apply and become members of professional associations (e.g., in the U.S., there is the National Association of Realtors). By virtue of their membership in these professional associations, they accept the association code of ethics. This becomes their "social contract." We will go over the National Association of Realtors' (NAR) code of ethics in meticulous detail in Chapter 13 of this book. However, for now consider the following case as shown in Ethics Box 12.1. This case illustrates how real estate agents and brokers submit to arbitration as a direct function of their membership in the NAR. As members they accept the NAR code of ethics and Article 17, which encourages members to arbitrate disputes through the local board instead of using the courts to adjudicate.

ETHICS BOX 12.1. REQUEST FOR ARBITRATION EXPENSES[i]

Two real estate brokers, Jim Sutherland (listing broker) and Tim Forham (cooperating broker), became embroiled in a heated dispute concerning who generated the sale of property listed (listed by Sutherland) and, therefore, who is entitled to part of the buyer representative commission. Both brokers found it very difficult to resolve the matter on their own. Therefore, as members of the National Realtors Association, they agreed to arbitrate based on Article 17 of the Code of Ethics. Article 17 reads as follows: In the event of contractual disputes or specific non-contractual disputes as defined in Standard of Practice 17–4 between REALTORS® (principals) associated with different firms, arising out of their relationship as REALTORS®, the REALTORS® shall submit the dispute to arbitration in accordance with the regulations of their Board or Boards rather than litigate the matter. Jim Sutherland made the request for arbitration to the local board, and the grievance committee of the board accepted the case and passed it on to the professional standards committee for a hearing. Five members from the board were appointed for the hearing panel.

Jim Sutherland learned that Tim Forham had practiced law before getting involved in real estate. Therefore, Sutherland decided to hire an attorney to ensure that he would not be at a disadvantage in presenting his case to the hearing. Sutherland also amended his arbitration request by asking that he be awarded not only the commission on the contested sale but also attorney's fees. At the hearing, Sutherland attended with his attorney.

The hearing panel stated that the arbitration process is provided to all members of the National Association of Realtors to minimize expenses (by avoiding high court costs), and the hiring of an attorney was Sutherland's own decision and not required. The hearing panel decided the commission dispute based strictly on the merits of the case presented but disallowed the request by Sutherland that he be awarded attorney's fees.

[i] *Request for Arbitration Expenses.* Accessed from www.realtor.org/2013-code-of-ethics-and-arbitration-manual/case-interpretations/related-to-article-17, on June 19, 2013.

THE MORAL REASONING TEST

The moral reasoning test is essentially a test based on Kohlberg's theory of moral develop-ment, which is displayed in Figure 12.1.[5] The theory holds that moral reasoning is developed in humans in terms of six stages, but these can be grouped into three levels of two stages each: pre-conventional, conventional, and post-conventional.

Level 1 (Pre-Conventional)

One can think of moral development in terms of age. The pre-conventional level of moral reasoning is commonly observed in children, although that is not to say that some adults do not think in pre-conventional terms. People in Level 1 of moral development reason that an action is ethical or not as a direct function of the self-consequences resulting from the action. In other words, if the consequences to the self are bad, then the action is unethical; conversely, if the consequences to the self are good, then the action is ethical. Specifically, the pre-conventional level consists of two stages that reflect self-interest. Typically, children focus largely on rewards and punishment (external consequences) that their actions bring. They also evaluate the ethicality of actions of other people similarly; that is, if a person's action is rewarded then it must be "good," and if it is punished it must be "bad." This is characteristic of Stage 1 (*obedience and punishment*), in which individuals focus on the direct consequences of their actions to themselves. For example, suppose a listing agent is think-ing of concealing certain defects in the property he is trying to sell to a prospective buyer. He reasons that this action is morally wrong because "I remember being punished for hav-ing done the same in the past." Or perhaps he has observed other real estate agents getting

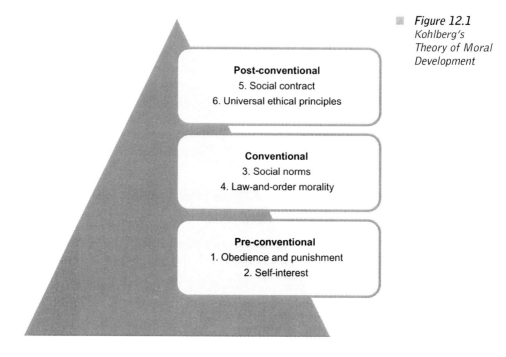

Figure 12.1
Kohlberg's
Theory of Moral
Development

punished for concealing certain defects. Note that his thinking is "egocentric," and there is deference to people who have power, mostly because the same people have the ability to provide reward and dispense punishment.

People in Stage 2 (*self-interest*) define what is right by whatever the individual believes to be in his best interest. People in this stage of moral development do not care about the needs of others. The right thing to do is to further one's own interests. They believe that reciprocity is the best rule: "You scratch my back, and I'll scratch yours." For example, a real estate broker at Stage 2 treats his agents based on his self-interest ("What can you do for me?"). If an agent performs well by generating high sales, then he is treated with respect. The broker invites him to social gatherings and periodically gives him expensive tickets to baseball events. In contrast, those agents that are faltering (not generating high sales) are treated with derision. (In one instance, a broker reasoning at Level 1 punished an agent for having a bad sales month by taking away his desk and giving it to an intern.)

Level 2 (Conventional)

From the perspective of physical age, the conventional level is typically observed among adolescents (but again, many adults may be arrested at this stage of development and reason morally like adolescents do). Conventional people judge the morality of actions in terms of what society expects (i.e., social norms and conventions). In other words, conventional morality is acceptance of society's conventions concerning right and wrong. People whose moral reasoning reflects conventional morality tend to obey rules even when there are no self-consequences for obedience or disobedience. They do things "by the book." That is, adherence to rules and conventions is very important to them. They think that people should do what they're told and what society expects of them: If we don't have rules, then we will have chaos. Thus, rules are very important. This level of moral development has two distinct stages: social norms and law-and-order morality.

In Stage 3 (social norms), people are accepting of social roles (e.g., a father is the bread winner; the mother stays home and takes care of the kids). Therefore, morality is based on adherence to well-established social norms. People who do not adhere to their prescribed roles are judged as "immoral." In other words, a "good boy" or a "good girl" is one who lives up to society's expectations. Hence, a person at Stage 3 judges the morality of an action by evaluating its consequences in terms of social approval and disapproval ("I want to be liked and thought well of; being ethical makes people like me and respect me." Consider the example of an agent who has an extramarital relationship with another agent in the same firm. The broker finds out and reasons that this is unethical conduct. As such, he reprimands both agents by shunning them and expressing his disapproval to other colleagues at the office. He believes that his action is justified on moral grounds, and that the agents' action is morally reprehensible. This reasoning captures the essence of Stage 3 moral development.

In Stage 4 (*law and order*), people think that obeying rules, laws, and other social conventions is very important because doing so helps regulate a well-functioning society. The underlying belief is "If one person violates a law, perhaps everyone will, and as such society suffers; thus, everyone has a duty to uphold society's expectations in terms of rules, laws, and other social conventions." Suppose a real estate agent fails to abide by a specific company policy (for real estate agents to keep tabs of the time they work by "punching the clock" every time they come in and go out). The broker believes this is a fair company policy because it provides evidence of "work." Suppose we have a real estate agent who seems to

261

do well (in terms of sales) but has failed to "punch the clock" regularly. As a matter of fact, that agent argues that this system of "punching the clock" is antiquated and demeaning. The broker, on the other hand, believes that the agent has violated company policy and this may be sufficient grounds for terminating the agent's employment with the firm. The broker believes that the agent is "not doing the right thing" because he violates company policy. Here the broker's moral reasoning reflects Stage 4 moral development.

Level 3 (Post-Conventional)

The post-conventional level is characterized by a growing realization that people have their own ideas of what is right and wrong and that sometimes a person's own rules of good conduct may take precedence over society's rules, laws, and other conventions. In other words, it is okay for people to disobey rules that are inconsistent with their own ethical principles (e.g., basic human rights as life, liberty, and justice). Rules, laws, and social conventions are not "etched in stone"; they are not absolute dictates that must be obeyed without question.

People at Stage 5 (*social contract*) view the world as multicultural. They realize that different people have different cultures, and as such they are likely to have different sets of values that guide their action. Therefore, Stage 5 individuals believe that laws are "social contracts," not rigid edicts. Laws that do not promote the general welfare, therefore, should be changed to maximize social welfare. This can be achieved through democratic action, participative management, negotiations, and inevitable compromise. Consider the previous example of "punching the clock." If the broker were reasoning at Stage 5, he would try to understand the point of view of the agent who is performing well but thinks that "punching the clock" is unfair. The agent wants to be judged by his sales output, not by how many hours he puts in at the office, as verified by the "clock punch." The broker would realize that such a company policy may work for some but not others. He would consider that the company policy can be changed to accommodate certain individuals in certain situations—that the company policy is essentially a social contract, not etched in stone. This contract can be negotiated and changed to meet the needs of the agents and the company, too.

Moral reasoning at Stage 6 (*universal ethical principles*) is based on universal ethical principles. Individuals at Stage 6 may reason that laws are valid only insofar as they are grounded in justice; as such, commitment to justice bears an obligation to disobey unjust laws. Stage 6 individuals tend to be highly empathic—this involves having to imagine what they would do if they are in another person's shoes. They may believe that action is never a means to an end, but always an end in itself. Their action reflects their own ethical principles; that is, they may believe that their action is "right" because it reflects a specific ethical principle that is immutable, not because the action is instrumental, expected, legal, or socially sanctioned. Consider the following situation. A broker is told by certain powerful people who hold a large share of the company stock that the company has an informal policy of not recruiting Mexican Americans as agents because they operate in a mostly white Anglo-Saxon community, and doing so is likely to adversely affect the competitive position of the company. The broker believes that all minorities should be provided an equal opportunity for employment as agents, and that their performance should be judged not by their ethnicity but by their sales performance. If the broker acts based on Stage 6 reasoning, what might he do? The broker should not be deterred; he or she should hire qualified Mexican Americans and justify his or her decision on moral grounds.

Applying the Moral Reasoning Test

Having learned a little about moral development and moral reasoning, how should we use this knowledge to help us make moral decisions in real estate marketing? In other words, how do we apply the "test"? The goal here is to emulate the moral reasoning of people who are in higher, rather than lower, stages of moral development. Obviously, to justify one's decisions based on obedience and punishment (Stage 1), self-interest (Stage 2), and social norms (Stage 3) is not advisable. We would not characterize these decisions as "ethical." We can do better using moral reasoning involved in Stages 4 and up. Law and order morality (Stage 4) is acceptable, but this type of ethics is a minimal standard. Remember the legal test we described earlier in the chapter. We can do better than law and order morality.

Moral reasoning based on social contract (Stage 5) is much better. The goal here is to go beyond existing rules, laws, and company policies. We should be able to have an appreciation for diversity. Certain rules, laws, and policies may apply in some situations, in some organizations, for some real estate agents and brokers, in some cultures. The catch is to be able to understand this complexity and apply moral standards with a certain degree of flexibility. Remember the illustration we used in describing the moral reasoning of a Stage 5 broker—the situation involving "punching the clock." This situation illustrates that company policies should be used as a guide, not as absolute dictum. The broker needs to better understand the needs of the agent, the conditions under which the policy of "punching the clock" works best, and make a decision to apply, not apply, or modify this policy to agents who do well without "punching the clock."

Of course, the most ideal application is to apply Stage 6 moral reasoning. That is, we can encourage real estate agents and brokers to develop their own set of moral principles that are well grounded in their own understanding of humanity and social welfare. But this may be too much to ask sometimes. We don't expect every real estate agent and broker to be a philosopher and moral thinker. If they embrace that role, that is "icing on the cake." Nevertheless, good leaders are ethical leaders, and as such their action should be guided by moral principles.

THE CONSUMER RIGHTS TEST

One way to appreciate the rights test in the context of real estate marketing is through understanding the consumer bill of rights. The consumer bill of rights traces its origins from a speech delivered on March 15, 1962 by President John F. Kennedy. The speech was delivered to the United States Congress. In that speech, President Kennedy extolled four basic rights, later titled "The Consumer Bill of Rights." The four consumer rights are: (1) the right to safety, (2) the right to be informed, (3) the right to choose, and (4) the right to be heard.[6] The United Nations has expanded these four consumer rights into eight—adding the following four: (5) the right to satisfaction of basic needs, (6) the right to redress, (7) the right to consumer education, and (8) the right to a healthful environment.[7] In addition to these rights, we will discuss the right to privacy—a new consumer right that has gained momentum in social and legal discourse over the last two decades.

The Right to Product Safety and to a Healthy Environment

The goal of asserting this right is to reduce consumer injuries caused by purchased goods and services. In other words, the right to safety makes companies that sell consumer goods and services responsible for safety, in that product safety should be an important standard in quality control. In the U.S., the Consumer Product Safety Commission (CPSC) has jurisdiction over a wide spectrum of commercial products (including housing-related products). The CPSC establishes performance standards, requires product testing, and oversees warning labels.

This particular consumer right applies more to real estate developers than real estate brokers and agents. Real estate developers ought to ensure that their building materials do not jeopardize the health and safety of the building inhabitants. This is not to say that real estate brokers and agents are off the hook when it comes to safety. In cases in which brokers and agents have specific knowledge of building hazards that are likely to jeopardize the health and safety of prospective buyers, then it is their ethical obligation to work closely with the seller to eliminate the sources of threat.

Akin to the right to product safety is the *right to a healthy environment*. That is, home owners have a right to live and work in an environment that is non-threatening to the well-being of present and future generations. Homeowners can sue neighboring real estate developers and other business entities that are proven to have polluted the environment. This can be in the form of land pollution, air pollution, water pollution, or noise pollution.

The Right to Be Informed and to Consumer Education

Consumers have the right to be informed about the full range of benefits and costs of the goods and services they purchase. Any business selling a good or service has an ethical and legal obligation to provide consumers with enough adequate and accurate information to help consumers make intelligent and informed purchase decisions. One may ask: Adequate and accurate information about what? About the quality and cost of the product! Quality should cover information about product benefits and cost should cover information about not only purchase price but also other costs, such as the cost of consumption, negative side effects related to consumption, and the cost of assembly and disposal.

This consumer right is fundamental to ethical capitalism. Consumers vote with their wallets by choosing those goods and services that are high in quality and low in cost. By doing so, consumers reward good businesses that innovate (develop and market high quality goods and services) and are efficient (reduce the cost of business and pass on these savings in the form of lower prices). By the same token, by not buying products that are low in quality and high in price consumers are weeding out bad businesses that do not innovate and are bloated with cost inefficiencies. In other words, a free and vibrant economy functions well when consumers vote with their wallets by purchasing high quality goods and services at low prices. For consumers to do this, they need information about product quality and price. If businesses are not truthful in their promotion of their products about quality and price, then consumers are handicapped in using their dollars to reward good companies and weed out bad companies. Therefore, companies are ethically and legally obligated to provide accurate and complete information about quality and price of their offerings. In doing so, they participate in ethical capitalism, ensuring the functioning of a free economy.

There are many laws in the U.S. that are specifically formulated to uphold consumers' right to information in the real estate industry. These include the Deceptive Trade Practices Act, the Interstate Land Sales Full Disclosure Act, the Real Estate Settlement Procedures Act, and the Truth in Lending Act. These are discussed at length in Chapter 11.

The Right to Choose and to Fair Trade

Consumers have the right of free choice among competing product offerings provided by different companies. In other words, consumers benefit from an economic environment in which competition thrives. Competition ensures that business firms compete to gain customers. To attract customers, businesses have to provide product offerings that are high in quality and low in price. The more competition there is, the greater the likelihood that business firms in a given industry will innovate (by developing higher quality products) and become efficient (by reducing the cost of doing business in the form of price savings to customers). Monopoly is the antithesis of competition. An industry riddled in monopoly is likely to have very few firms that are not innovative and inefficient. Thus, monopoly does not serve consumers well, and does not serve society at large.

Many of the laws in the free economies of the world have specific anti-trust laws to discourage monopolistic business practices. Consider the Robinson-Patman Act.[8] This is a major anti-trust law in the U.S. The law prohibits anticompetitive practices. Example of anti-competitive practices include dumping (a company sells a product below cost in an effort to drive out the competition), exclusive dealing (where a seller forces a buyer to sign a contract obligating the buyer to purchase exclusively from the seller and only the seller), price fixing (when two or more companies collude to fix price at a high level, thus negating the price savings that derive from competition), refusal to deal (when two or more companies collude to shut out a certain vendor by refusing to do business with that vendor), dividing territories (when companies collude in dividing the market into designated territories and agreeing to stay out of each other's way), and tying (when products are bundled together in a way that buyers are forced to accept a less attractive product because they need the more attractive one).[9]

The same can be said in the real estate industry. Competition serves individual consumers well by providing a variety of housing options for renters and buyers. Imagine a community dominated by one real estate developer. This developer has a monopoly on the vast majority of the homes for sale and rent. By monopolizing the market, this developer can set high prices and provide less-than-optimal housing. Buyer and renters do not have much to say because they are "stuck" dealing with this one giant real estate company. Buyers in a monopolistic market have very little bargaining power. Much of the power is wielded by the monopolistic real estate company. Similarly, competition serves business well by providing a variety of commercial real estate offering for corporate renters and buyers. Ethics Box 12.2 describes an interesting case involving short sales that can be considered unfair trade.

ETHICS BOX 12.2. REAL ESTATE: SHORT SALES POSE RISKS TO LENDERS[i]

What is a "short sale"? When a homebuyer is in trouble for failing to pay the monthly mortgage, he risks foreclosure (i.e., the mortgage lender repossessing the house). To mitigate this risk many homebuyers opt for a short sale. This is the process of selling the house in the open market at a deflated price. The mortgage lender normally agrees to the short sale to avoid the higher cost and public stigma associated with evicting the homeowner.

Some real estate agents have colluded with investors to take advantage of the homebuyer's predicament to make extra money. This is how it works. The real estate agent obtains an appraisal that reflects a significant devaluation (a much lower price than the original appraisal). He turns in the appraisal to the investor, who in turn uses the new appraisal and submits to the mortgage lender two dummy offers for the listed property that are much lower than they could get in the open market. Then the investor submits a formal "low ball" offer, which is then accepted by the lender. The investor then asks the real estate agent to list the property at a significantly higher price and gets it resold at a sizeable profit. The investor hides the ultra-low price on the purchased property by going through an out-of-state lawyer to close the deal. Doing so allows the investor to hide the purchase price because the out-of-state closing is not recorded in the local multiple listings. Therefore, the original homeowner, the lender, and the second buyer would not be privy to the fact that the house's new price increased significantly overnight.

[i] *Real Estate: Short Sales Pose Risks to Lenders.* Accessed from www.pe.com/business/business-headlines/20120504-real-estate-short-sales-pose-risks-to-lenders.ece, on July 3, 2013.

The Right to Be Heard and to Redress

Consumers have a right to voice their concerns about product malfunctions, unsafe products, shoddy products, and other mistreatment; this is their *right to be heard*. There are consumer advocacy institutions whose purpose is to document consumer complaints and make them public. Doing so allows other consumers to refrain from doing business with companies that do not deliver on their promises. For example, there is the Better Business Bureau (BBB) in the U.S. This is a consumer advocacy group found in almost every community in every state. Consumers register their complaints with the BBB. When a consumer is deciding whether to do business with a company, he can check the company's record of customer complaints at the BBB. Companies that have a good record are approached and companies with a poor record are avoided. Furthermore, state and federal Attorney Generals in the U.S. provide aid to customers who were provided a good or service in a manner that violates an applicable law.

Consumers of real estate in the U.S. have outlets in the form of the Real Estate Board (an agency at the state level that hears complaints and makes attempts to restore justice through the threat of revoking real estate licenses). Buyers and sellers of real estate can file complaints with the Real Estate Board and hearing panels are assembled at the local level. The hearing panel adjudicates cases.

Akin to the right to be heard is consumers' *right to redress*. That is, consumers have a right to seek a fair settlement of just claims related to shoddy goods and unsatisfactory services in the form of compensation through the courts. Suppose a home buyer is deceived regarding certain housing defects. This buyer has a right to sue the seller through the judicial system (i.e., the courts). If the court finds the home seller (defendant) guilty as charged, the home buyer may be compensated not only for damages (to recover the costs related to housing repairs), but also for punitive damages (as means to punish the home seller for deceiving the home buyer into believing that the property is in good shape).

The Right to Privacy

People have the right to be "let alone." They have the right to enjoy the comfort of their homes without being badgered by real estate telemarketers or other real estate sales agents. They do not want their personal lives to be invaded by a barrage of real estate advertising and solicitations. Privacy rights are inherently intertwined with information technology. That is, many real estate marketers have managed to conduct intrusive promotion campaigns using information technology in the form of telephone solicitations, e-mail promotion campaigns, texting campaigns, and the use of social media (e.g., Facebook). When prospective real estate buyers and sellers visit a website of some real estate firm, they may be asked to "sign in." The sign-in process involves disclosing personal information that some firms sell to other firms for prospecting. Many consumers find this practice to be highly unethical because their personal information is used by prospecting agents and firms to pressure them into adopting real estate products that may be unwanted. This is not to mention the fact that the sales techniques used by real estate firms and agents can be highly intrusive (i.e., the techniques violate one's sense of privacy).

The right to privacy is explicitly incorporated in the Universal Declaration of Human Rights. It states that no one shall be subjected to arbitrary interference with his privacy, family, home, or correspondence, nor to attacks upon his honor and reputation. Everyone has the right to the protection of the law against such interference or attacks.[10] Most states in the U.S. recognize the right to privacy as law.

THE SOCIAL JUSTICE TEST

The social justice test asserts that an action is judged to be unethical if it discriminates against any group of people using criteria such as religion, race, ethnicity, national origin, gender, sexual orientation, familial status, or age. Social justice theory also asserts that if discrimination is to occur, discrimination has to be in favor of groups considered to be vulnerable or culturally disadvantaged. For example, based on the principle of social justice, an agent should not only refrain from discriminating against a minority buyer, but should also assist that buyer in ways above and beyond what the agent might do for other regular clients.

The concept of "social justice" has been attributed to a Jesuit priest by the name of Luigi Taparelli in 1840, based on the teachings of the great philosopher and ethicist, St. Thomas Aquinas.[11] However, social justice is now considered a secular concept, not affiliated with any particular religion. The teachings related to social justice have been influenced heavily by the famous modern political philosopher, John Rawls.[12] Rawls argued that any practice that discriminates against people unfairly is unjust.

For example, the fair housing laws in the U.S. are based on the concept of social justice. Fair housing laws make it illegal to discriminate in all types of housing-related transactions, such as advertising, mortgage lending, homeowner's insurance, and zoning. For example, the law makes it unlawful to discriminate based on race, color, religion, sex, familial status, or national origin in the terms and conditions of the sale or rental of the residential property.[13]

Rawls also advocates that if discrimination is to occur, it should be done in favor of groups who are considered to be vulnerable or culturally disadvantaged. The rationale of affirmative action in the U.S. is based on Rawls second principle of social justice. Affirmative action is also known as "positive discrimination" (e.g., in the United Kingdom). It involves employment equity and refers to policies that take factors including race, color, religion, sex, or national origin into consideration in order to benefit underrepresented groups (considered culturally disadvantaged or vulnerable) in areas of employment, education, and business.[14]

For example, if a real estate developer is contracted to assist in the development of a public facility, such as a government building, the developer has to abide by affirmative action rules and regulations in the personnel recruitment. Affirmative action recruitment may require that the real estate developer hire a certain percentage of the workforce from certain disadvantaged groups, such as African Americans, Latinos, and Native Americans.

THE STAKEHOLDERS TEST

Some business ethicists view ethical conduct in business as meeting the needs and demands of the firm's various stakeholders.[15] Stakeholders are groups external to the organization that influence (and/or are influenced by) the organization. For example, a real estate development firm has many stakeholder groups, such as the shareholders (i.e., the owners of the firm), the employees, the customers, the suppliers, the real estate sales organizations supporting the firm, the local community, and those concerned about the environment. Some of these stakeholders may be considered by management to be more important than others (e.g., customers are often considered much more important than suppliers, the local community, or the environment).

Much of what we call "corporate social responsibility," or CSR, is based on the concept of stakeholders' ethics. That is, the firm is deemed socially responsible if it caters not only to the desires and whims of the stockholders (i.e., the owners of the firm) but also to its many and diverse stakeholder groups, such as its employees, customers, and the local community.[16] Note that the definition of *marketing* as set by the American Marketing Association takes into account the various stakeholder groups:

> Marketing is an activity, set of institutions, and processes for creating, communicating, delivering, and exchanging offerings that have value for customers, clients, partners, and society at large.[17]

A better definition was recently provided by marketing ethicists, Professors Patrick E. Murphy, Magalena Oberseder, and Gene R. Lacniak:

> Corporate societal responsibility in marketing encompasses balancing the legitimate demands of stakeholders (namely customers, employees, environment, suppliers/dealers, regulators, local community and shareholders); accepting accountability for marketing decisions; and integrating ethical and societal obligations into the firm's marketing activities.[18]

Examine Table 12.1. This table spells out responsibilities of real estate marketing professionals in major real estate development firms. The table shows that ethical responsibilities are directed toward different stakeholder groups: customers/end users, employees, suppliers, the local community, shareholders, and the environment. The responsibilities of marketing professionals are also divided in terms of three major categories: product, pricing, and promotion.

Table 12.1 *Corporate Societal Responsibility as Applied to the Marketing Function in a Real Estate Development Firm*

Domain	Product Responsibility	Pricing Responsibility	Promotion Responsibility
Customers and End Users	Ensuring healthy and safe real estate structures housing end users Delivering the final offering to the customer as promised	Providing fair pricing of the real estate offerings to customers and end users Making price as transparent as possible to customers and end users	Disseminating accurate information to customers and end users about the full range of costs and benefits of the firm's offerings
Employees	Ensuring the health and safety of all employees of the real estate development firm	Paying employees a fair and decent living wage	Ensuring that the sales force is providing the customer and end users with accurate information about the full range of costs and benefits of the firm's offerings
Suppliers	Encouraging healthy and safe working conditions of sub-contractors and other suppliers of building products	Paying suppliers and sub-contractors fair and reasonable prices for their offerings	Ensuring that all communications directed to suppliers and subcontractors are accurate
Local Community	Ensuring that the building process does not in any shape or form harm local residents and their environment Encouraging employees to volunteer their building construction services to the local community pro bono	Recruiting employees and subcontractors (as well as purchasing supplies) from the local community	Disclosing information without bias about negative incidents involving the firm's employees or sub-contractors being responsible for harm to the local residents
Shareholders	Ensuring that the offering is of high enough quality to bestow a positive image on the company, ultimately enhancing long-term profitability of the firm	Ensuring that the price of the offerings will generate a fair return on investment to the shareholders	Communicating and sharing with shareholders information not only about the financial health of the firm but also about the health and well-being of the firm's stakeholder groups
Environment	Ensuring the safe disposal of building by-products Using environmentally friendly/green building products	Ensuring that the price of the offering is not too low to the point where the firm will have to engage in environmentally unfriendly practices to cut costs	Communicating and sharing information with environmental stewards about ways that the firm acts toward the environment

Source: Adapted from Murphy, P. E., Oberseder, M., & Laczniak, G. R. (2013). Corporate societal responsibility in marketing: Normatively broadening the concept. *AMS Review, 3*, 98.

Product Responsibilities

With respect to product responsibilities, the marketing professional should ensure that the structures the firm builds are physically safe for the end users—the families who will end up living in that space (as in a residential housing structure) or the employees who will end up working in that space (as in commercial structures). A distinction is made between customers and end users in this case because in commercial real estate the commercial space is designed to meet customer's specifications (the corporate client), but the end users in this case are employees and other people who will end up using the space for commercial reasons. Hence, the marketing professional is responsible not only for meeting the building specifications of the customer (buyer) but also for ensuring the health and safety of the people who will use the building.

The marketing professional is also responsible for confirming the health and safety of the workers who are involved in the process of building and construction. Those involved in marketing can create incentives and disincentives for suppliers of building and construction materials to meet suppliers' social responsibility standards. The same can be said for subcontractors. Suppose, for example, the marketing manager of the real estate development firm finds out that one of the subcontractors the firm is using has repeatedly violated fair labor standards by failing to protect its workers from unsafe working conditions. The marketing professional, in this case, cannot simply turn a blind eye. He has to take action by applying incentives or disincentives to induce the subcontractor to take measures designed to protect his workers. The same can be said about the marketing manager taking action within the company to "right a moral wrong."

With respect to the local community, the process involved in building and construction has to be safe for community residents. Suppose that the workers involved in building and construction become embroiled in a physical quarrel with local residents, resulting in destruction of local property. The marketing professional has an ethical obligation to discipline these workers and pay for damages to compensate the local community for the damaged property.

The marketing professional is also responsible to the firm's shareholders. He meets this responsibility by ensuring that the constructed building reflects highly on the firm. Doing so enhances the overall image and reputation of the firm, which in the long run should affect the firm's bottom line.

Finally, the marketing professional should ensure that the disposal of building by-products, such as hazardous waste, is carried out in ways that would not degrade the environment. Doing so helps the firm meet its ethical obligation toward the environment.

Pricing Responsibilities

The pricing responsibilities of the marketing professional can also be discussed in terms of the various stakeholder groups. With respect to customers as a key stakeholder group, the marketing professional has an ethical obligation to price real estate offerings fairly, guided by well-established and accepted pricing practices such as comparative market analysis (commonly used in residential housing), the cost approach (commonly used in the construction of new buildings, especially commercial real estate), and the income method (commonly used in rental properties). The price of the real estate offerings has to be transparent, in the sense that it can be justified based on objective criteria (i.e., comparative market analysis).

With respect to employees as another major stakeholder group, the marketing professional has an ethical obligation to ensure that the workers involved in the building and construction of a given real estate offering are paid fairly. For example, the marketing professional should ensure that there is no pay discrimination against a certain class of workers (e.g., minority workers such as Latinos and African Americans in the U.S.) or preferential treatment in favor of certain class of workers (e.g., workers of a certain religious faith, such Southern Baptists). Also, the marketing professional should ensure that workers are compensated adequately to meet the increasing cost of living and living standards. The same can be said in relation to suppliers and subcontractors. These suppliers have to be paid fairly and on time.

The ethical marketing professional also recognizes that the real estate firm has an obligation to support the local community. This can be done by recruiting workers to staff a given building and construction project from the local community. Similarly, many of the building supplies can be purchased from local stores, providing economic support to the retailing establishments in the community. The marketing professional has a responsibility to the shareholders, too, in that the price has to be high enough to ensure that the real estate project results in a satisfactory profit margin.

Finally, pricing decisions have to be made with the environment in mind. We see on many occasions instances in which the price of an offering is low because of fierceness of the competitive landscape. Real estate firms have to present bids that reflect low prices to win the bid. However, by doing so, they cut costs in other, "less important," areas, such as the efficient use of resources and using green building supplies that sometimes cost more. Such practices undermine the environment. Ethically speaking, such practices should be avoided as much as possible.

Promotion Responsibilities

The marketing professional engaged in promotion has obligations in relation to the same stakeholder groups. She is ethically obligated to disseminate accurate information to customers and end users about the full range of costs and benefits of the firm's offerings. For example, in the bidding process, the marketing professional has to resist the temptation of embellishing the proposed benefits and costs for the single purpose of winning the bid. There are many instances in which managers submit bids with a low price that cannot be implemented once the bid is accepted. Once the bid is accepted, the real estate firm submits additional invoices to the client that reflect cost overruns. Such practice of deflating the price in the bid is considered deceptive promotion and should be avoided by all means. There are also other instances in which real estate firms submit a bid by misrepresenting their capabilities. Again, information has been provided to prospective clients couched in misrepresentation and puffery for the single purpose of winning the bid.

With respect to the employees' stakeholder group, the marketing professional is obligated to ensure that the firms' sales force is promoting the real estate offering using accurate information about the offering. The sales force in real estate is typically compensated using commission. Thus, the pressure to sell is enormous, given the fact that each agent's livelihood is directly dependent on making a sale. Such pressure induces salespeople to distort information by inflating the benefits of an offering to prospective clients and deflating the costs. Misrepresentation should be avoided, and the marketing professional has an obligation to ensure that the sales force uses accurate information in the promotion of the firms' offerings.

The marketing professional also has an obligation to ensure that all forms of communications directed to suppliers and subcontractors are accurate. For example, there are instances in which subcontractors are hired and presented with inaccurate information about the hiring company (e.g., the company's website says "we have a history of rewarding our subcontractors who complete their assigned projects on time," while in actuality, there is no such practice).

With respect to the local community, the marketing professional should be forthcoming in disclosing information without bias to the public about incidents that may tarnish the image of the firm. For example, consider an incident in which the firm's employees may have been complicit in, perhaps, a major construction accident due to a case of workers' intoxication on the job. Such publicity can adversely affect the company's reputation. However, making any attempt to conceal and not disclose pertinent facts about the intoxicated workers is not ethical. The ethical course of action in this case is to be forthcoming with the information and avoid concealment.

The marketing professional is ethically obligated to share information with the shareholders about all aspects of the firm's conditions and actions—both positive and negative information. Presenting only financial information and information that sheds positive light on the firm is unethical because it can be misleading by omitting vital information to the shareholders about the future prospects of the company and the extent to which the shareholders should continue to invest in the company. The ethical thing to do is to present information to the shareholders that reveals the good, the bad, and the ugly of the firms' conditions and actions.

In addition, communicating honestly and with integrity with the stewards of the environment is the ethical thing to do. In other words, the marketing professional should make known facts about what the firm is doing (and not doing) to protect the environment. In case of an environmental disaster in which the company is culpable, the marketing professional should not conceal any information that reflects the true facts surrounding the incident. The marketing professional should work cooperatively with the stewards of the environment to disclose all the facts that may assist in remediation efforts.

THE DUTIES TEST

The duties test is grounded in what ethicists refer to as deontological ethics. Deontological ethics is a school of thought in professional ethics that focuses on codes of conduct. The central tenet of deontological ethics asserts that morality of an action can be judged based on the action's adherence to a rule or rules.[19] Deontological ethics is sometimes referred to "duty" or "rule"-based ethics, and, as mentioned earlier in this chapter, it is commonly contrasted with teleological ethics or consequentialism (utilitarian ethics). Codes of ethics that are derived from religious scriptures are a good example of deontological ethics. The Ten Commandments, well known to Jews, Christians, and Muslims, are considered to be a set of divine rules or duties (e.g., honor your father and your mother, you shall not murder, you shall not commit adultery, you shall not steal, you shall not bear false witness against your neighbor, you shall not covet your neighbor's house, and you shall not covet your neighbor's wife).

Every professional association has its own code of ethics. To become a member in a professional society, the applicant has to read the society's code of ethics and agree to subscribe to its tenets. For example, in the U.S., marketing executives belonging to the American Marketing Association subscribe to their society's code of ethics,[20] advertising professionals subscribe to the American Association of Advertising Agencies' code of ethics,[21] marketing research professionals subscribe to the Marketing Research Association code of marketing research standards,[22] sales and marketing professionals subscribe to the code of ethics of Sales & Marketing Executives International,[23] real estate appraisers subscribe to the Appraisal Institute's code of ethics,[24] home builders subscribe to the code of ethics from the National Home Builders Association,[25] and real estate agents and brokers subscribe to the code of ethics of the National Association of Realtors,[26] among others. The next chapter will describe the National Association of Realtors' code of ethics in some detail.

SUMMARY

The goal of this chapter is to sensitize the student to ethical issues in real estate marketing. We covered a series of ethics tests and discussed their use in helping real estate marketing professionals make ethical decisions. These tests include the legal test, the consequence and utilitarian test, the social contract test, the moral reasoning test, the consumer rights test, the social justice test, the stakeholders test, and the duties test.

The legal test is essentially following the law. The real estate marketing professional should be familiar with all real estate laws and make a concerted effort to abide by these laws. Among the popular laws in the U.S. are the Deceptive Trade Practices Act, the Interstate Land Sales Full Disclosure Act, the Real Estate Settlement Procedures Act, the Equal Credit Opportunity Act, the Truth in Lending Act, the Fair Housing Law, the National Environmental Policy Act, the Flood Disaster Protection Act, the Clean Air Amendment, the Federal Water Pollution Control Act, the Noise Pollution and Abatement Act, and the Coastal Zone Management Act.

The consequence and utilitarian test focuses on assessing the consequences of one's action on affected stakeholder groups. An action that generates the greatest good is the one deemed most ethical. The social contract test focuses on community norms. "Do in Rome what the Romans do" is the motto. In other words, real estate marketing professionals should ensure that their decisions are in accord with community norms.

The moral reasoning test is based on a moral development theory. The theory asserts that people go through three major levels of moral development: pre-conventional, conventional, and post-conventional. The pre-conventional level is divided into two stages: obedience and punishment and self-interest. The conventional level is broken down into two stages: social norms and law and order morality. The post-conventional level has two stages: social contract and universal ethical principles. The goal here is to emulate the moral reasoning of people who are in higher, rather than lower, stages of moral development.

The rights test focuses on consumer rights established in the U.S. and internationally: (1) the right to product safety and to a healthy environment, (2) the right to be informed and to consumer education, (3) the right to choose and to fair trade, and (4) the right to be heard and to redress, and (5) the right to privacy.

The social justice test focuses on avoiding discrimination and making preferential treatment decisions. If the real estate marketing professional is in a position to discriminate, then discrimination has to occur in favor of the disadvantaged. We also briefly mentioned the duties test, which will be described in greater detail in the following chapter.

The stakeholder test posits that a real estate firm is likely to survive and prosper in the long run by meeting the demand of its primary stakeholder groups (customers and end users, employees, suppliers and subcontractors, the local community, the environment, and last, but not least, the shareholders).

DISCUSSION QUESTIONS

1. What is the legal test? Can you distinguish between what is legal and what is ethical in real estate marketing?

2. The ethicality of an action taken by a real estate marketing professional can be judged by the consequences to the self, the firm, and others. Please explain.

3. "Do in Rome what the Romans do" is the motto of social contract ethicists. Please explain.

4. What are the six different stages of the theory of moral development? How can this theory be applied to help real estate marketing professionals make ethical decisions?

5. There are the important consumer rights in real estate marketing. What are they? How do real estate marketing professionals use them to guide ethical decision-making?

6. What is the social justice test? How do real estate marketing professionals use this test to judge the ethicality of their decisions and actions?

7. The stakeholder theory of business ethics is popular. How does it apply to real estate marketing?

NOTES

1. Consequentialism. *Wikipedia*. Accessed from http://en.wikipedia.org/wiki/Teleological_ethics, on June 19, 2013.
2. Harrison, R. (1995). Jeremy Bentham. In T. Honderich (Ed.), *The Oxford Companion to Philosophy* (pp. 85–88). Oxford: Oxford University Press.
3. Krantz, S. F. (2002). *Refuting Peter Singer's ethical theory: The importance of human dignity*. Westport, CT: Greenwood Publishing Group (pp. 28–29).
4. Rousseau, J.-J. (1920). The social contract and discourses (No. 660). London: JM Dent & Sons.
5. Kohlberg, L. (1973). The claim to moral adequacy of a highest stage of moral judgment. *Journal of Philosophy, 70,* 630–646. Kohlberg, L. (1981). *Essays on moral development, Vol. I: The philosophy of moral development.* San Francisco, CA: Harper & Row.
6. *John F. Kennedy: Special message to the Congress on protecting the consumer interest.* Accessed from www.presidency.ucsb.edu, on June 19, 2013.
7. Consumer Bill of Rights. *Wikipedia*. Accessed from http://en.wikipedia.org/wiki/Consumer_bill_of_rights, on June 19, 2013.
8. Robinson-Patman Act. *Wikipedia*. Accessed from http://en.wikipedia.org/wiki/Robinson%E2%80%93Patman_Act, on June 21, 2013.
9. Anti-competitive practices. *Wikipedia*. Accessed from http://en.wikipedia.org/wiki/Anti-competitive_practices, on June 21, 2013.
10. Universal Declaration of Human Rights. *Wikipedia*. Accessed from www.un.org/Overview/rights.html, on July 2, 2013.
11. Zajda, J. I., Majhanovich, S., & Rust, V. D. (Eds.) (2006). *Education and social justice*. New York: Springer.
12. Rawls, J. (2003). *Political liberalism*. New York: Columbia University Press.
13. Fair housing (United States). *Wikipedia*. Accessed from http://en.wikipedia.org/wiki/Fair_housing_(United_States), on June 9, 2013.

14. *Affirmative Action: History and Rationale*. Clinton Administration's *Affirmative Action Review: Report to the President*. Accessed from http://clinton2.nara.gov/WH/EOP/OP/html/aa/aa02.html, on July 2, 2013.
15. Sirgy, M. J. (2002). Measuring corporate performance by building on the stakeholders' model of business ethics. *Journal of Business Ethics*, 35, 143–162.
16. Freeman, R. E. (1984). *Strategic management: A stakeholder approach*. Boston: Pittman.
17. Definition of marketing. *American Marketing Association, 2007*. Accessed from www.marketing. power.com/AboutAMA/Pages/DefinitionofMarketing.aspx?sq=definition+of+marketing, on January 2, 2010.
18. Murphy, P. E., Oberseder, M., & Laczniak, G. R. (2013). Corporate societal responsibility in marketing: Normatively broadening the concept. *AMS Review*, 3, 92.
19. Beauchamp, T. L. (1991). *Philosophical ethics: An introduction to moral philosophy* (2nd ed.). New York: McGraw Hill.
20. Statement of ethics. *American Marketing Association*. Accessed from www.marketingpower.com/ aboutama/pages/statement%20of%20ethics.aspx, on July 3, 2013.
21. Standards of practice of the American Association of Advertising Agencies. *American Association of Advertising Agencies* Accessed from https://ams.aaaa.org/eweb/upload/inside/standards.pdf, on July 3, 2013.
22. The code of marketing research standards. *Marketing Research Association*. Accessed from www. mra-net.org/images/documents/expanded_code.pdf, on July 3, 2013.
23. Sales & marketing creed: The international code of ethics for sales and marketing. *Sales and Marketing Executives International*. Accessed from www.smei.org/displaycommon.cfm?an=1&subarticlenbr= 16, on July 3, 2013.
24. Code of professional ethics of the Appraisal Institute. *Appraisal Institute*. Accessed from www. appraisalinstitute.org/airesources/downloads/cpe/CPE.pdf, on July 3, 2013.
25. Code of ethics of the National Association of Home Builders. *National Association of Home Builders*. Accessed from www.hbacentralmo.com/hba.nsf/ContentPage.xsp?action=openDocument& documentId=20698D045C5A37C08625769A005CF086, on July 3, 2013.
26. Real estate resources: 2012 code of ethics and standards of practice. *NAR*. Accessed from www. realtor.org/mempolweb.nsf/pages/code, on July 3, 2013.

Chapter 13

A Code of Ethics for Real Estate Marketing Professionals

LEARNING OBJECTIVES

As were the previous two chapters, this chapter is also designed to sensitize the student to ethical issues in real estate marketing. However, the goal of this chapter is to introduce a specific code of ethics from the U.S. National Association of Realtors. We will discuss the use of this code and make reference to specific case studies to help students appreciate the significance of the various articles covered in the code. Specifically, this chapter is designed to help students learn:

■ the ethics and professional duties of the real estate agent and broker to clients and customers
■ the ethical and professional duties of the real estate agent and broker to the public
■ the ethical and professional duties of the real estate agent and broker to other real estate agents and brokers

THE NAR CODE OF ETHICS

The National Association for Realtors (NAR) Code of Ethics is divided into three major sets of articles and standards of practice: (1) duties to clients and customers, (2) duties to the public, and (3) duties to realtors.

Duties to Clients and Customers

Duties to clients and customers are covered by the first nine articles of the code (Articles 1–9). See a summary of Articles 1–9 pertaining to duties to clients and customers in Table 13.1.

Article 1

The article states that when real estate agents and brokers represent a client (e.g., buyer, seller, landlord, or tenant), they are committed to protect and promote the interests of their clients. This obligation does not relieve them of their obligation to treat all parties honestly.[1]

Here is a case to help you better understand this article in the code. Jim Sutherland agreed to be the listing agent for Janet Smith, who wanted to sell her home. One of the important features of the home was the fact that it was located near a bus stop. Immediately following the agreement, Sutherland advertised it as such (located near a bus stop). A prospective

Table 13.1 *Articles in the NAR Code of Ethics Related to Duties to Clients and Customers*

Article	Description
Article 1	Article 1 states that when real estate agents and brokers represent a client (e.g., buyer, seller, landlord, or tenant), they are committed to protect and promote the interests of their clients. This obligation does not relieve them of their obligation to treat all parties honestly.
Article 2	Article 2 states that real estate agents and brokers avoid exaggeration, misrepresentation, or concealment of pertinent facts relating to any aspect of the transaction. However, they are not obligated to disclose defects about the property they may not have knowledge of. They do not advise on matters outside the scope of their license.
Article 3	Article 3 states that real estate agents and brokers usually cooperate with other agents/brokers except when such action may not be in the best interest of their clients. Of course, the obligation to cooperate does not extend to sharing commissions, fees, or forms of compensation.
Article 4	Article 4 states that real estate agents/brokers do not buy or present offers from themselves (or any member of their immediate families or their firms) any real property without making their true position known to the owner (or the owner's agent or broker). Conversely, in selling property they own (or partially own), agents/brokers reveal their ownership in writing to the buyer (or the buyer's representative).
Article 5	Article 5 states that real estate agents and brokers do not provide professional services concerning a property or its value where they have an invested interest. However, they can do so if they disclose their interest to all affected parties.
Article 6	Article 6 states that real estate agents and brokers do not accept any form of compensation made for their client without the client's direct knowledge and approval. Agents and brokers disclose to their clients any financial interest they may have with real estate products (e.g., homeowner's insurance, warranty programs, mortgage financing, title insurance, etc.) when recommending these products to their clients.
Article 7	Article 7 states that real estate agents and brokers disclose the exact compensation they receive from any real estate transaction to all parties concerned.
Article 8	Article 8 states that real estate agents and brokers keep funds given to them in trust of their clients in a special account (e.g., escrow) in appropriate financial institution, separated from their own financial accounts.
Article 9	Article 9 states that real estate agents and brokers try their best to ensure that all agreements related to real estate transactions (e.g., listing and representation agreements, purchase contracts, and leases) are in writing in clear and understandable language spelling out the specific terms, conditions, and obligations of the parties. A copy of each agreement is provided to each party upon their signing or initialing.

Source: Adapted from Real estate resources: 2012 code of ethics and standards of practice. *NAR*. Accessed from www.realtor.org/mempolweb.nsf/pages/code, on June 9, 2013.

buyer by the name of Hughes Jones was shown the property; he liked it and made a deposit. Two days later, Sutherland read a notice that the bus line running near Janet Smith's house was being discontinued. He was concerned about this because Jones had explained that his daily schedule required that he takes the bus to work, and that a bus stop needed to be close to the property. Sutherland followed up by notifying Jones of this, who responded by saying that he was no longer interested in the property because the availability of bus transportation is very important to him. Sutherland contacted his client (Smith) and recommended that the deposit be returned. Smith complied. However, Smith was angry about this situation and filed a complaint with the local board of realtors that Sutherland had failed to promote her interests by disclosing to Jones the bus route change. Sutherland countered before a hearing panel of the board's professional standards committee that given the fact that the proximity of the property to a bus stop had been prominently advertised, and given the fact that he knew that Jones' physical disability necessitated a home near a bus stop, he felt that he should inform Jones—because he is obliged to treat all parties honestly—and in doing so, he was not violating his obligation to his client, Smith. The hearing panel concluded that Jim Sutherland had not violated Article 1; he had acted ethically.[2]

Article 2

Article 2 states that real estate agents and brokers avoid exaggeration, misrepresentation, or concealment of pertinent facts relating to any aspect of the transaction. However, they not obligated to disclose defects about the property they may not have knowledge of. They do not advise on matters outside the scope of their license.[3]

Here is a case that brings Article 2 to life. Bob Moore, a real estate agent acting as a management agent, presented a housing unit for rent to Elizabeth Miller, a prospective tenant. When he showed the unit to Miller, he stated that the unit was in good condition. Miller liked the unit, signed the lease, and moved into the unit shortly thereafter. However, after a couple of months in the house, she filed a complaint against Moore with the local board. She charged Moore with misrepresentation. She discovered a clogged sewer line and a defective heater. She was very upset with Moore because he had assured her that the house was in good condition.

At the hearing, Moore admitted that he had assured Miller that "the house was in good condition," and that as soon as he found out about the clogged sewer line and the defective heater, he responded immediately by calling a plumber and a repairman for the heater to remedy the situation. Moore stated that he had no prior knowledge of these problems, and that he had acted promptly and responsibly to address them. The hearing panel decided that Bob Moore was not in violation of Article 2.[4]

Article 3

This article states that real estate agents and brokers usually cooperate with other agents/ brokers except when such action may not be in the best interest of their clients. Of course, the obligation to cooperate does not extend to sharing commissions, fees, or forms of compensation.[5]

Consider the following case. Real estate agent Matt Little complained to his local board that a particular procedure of the board's Multiple Listing Service (MLS) was in conflict with Article 3 of the Code of Ethics. Specifically, this procedure allowed real estate agents/ brokers participating in the MLS to ignore their duties and obligations spelled out under Article 3 (real estate agents/brokers cooperating with one another). He complained that, as exclusive agent for a client (Betsy Fields), he had filed the client's property in the MLS.

Other real estate agents/brokers participating in the MLS had contacted his client, Fields, directly—and made appointments to show the property and transmitted purchase—without his (Little's) knowledge or consent. In essence, the MLS procedure encouraged participant agents/brokers to deal with the seller directly without having to cooperate with the seller's agent/broker.

The administrative staff of the MLS argued that the MLS rule is legitimate because listing the property with the MLS authorizes all MLS participants to deal directly with the sellers. Little argued that this MLS procedural rule should be changed to avoid conflict with Article 3.

The directors of the local board summoned the chairperson of the board's MLS committee to a special directors' meeting to discuss this conflict. It was argued that maintaining this procedural rule undermines the nature and purpose of the MLS itself, because the MLS does not provide brokerage services and cannot function as an agent of sellers. The MLS was ordered to change the procedural rule that permitted any participant of the MLS to undermine the agency status of the agent/broker holding an exclusive listing.[6]

Article 4

This article states that real estate agents/brokers do not buy or present offers from themselves (or any member of their immediate families or their firms)any real property without making their true position known to the owner (or the owner's agent or broker). Conversely, in selling property they own (or partially own), agents/brokers reveal their ownership in writing to the buyer (or the buyer's representative).[7]

The following case brings Article 4 to life. Pam Johnson, owner of a lot zoned for commercial use, consulted Hank Taylor about the value of this lot. Johnson indicated to Taylor that she was planning to leave town soon, and she would like to sell the lot. Taylor suggested that the lot be appraised by a professional appraiser, which resulted in a valuation of $130,000. Johnson hired Taylor as the listing agent, and he, in turn, listed the lot at that price. Shortly thereafter, Taylor received an offer of $122,000. He communicated that offer to Johnson, who rejected the offer. No other offers were made during the next four months. At that time, Johnson asked Taylor if he would be willing to buy the lot himself. Taylor made Johnson an offer of $118,000. Johnson accepted that offer. Several months later, Johnson returned and discovered that Taylor had sold the lot for $125,000, three weeks after he had purchased it from Johnson for $118,000.

Johnson filed a complaint to the local board charging that Taylor took advantage of her. She stated that she could not understand Taylor's inability to sell the lot at $122,000 in four months, given that he managed to sell the lot at $125,000 only three weeks after he took ownership of it.

At the hearing, Taylor introduced evidence that he indeed worked hard to sell the lot for $122,000. He introduced several letters from prospects while the property was listed with him. These prospects expressed the opinion that the lot was overpriced. The buyer who ended up purchasing the lot for $125,000 showed up at the hearing as a witness. He stated that he had never met Taylor prior to the date of the actual purchase. The panel concluded that because Johnson's suspicion of duplicity was unfounded, Taylor had not violated Article 4 of the Code of Ethics.[8]

Article 5

This article states that real estate agents and brokers do not provide professional services concerning a property or its value where they have an invested interest. However, they can do so if they disclose their interest to all affected parties.[9]

The following case illustrates the significance of Article 5. Melissa Stanton, owner of an apartment building, was negotiating with Hugh Jones, a prospective buyer, in the sale of the building. Stanton and Jones could not agree on a price that was mutually satisfactory for both parties. To break the stalemate, they agreed that each would hire an appraiser and they would accept the average of the two appraisals as a fair price. Melissa Stanton hired Eric Donahue as her appraiser, and Hugh Jones hired Tom Fordham. The two appraisers submitted their reports, and the average of the two was determined as "fair price." As such, the transaction was completed.

Six months later, Melissa Stanton came to understand that Eric Donahue (the appraiser that she hired) had managed five buildings owned by Hugh Jones, and that he had been Jones' property manager at the time Stanton hired him to appraise the property. At this point, Melissa Stanton hired Christine Tennessee, another agent, to make another appraisal of the property she had sold to Jones. Tennessee's valuation turned out to be 30% higher than Donahue's.

Melissa Stanton filed a complaint against Eric Donahue to the local board. She charged Donahue of unethical behavior based on Article 5 of the code of ethics. She argued that because Donahue managed all of Jones' properties and because he had not disclosed that relationship to her, he had acted unethically by coming up with an excessively low valuation on the property he had appraised for her. At the hearing, Stanton used a witness who testified that he had heard Hugh Jones admit that he had made a good buy in purchasing Stanton's building because Stanton's appraiser was his property manager. Hugh Jones, appearing as a witness for Eric Donahue, disputed this fact and argued that that average price of the two original appraisals was a fair price.

The hearing panel concluded that Eric Donahue had a contemplated interest in this transaction, having served as property manager for Hugh Jones. Given this contemplated interest, he should have disclosed this interest to Melissa Stanton. He was found to be in violation of Article 5 because he had failed to disclose this relationship to the seller.[10]

Article 6
This article states that real estate agents and brokers do not accept any form of compensation made for their client without the client's direct knowledge and approval. Agents and brokers disclose to their clients any financial interest they may have with real estate products (e.g., homeowner's insurance, warranty programs, mortgage financing, title insurance, etc.) when recommending these products to their clients.[11]

The following case reflects the essence of Article 6. Joe Estes, a real estate agent, managed a property owned by Tina Holloway. In doing so, he bought janitorial supplies at wholesale prices and billed Holloway at retail level, thereby making a significant profit on the sale of these supplies.

Tina Holloway came to notice Estes' practice and filed a complaint with the local board charging Sutherland with unethical conduct—in reference to Article 6 of the NAR Code of Ethics. During the hearing, Estes tried to argue that the prices of the supplies he billed to Holloway were not significantly higher than the prices Holloway had been paying before he took over the management of Holloway's property. The panel noted that Estes failed to disclose to Holloway the profit he made on the supplies. Estes was found to be in violation of Article 6.[12]

Article 7
This article states that real estate agents and brokers disclose the exact compensation they receive from any real estate transaction to all parties concerned.[13]

Consider the following case. Michelle Sowers hired a real estate agent by the name of Jonathon Fibbs to help her buy a small commercial property. Sowers offered Fibbs an incentive: If Fibbs could find a property that met her exact specifications and price, she would pay a finder's fee.

After several weeks, Fibbs notified Sowers that a Matthew Jones had just listed a property with him that seemed to meet Sowers' requirements with one exception: The listed price was somewhat higher than Sowers' desired price. Sowers examined the property and insisted on paying only her original specified price. A couple of days later, Fibbs contacted Sowers, saying that he had persuaded Jones to accept Sowers' offer. The transaction was completed and Fibbs collected a commission from the seller (Matthew Jones) and a finder's fee from the buyer (Michelle Sowers). The finder's fee was not disclosed to Jones.

A few weeks later, Jones learned about the finder's fee and was very upset, feeling that he was duped by Fibbs. He filed a complaint with the local board in which he charged Fibbs with unethical conduct. In the complaint, Jones argued that when Fibbs had presented Sowers' offer, he was reluctant to accept it. However, Fibbs made a case that the offer was fair and that he was not likely to receive a better offer. The complaint also specified that Jones (seller) was not aware that Fibbs (agent) was also representing Sowers (buyer).

At the hearing, Jonathon Fibbs argued that he did his best to serve both buyer and seller; that in his judgment the listed price was excessive and the final price that was accepted was indeed a fair price. The hearing panel found Fibbs guilty in violation of Article 7 of the Code of Ethics. Given that he represented both buyer and seller, he should have fully disclosed his dual agency role to both parties.[14]

Article 8

This article states that real estate agents and brokers keep funds given to them in trust of their clients in a special account (e.g., escrow) in appropriate financial institution, separated from their own financial accounts.[15]

The following case well illustrates Article 8. Tom Newton, a listing broker, received an offer from Martha Heminway to purchase one of his listings with a check for $5,000 as an earnest money deposit. Heminway's offer was subject to the sale of the house in which she was, at that time, living. Newton presented the offer to the seller (Abraham Jones), and the offer was accepted. Newton then made a mistake. Instead of depositing the $5,000 earnest money in the brokerage escrow account, he deposited the check in his personal checking account. Because Heminway's offer was contingent on the sale of her house, Jones' house remained on the market. A few days later, Newton received another offer to purchase Jones' house from another broker and presented it to Jones as an alternative offer. Jones (seller) asked Newton to see if he could expedite the sale by pressuring Heminway (the first prospective buyer) to waive the sale contingency and proceed to complete the transaction. Heminway felt very reluctant to waive the sale contingency because of the risk imposed (the possibility of buying the new house without being able to sell the current house). Heminway then decided to withdraw her offer and asked Newton for her $5,000 check back. Newton explained that he had inadvertently deposited her check in his personal bank account which then became "attached" (because of a loss in a civil suit that was filed against him). In other words, he was temporarily unable to refund the deposit to Heminway.

Martha Newton filed a complaint with the real estate board against Tom Newton. At the hearing, Newton explained that his bank account had been attached because of the loss of a civil suit, that he was appealing, and that the situation would be straightened out in a matter of days. The hearing panel concluded that Tom Newton was in violation of Article 8 of the Code of Ethics for having failed to place the earnest money deposit in a business escrow account separate from his own personal account.[16]

Article 9

This article states that real estate agents and brokers try their best to ensure that all agreements related to real estate transactions (e.g., listing and representation agreements, purchase contracts, and leases) are in writing in clear and understandable language spelling out the specific terms, conditions, and obligations of the parties. A copy of each agreement is provided to each party upon their signing or initialing.[17]

Consider the following situation. Tyson Grey, a real estate agent representing a buyer, Misty Harvey, identified a house that met the needs of Harvey and her family. He accompanied Harvey to the house; she liked the house very much and asked Grey to submit an offer contingent upon the owner (Tom Jones) fixing a few things around the house (e.g., repair several cracks in the ceiling, replace a cracked window, repair a small plumbing leak in the kitchen). Grey discussed these fix-ups with Jones, and Jones agreed to do them. They shook hands. Grey proceeded to close the deal.

After the sale, Misty Harvey and her family moved into the house and immediately noticed that the owner (Tom Jones) had failed to do the needed repairs—the promised fix-ups. Harvey then filed a complaint to the local real estate board charging that Tyson Grey failed to perform his duties by ensuring that the repair work was done. At the hearing panel, it was noted that Tyson Grey failed to get Jones (the seller) to sign a contingency clause specifying these repairs. Grey testified that he assumed that Jones would follow up on the repairs and that they had shaken hands on that promise. Jones had reneged on that promise, and Grey argued that Jones was to be blamed, not he. The hearing panel found Grey to be in violation of Article 9 of the NAR Code of Ethics.

Duties to the Public

Duties to the public are covered by the second set of articles of the code: Articles 10–14. See a summary of Articles 10–14 pertaining to duties to the public in Table 13.2.

Article 10

This article states that real estate agents and brokers do not deny equal professional services to any person for reasons of race, color, religion, sex, handicap, familial status, national origin, or sexual orientation. Agents and brokers do not discriminate against a person or persons on the basis aforementioned social groups.[18]

The following case illustrates the spirit of the article. An African-American couple called on a real estate agent, Aaron Butterfield, and expressed interest in buying a home. Their specs were: $130,000–$145,000 price range, at least three bedrooms, a large lot, and located in the Cedar Orchard area of town. Butterfield explained that homes in the Cedar Orchard neighborhood were generally sold in the $180,000–$220,000 range. After assessing the couple's financial situation, Butterfield concluded that the family could not afford a home priced over $137,500. Butterfield proceeded to show the couple homes below the maximum of $137,000. However, the couple did not like any of the properties Butterfield showed them.

The minority couple then filed a complaint with the local real estate board charging Aaron Butterfield with a violation of Article 10 of the Code Ethics. At the hearing, the minority couple argued that Aaron Butterfield engaged in racial steering; they had specifically expressed an interest in buying a home in the Cedar Orchard neighborhood, but he

Table 13.2 Articles in the NAR Code of Ethics Related to Duties to the Public

Article	Description
Article 10	Article 10 states that real estate agents and brokers do not deny equal professional services to any person for reasons of race, color, religion, sex, handicap, familial status, national origin, or sexual orientation. Agents and brokers do not discriminate against a person or persons on the basis aforementioned social groups.
Article 11	Article 11 states that the services that real estate agents and brokers provide to their clients and customers are expected to conform to the standards of practice and competence that are reasonably expected in these professions (e.g., residential real estate brokerage, real property management, commercial and industrial real estate brokerage, land brokerage, real estate appraisal, real estate counseling, real estate syndication, real estate auction, and international real estate). Agents and brokers do not provide professional services outside their field of specialty unless they do this with the assistance professionals specializing in these services, or unless the facts are fully disclosed to the client.
Article 12	Article 12 states that real estate agents and brokers are honest and truthful in their marketing communications. That is, they present the facts in their advertising, and other promotions. Agents and brokers do their best to ensure that their status as real estate professionals is readily apparent in their advertising and promotion.
Article 13	Article 13 states that real estate agents and brokers do not engage in law-related activities (e.g., legal counsel).
Article 14	Article 14 states that if charged with unethical practice (or requested to participate and cooperate), in any professional standards proceeding or investigation, real estate agents and brokers present all pertinent facts to the proper tribunals of the member board (or affiliated institute, society, or council in which membership is held). Furthermore, agents and brokers do not take action to disrupt or obstruct such proceedings or investigations.

Source: Adapted from Real estate resources: 2012 code of ethics and standards of practice. *NAR*. Accessed from www.realtor.org/mempolweb.nsf/pages/code, on June 9, 2013.

had steered them away from this neighborhood because they were African American and the vast majority of the families in that neighborhood were Caucasian. Butterfield responded by producing documentation indicating the couple's housing preference, including price range. Additional documentation was provided indicating the exact homes they were shown based on their housing preference and price affordability, although admittedly, none of the homes shown were located in Cedar Orchard, their preferred neighborhood. Further, Butterfield was able to produce listing and sales information of homes in the Cedar Orchard neighborhood, confirming that no homes were available below $145,000. Butterfield stated, "If there were listings in Cedar Orchard in the $130,000–$145,000 price range with three bedrooms and a large lot, and I had refused to show them these homes, I should be charged. But the fact of the matter is, there were no such homes available then. How could I show them homes that met their criteria that did not exist?" The hearing panel concluded that Aaron Butterfield did not engage in racial steering, and therefore he is not in violation of Article 10.[19]

Article 11

This article states that the services that real estate agents and brokers provide to their clients and customers are expected to conform to the standards of practice and competence that are reasonably expected in these professions (e.g., residential real estate brokerage, real property management, commercial and industrial real estate brokerage, land brokerage, real estate appraisal, real estate counseling, real estate syndication, real estate auction, and international real estate). Agents and brokers do not provide professional services outside their field of specialty unless they do this with the assistance professionals specializing in these services, or unless the facts are fully disclosed to the client.[20]

Here is a case illustration. Jim Fowler, a real estate agent, sold a light industrial property to David Jones, a laundry operator. Several months later, Jones (the buyer) hired Jim Fowler to do another job: appraise the property for use in possible merger with another laundry. Fowler happily accepted this new assignment and performed the appraisal. David Jones (former buyer, now client) was very unhappy with Fowler's report because he felt that the valuation was excessively low. David Jones then hired another appraiser, George Fordham, specializing in commercial real estate. As expected, Fordham's valuation was significantly higher than Fowler's. This made David Jones more upset, after which he filed a complaint with the local real estate board charging Jim Fowler with incompetent and unprofessional service.

At the hearing, Jim Fowler admitted that he had no professional credentials or experience in commercial real estate appraisal, and that his appraisal experience had been limited mostly to residential real estate. He also admitted that he had readily accepted the assignment and did not disclose to David Jones his inexperience and lack of credentials in this area. Jim Fowler was found by to be in violation of Article 11.[21]

Article 12

This article states that real estate agents and brokers are honest and truthful in their marketing communications. That is, they present the facts in their advertising, and other promotions. Agents and brokers do their best to ensure that their status as real estate professionals is readily apparent in their advertising and promotion.[22]

Here is an illustration. Page Smith, a prospective real estate buyer, observed a sign on a vacant lot that said: "For Sale—Call 552–5315." She inferred from the sign that this was a For-Sale-by-Owner type of situation. This was a desirable situation for Smith because she hoped that in buying the lot directly from the owner, she would save a significant chunk of money. So she called the number on the sign and was surprised to learn that the lot was exclusively listed by Cory Gutherland, a real estate agent. Page Smith was very upset about this, prompting her to file a complaint against Cory Gutherland charging misrepresentation (a violation of Article 12 of the Code of Ethics).

At the hearing, Gutherland stated that the sign was not a "formal" advertisement, such as a newspaper advertisement, business card, or billboard. Therefore, Article 12 should not apply in this case. The hearing panel determined that the sign was indeed an advertisement, and that Gutherland had violated Article 12 of the Code of Ethics.[23]

Article 13

This article states that real estate agents and brokers do not engage in law-related activities (e.g., legal counsel).[24]

Think about the following case. Joseph Jones had recently hired Jim Cassey, a real estate agent, to help Jones sell a piece of commercial property his company owned. Jones said the company was sending him on business to Latin America and that he would be away for some time. As such, he would like to leave a power of attorney with his wife while he was gone to handle urgent situations. Jones asked Cassey to prepare a power of attorney for him, and Cassey obliged. This action came to the attention of one person, Christine Tennessee, who had involvement in the Grievance Committee of the local board. She asked the board to look into this and charged Jim Cassey with unprofessional conduct.

At the hearing, Jim Cassey claimed that Jones' request was essentially related to real estate service, not necessarily legal in nature. He had inferred that the reason for Jones asking him (rather than an attorney) to prepare a power of attorney was because Jones wanted his wife to handle his real estate business in case of an emergency when he was overseas. He stated that Jones was a personal friend, and he knew a lot about Jones' personal affairs. Given that the power of attorney was directly related to real estate, he felt that he had rendered a real estate service to Jones, not a legal service. However, the hearing panel did not agree with Jim Cassey and found him to be in violation of Article 13 of the Code.[25]

Article 14

This article states that if charged with unethical practice (or requested to participate and cooperate), in any professional standards proceeding or investigation, real estate agents and brokers present all pertinent facts to the proper tribunals of the member board (or affiliated institute, society, or council in which membership is held). Furthermore, agents and brokers do not take action to disrupt or obstruct such proceedings or investagations.[26]

The following case does a good job illustrating this article. A local real estate board received a complaint from a Kimberly Smith, client of Robert Sutherland, about a possible violation of Article 1 of the Code of Ethics. The complaint was referred to the chairperson of the grievance committee of the board, who sent a copy of it to Robert Sutherland, requesting that he respond by sharing a specific document and indicate his willingness to appear in front of a hearing panel. Robert Sutherland responded by denying the charge. He also indicated that he was willing to appear at a hearing and establish his innocence by presenting all of the pertinent facts. However, he also indicated that his attorney had advised him not provide the requested document to the committee in advance of the hearing.

The grievance committee found his unwillingness to share the requested document to be in violation of Article 14. The board asserted that the grievance committee had a right to examine the requested documents before the hearing, and that he was in violation of Article 13—and possibly Article 1, too (which was yet to be determined).[27]

Duties to Other Real Estate Agents and Brokers

Duties to the public are covered by the last three articles of the code (Articles 15–17). See a summary of Articles 15–17 pertaining to duties to other real estate agents and brokers in Table 13.3.

Article 15

This article states that real estate agents and brokers do not make false or misleading statements about other real estate professionals and affiliated business practices.[28]

Here is a case illustration. Richard Fondue managed a residential brokerage firm in a market that was highly competitive. He frequently used information from the MLS to keep

Table 13.3 *Articles in the NAR Code of Ethics Related to Duties to Other Real Estate Agents and Brokers*

Article	Description
Article 15	Article 15 states that real estate agents and brokers do not make false or misleading statements about other real estate professionals and affiliated business practices.
Article 16	Article 16 states that real estate agents and brokers do not engage in any practice that counters the exclusive representation or brokerage relationship agreements that other agents and brokers have with clients.
Article 17	Article 17 states that in the event of disputes (contractual or specific non-contractual) between real estate agents or brokers associated with different firms, the disputing parties accept mediation if the board deems mediation as necessary. If the dispute is not resolved through mediation (or if mediation is not required), the disputing parties accept arbitration rather than litigation. The disputing parties are bound by any resulting agreement or award.

Source: Adapted from Real estate resources: 2012 code of ethics and standards of practice. *NAR*. Accessed from www.realtor.org/mempolweb.nsf/pages/code, on June 9, 2013.

close track of his listing and sales activity vis-à-vis the competition. One day, an ad from a competitor brokerage house (managed by Tom Sordham) came to his attention. The ad showed a figure in which his firm (Fondue's firm) had 10% market share of the properties in the MLS over the past three months. Fondue thought this was somewhat "low." His own tabulation of the MLS data showed his market share to be 11%. Fondue filed an ethics complaint against Sordham, citing Article 15 of the Code of Ethics, in that Sordham's brokerage house made a misleading statement about competitors. The grievance committee of the local board determined that an ethics hearing was warranted.

At the hearing, Tom Sordham indicated his that market share claim was based in data but that a glitch in the system had resulted in miscalculation. Specifically, he produced documentation from the Board's MLS administrator indicating a programming error and that this error resulted in miscalculations. Adjusting for this error, the actual market share of Fondue's firm turned out to be 10.9%.

The Hearing Panel accepted Sordham's testimony. The panel members reasoned that the error had been small and did not seem to have a malicious intent. The panel concluded that Sordham's comparison with his competitors, while slightly inaccurate, was made in good faith. Therefore, Sordham was found not to be in violation of Article 15.[29]

Article 16

This article states that real estate agents and brokers do not engage in any practice that counters the exclusive representation or brokerage relationship agreements that other agents and brokers have with clients.[30]

Here is a case illustration. Hope Fuss (client) listed her home for sale with Richard Garrison. In doing so, she gave specific instructions to Garrison: The sale was to be handled without advertising and without attracting any attention. This meant that she did not want the house filed with the MLS, advertised, or promoted publicly in any way. She also emphasized that Garrison should make sure that other real estate agents who might become involved in the transaction maintained secrecy.

Richard Garrison approached another real estate agent (Tom Jordan), whom Garrison knew to be in touch with several prospective buyers. Garrison explained to Jordan his client's wish that this transaction stay "under the radar." Garrison refrained from promoting the property further. However, shortly after he approached Jordan Garrison found out that Christine Tennley, another real estate agent, was discussing the property with several prospective buyers. Garrison was very upset by this revelation; after questioning Jordan, Garrison found out that Jordan had approached Tennley to help with the sale of the property.

Richard Garrison then charged Tom Jordan with unethical conduct (violation of Article 16) in a complaint to the local board, stating that Jordan's breach of confidence was a failure to respect his, Garrison's, exclusive agency, and that Jordan's action placed his relationship with his client (Hope Fuss) at risk. At the hearing, Tom Jordan defended himself by saying that he had indeed discussed this situation with Christine Tennley, but he had not formally invited Tennley to help with the sale. Therefore, he had not violated Article 16. The hearing panel did not accept his excuse and found him guilty as charged.[31]

Article 17

This article states that in the event of disputes (contractual or specific non-contractual) between real estate agents or brokers associated with different firms, the disputing parties accept mediation if the board deems mediation as necessary. If the dispute is not resolved through mediation (or if mediation is not required), the disputing parties accept arbitration rather than litigation. The disputing parties are bound by any resulting agreement or award.[32]

Consider the following case. Two real estate agents, Terry Blanchard and Kent Henderson, had been involved in a transaction that resulted in a dispute about compensation. Instead of using the local board to help mediate or arbitrate this dispute, Terry Blanchard filed suit against Kent Henderson for payment of compensation that he felt Henderson owed him. Instead of hiring an attorney and responding in kind, Kent Henderson filed a request for arbitration with the local board. The local board's grievance committee reviewed the case and determined that this case necessitated arbitration. However, Terry Blanchard refused to withdraw his lawsuit against Henderson. As such, Kent Henderson filed a complaint with the board charging Terry Blanchard with a violation of Article 17.

The board directed Terry Blanchard to attend a scheduled hearing. Kent Henderson presented evidence demonstrating that he had asked Terry Blanchard to submit the dispute to arbitration by the board. Terry Blanchard did attend the hearing and said that the board had no authority to bar his access to the judicial system or to require him to arbitrate his dispute with Henderson. The board concluded that Henderson was correct as to his legal rights. However, the board members explained to Henderson that his membership in the local chapter of the realty association assumes that all members accept certain obligations as specified by the association code of ethics, and that if he wished to continue as a member of the local chapter, he would be obliged to adhere to the code of ethics, and therefore to arbitration.

Terry Blanchard did not accept the board's explanation and failed to withdraw the lawsuit and submit to arbitration by the board. As a result of Blanchard's action, the board concluded that Terry Blanchard was in violation of Article 17. However, it was also noted that if the court remanded the case to the local board, the finding of a violation of Article 17 would be dropped because the Board's arbitration process would be the final outcome.[33]

SUMMARY

This chapter was designed to sensitize the student to ethical issues in real estate marketing through the NAR code of ethics. This code is divided into three major sets of articles and standards of practice: (1) duties to clients and customers (Articles 1–9), (2) duties to the public (articles 10–14), and (3) duties to real estate agents and brokers (Articles 15–17).

Article 1 states that when real estate agents and brokers represent a client (e.g., buyer, seller, landlord, or tenant), they are committed to protect and promote the interests of their clients. This obligation does not relieve them of their obligation to treat all parties honestly. Article 2 states that real estate agents and brokers avoid exaggeration, misrepresentation, or concealment of pertinent facts relating to any aspect of the transaction. However, they are not obligated to disclose defects about the property they may not have knowledge of. They do not advise on matters outside the scope of their license. Article 3 states that real estate agents and brokers usually cooperate with other agents/brokers except when such action may not be in the best interest of their clients. Of course, the obligation to cooperate does not extend to sharing commissions, fees, or forms of compensation. Article 4 states that real estate agents/brokers do not buy or present offers from themselves (or any member of their immediate families or their firms)any real property without making their true position known to the owner (or the owner's agent or broker). Conversely, in selling property they own (or partially own), agents/brokers reveal their ownership in writing to the buyer (or the buyer's representative). Article 5 states that real estate agents and brokers do not provide professional services concerning a property or its value where they have an invested interest. However, they can do so if they disclose their interest to all affected parties. Article 6 states that real estate agents and brokers do not accept any form of compensation made for their client without the client's direct knowledge and approval. Agents and brokers disclose to their clients any financial interest they may have with real estate products (e.g., homeowner's insurance, warranty programs, mortgage financing, title insurance, etc.) when recommending these products to their clients. Article 7 states that real estate agents and brokers disclose the exact compensation they receive from any real estate transaction to all parties concerned. Article 8 states that real estate agents and brokers keep funds given to them in trust of their clients in a special account (e.g., escrow) in appropriate financial institution, separated from their own financial accounts. Article 9 states that real estate agents and brokers try their best to ensure that all agreements related to real estate transactions (e.g., listing and representation agreements, purchase contracts, and leases) are in writing in clear and understandable language spelling out the specific terms, conditions, and obligations of the parties. A copy of each agreement is provided to each party upon their signing or initialing. Duties to the public are covered by the second set of articles of the code: Articles 10–14.

Article 10 states that real estate agents and brokers do not deny equal professional services to any person for reasons of race, color, religion, sex, handicap, familial status, national origin, or sexual orientation. Agents and brokers do not discriminate against a person or persons on the basis aforementioned social groups. Article 11 states that the services that real estate agents and brokers provide to their clients and customers are expected to conform to the standards of practice and competence that are reasonably expected in these professions (e.g., residential real estate brokerage, real property management, commercial and industrial real estate brokerage, land brokerage, real estate appraisal, real estate counseling, real estate syndication, real estate auction, and international real estate). Agents and brokers do not provide professional services outside their field of specialty unless they do this with the assistance professionals specializing in these services, or unless the facts are fully disclosed to the client. Article 12 states that real estate agents and brokers are honest and truthful in their marketing communications. That is, they present the facts in their advertising, and other promotions. Agents and brokers do their best to ensure that their status as real estate professionals is readily apparent in their advertising and promotion. Article 13 states that real estate agents and brokers do not engage in law-related activities (e.g., legal counsel). Article 14 states that if charged with unethical practice (or requested to participate and cooperate), in any professional standards proceeding or investigation, real estate agents and brokers present all pertinent facts to the proper tribunals of the member board (or affiliated institute, society, or council in which membership is held). Furthermore, agents and brokers do not take action to disrupt or obstruct such proceedings or investigations.

Duties to the public are covered by the last three articles of the code (Articles 15–17). Article 15 states that real estate agents and brokers do not make false or misleading statements about other real estate professionals and affiliated business practices. Article 16 states that real estate agents and brokers do not engage in any practice that counters the exclusive representation or brokerage relationship agreements that other agents and brokers have with clients. Article 17 states that in the event of disputes (contractual or specific non-contractual) between real estate agents or brokers associated with different firms, the disputing parties accept mediation if the board deems mediation as necessary. If the dispute is not resolved through mediation (or if mediation is not required), the disputing parties accept arbitration rather than litigation. The disputing parties are bound by any resulting agreement or award.

DISCUSSION QUESTIONS

1. Describe one major article related to duties to clients and customers. Think of a scenario that would illustrate this important article.

2. Do the same in relation to one article related to the public.

3. Repeat this exercise in relation to one article involving duties to real estate agents and brokers.

NOTES

1. Real estate resources: 2012 code of ethics and standards of practice. *NAR*. Accessed from www.realtor.org/mempolweb.nsf/pages/code, on June 9, 2013.
2. Case #1–1: Fidelity to client. *Realtor*. Accessed from www.realtor.org/2013-code-of-ethics-and-arbitration-manual/case-interpretations/related-to-article-1, on June 9, 2013.
3. Real estate resources: 2012 code of ethics and standards of practice. *NAR*. Accessed from www.realtor.org/mempolweb.nsf/pages/code, on June 9, 2013.
4. Case #2–1: Disclosure of pertinent facts. *Realtor*. Accessed from www.realtor.org/2013-code-of-ethics-and-arbitration-manual/case-interpretations/related-to-article-2, on June 9, 2013.
5. Real estate resources: 2012 code of ethics and standards of practice. *NAR*. Accessed from www.realtor.org/mempolweb.nsf/pages/code, on June 9, 2013.
6. Case #3–1: Rules of MLS may not circumvent code. *Realtor*. Accessed from www.realtor.org/2013-code-of-ethics-and-arbitration-manual/case-interpretations/related-to-article-3, on June 9, 2013.
7. Real estate resources: 2012 code of ethics and standards of practice. *NAR*. Accessed from www.realtor.org/mempolweb.nsf/pages/code, on June 9, 2013.
8. Case #4–1: Disclosure when buying on own accord. *Realtor*. Accessed from www.realtor.org/2013-code-of-ethics-and-arbitration-manual/case-interpretations/related-to-article-4, on June 9, 2013.

9. Real estate resources: 2012 code of ethics and standards of practice. *NAR*. Accessed from www.realtor.org/mempolweb.nsf/pages/code, on June 9, 2013.

10. Case #5–1: Contemplated interest in property appraised. *Realtor*. Accessed from www.realtor.org/2013-code-of-ethics-and-arbitration-manual/case-interpretations/related-to-article-5, on June 9, 2013.

11. Real estate resources: 2012 code of ethics and standards of practice. *NAR*. Accessed from www.realtor.org/mempolweb.nsf/pages/code, on June 9, 2013.

12. Case #6–1: Profit on supplies used in property management. *Realtor*. Accessed from www.realtor.org/2013-code-of-ethics-and-arbitration-manual/case-interpretations/related-to-article-6, on June 9, 2013.

13. Real estate resources: 2012 code of ethics and standards of practice. *NAR*. Accessed from www.realtor.org/mempolweb.nsf/pages/code, on June 9, 2013.

14. Case #7–1: Acceptance of compensation from buyer and seller. *Realtor*. Accessed from www.realtor.org/2013-code-of-ethics-and-arbitration-manual/case-interpretations/related-to-article-7, on June 9, 2013.

15. Real estate resources: 2012 code of ethics and standards of practice. *NAR*. Accessed from www.realtor.org/mempolweb.nsf/pages/code on June 9, 2013.

16. Case #8–1: Failure to put deposit in separate account. *Realtor*. Accessed from www.realtor.org/2013-code-of-ethics-and-arbitration-manual/case-interpretations/related-to-article-8, on June 9, 2013.

17. Real estate resources: 2012 code of ethics and standards of practice. *NAR*. Accessed from www.realtor.org/mempolweb.nsf/pages/code, on June 9, 2013.

18. Real estate resources: 2012 code of ethics and standards of practice. *NAR*. Accessed from www.realtor.org/mempolweb.nsf/pages/code, on June 9, 2013.

19. Case #10–1: Equal professional services by realtors. *Realtor*. Accessed from www.realtor.org/2013-code-of-ethics-and-arbitration-manual/case-interpretations/related-to-article-10, on June 9, 2013.

20. Real estate resources: 2012 code of ethics and standards of practice. *NAR*. Accessed from www.realtor.org/mempolweb.nsf/pages/code, on June 9, 2013.

21. Case #11–1: Appraiser's competence for assignment. *Realtor*. Accessed from www.realtor.org/2013-code-of-ethics-and-arbitration-manual/case-interpretations/related-to-article-11, on June 9, 2013.

22. Real estate resources: 2012 code of ethics and standards of practice. *NAR*. Accessed from www.realtor.org/mempolweb.nsf/pages/code, on June 9, 2013.

23. Case #12–1: Absence of name on sign. *Realtor*. Accessed from www.realtor.org/2013-code-of-ethics-and-arbitration-manual/case-interpretations/related-to-article-12, on June 9, 2013.

24. Real estate resources: 2012 code of ethics and standards of practice. *NAR*. Accessed from www.realtor.org/mempolweb.nsf/pages/code, on June 9, 2013.

25. Case #13–1: Preparation of instrument unrelated to real estate transaction. *Realtor*. Accessed from www.realtor.org/2013-code-of-ethics-and-arbitration-manual/case-interpretations/related-to-article-13, on June 9, 2013.

26. Real estate resources: 2012 code of ethics and standards of practice. *NAR*. Accessed from www.realtor.org/mempolweb.nsf/pages/code, on June 9, 2013.

27. Case #14–1: Establishing procedure to be followed in handling complaints. *Realtor*. Accessed from www.realtor.org/2013-code-of-ethics-and-arbitration-manual/case-interpretations/related-to-article-14, on June 9, 2013.

28. Real estate resources: 2012 code of ethics and standards of practice. *NAR*. Accessed from www.realtor.org/mempolweb.nsf/pages/code, on June 9, 2013.

29. Case #13–1: Knowing or reckless false statements about competitors. *Realtor*. Accessed from www.realtor.org/2013-code-of-ethics-and-arbitration-manual/case-interpretations/related-to-article-15, on June 9, 2013.

30. Real estate resources: 2012 code of ethics and standards of practice. *NAR*. Accessed from www.realtor.org/mempolweb.nsf/pages/code, on June 9, 2013.

31. Case #16–1: Confidentiality of cooperating realtor's participation. *Realtor*. Accessed from www.realtor.org/2013-code-of-ethics-and-arbitration-manual/case-interpretations/related-to-article-16, on June 9, 2013.

32. Real estate resources: 2012 code of ethics and standards of practice. *NAR*. Accessed from www.realtor.org/mempolweb.nsf/pages/code, on June 9, 2013.

33. Case #17–1: Obligation to submit to arbitration. *Realtor*. Accessed from www.realtor.org/2013-code-of-ethics-and-arbitration-manual/case-interpretations/related-to-article-17, on June 9, 2013.

Glossary

Absenteeism   Absenteeism refers to the situation in which employees (e.g., real estate agents) are taking too much time off work.

Advertorial columns   Advertorial columns are editorial columns written by journalists on issues and events related to the real estate industry. These editorials are then disseminated in the form of digital blogs. It is considered to be a form of publicity by the real estate development firm because real estate marketers provide information to the columnists that may serve as good publicity for the firm.

Affirmative action   Affirmative action refers to policies that take factors (e.g., race, color, sex, religion, and national origin) into account to benefit an underrepresented group in employment and education.

Age Discrimination in Employment Act   The Age Discrimination in Employment Act (of 1967) is a law that forbids employment discrimination against anyone at least 40 years of age. The law prohibits discrimination in hiring, promotion, compensation, and termination of employment.

Americans with Disabilities Act   The Americans with Disabilities Act (of 1990) is a law that prohibits discrimination based on disability.

Appraisal fees   Appraisal fees are fees paid to a real estate appraiser for determining the market value of the property.

Assumptive close   The assumptive close occurs when the buyer agent assumes that the client is comfortable with the target property and would like to move to closing.

Audience selection   Audience selection is the first step in promotion planning, in which the real estate marketer identifies the exact audience that will be the target of the marketing communication campaign.

Autocratic leadership style   Autocratic leadership (also known as "authoritarian leadership") refers to the situation in which the sales manager controls all important decisions of the sales organization, including all the selling-related tasks.

Balance-sheet closing technique   This closing technique refers to the buyer agent summarizing advantages as well as disadvantages of the property by comparing the "assets" and "liabilities" of those properties that have been presented to the client.

Best Alternative to a Negotiated Agreement (BATNA)   As the name suggests, BATNA is a best alternative to a negotiated agreement, which real estate negotiators use as leverage. That is, a real estate negotiator who has a BATNA is in a better position to exert influence on the other party and extract concessions.

Broadcast advertising   Broadcast advertising involves real estate advertising through television and radio.

Broker's fee   The broker's fee is the fee of commission that the seller pays the brokerage firm of the listing agent for services rendered.

Building-code requirements   Building code requirements are standards that a builder has to meet before a certificate of occupancy can be issued, allowing residents to occupy the structure.

Canvassing   Canvassing refers to door-to-door solicitations. This is typically done when the real estate agent has served one or more sellers in the neighborhood. The agent knocks on the neighbor's door, introduces himself to the neighbor, informs the neighbor that he is representing a neighbor as a client, and requests that the neighbor keep him in mind in the event that the neighbor or other family members enter the market for real estate.

Caravan tour   This is a van (or bus) tour event organized by the manager of a real estate brokerage house in which the real estate agents at the brokerage house tour through available listings.

Civil Rights Act   The Civil Rights Act (of 1964) is a landmark piece of legislation in the U.S. that outlawed major forms of discrimination against racial, ethnic, national and religious minorities, and women.

Clean Air Act   This is legislation governing the emission of air pollutants from human sources into the atmosphere.

Clean Water Act (CWA)   CWA of 1972 is the national law in the United States related to water pollution. The goal is to eliminate releases of high amounts of toxic substances into water systems and to ensure that surface water (e.g., rivers and lakes) meets standards necessary for human sports and recreation.

Closing   Closing is the process in which the buyer representative attempts to wrap up the real estate deal by getting the buyer to agree and sign the paperwork that makes the deal legal.

Coastal Zone Management Act   This law was passed in 1972 in the United States. It is relevant to land use and water development in 30 states that have coastal waters. The law is designed to preserve, protect, and develop coastal resources.

Collateral   Collateral is a financial criterion used by real estate agents and mortgage lenders to qualify prospective buyers. Collateral refers to specific assets such as property, saving and investments, and valuable material possessions that can secure the mortgage loan.

Comparative market analysis (CMA)   A CMA is a market valuation method commonly used in residential real estate. This method determines the price of the property based on comparable properties recently sold.

Competitive analysis   Competitive analysis involves data collection about the marketing behavior of the key competitors of the real estate development firm. This analysis assists the marketing manager with target marketing, product strategy, pricing strategy, place strategy, and promotion strategy.

Consequentialism   Consequentialism is a group of ethical theories asserting that the consequences of one's action should be the basis for any judgment about the rightness or wrongness of that action.

Corporate societal responsibility   Corporate societal responsibility in marketing encompasses balancing the legitimate demands of stakeholders (namely, customers, employees, environment, suppliers/dealers, regulators, local community and shareholders); accepting accountability for marketing decisions; and integrating ethical and societal obligations into the firm's marketing activities.

Corporate sponsorship   Corporate sponsorship is a form of public relations in which the real estate development firm sponsors sports and other community events; in return, the company's name, logo, and slogan appears using a variety of media (e.g., banners) directed to the audience of these events.

Costing approach   This is a method of determining market value of real estate commonly used by builders and real estate developers. The method involves determining the price by adding the total cost of the project plus a profit margin.

Customer analysis   Customer analysis focuses on an assessment of the housing (or property) needs and satisfaction of current customers of a real estate development firm.

Deceptive Trade Practices Act (DTPA)   The DTPA is a U.S. law providing protection for consumers from unscrupulous business practices. Under this law, business firms, including real estate firms, can be prosecuted for false, misleading, or deceptive practices.

Deed restrictions   Deed restrictions are legal demands typically made by the real estate developer that restrict the use of the developed land in certain ways (e.g., land can only be used for a single-family residence containing at least 1,200 square feet of living area).

Democratic leadership style   Democratic leadership (sometimes known as "participative leadership") refers to situations in which the sales manager leads the sales force by encouraging the sales associates to participate in many decisions involving the sales organization (recruiting new sales associates, deciding on who needs training and in which topics, determining who should be given a sales achievement, etc.).

Deontological ethics   Deontological ethics assert that the rightness or wrongness of one's action is judged based on rules or conventions related to that action.

Digital advertising   Digital advertising is a form of real estate advertising involving websites, banner ads on related websites, e-mail messages, and video marketing.

Direct-response advertising   Direct-response advertising is a passive prospecting method a real estate agent uses to attract prospective buyers. A typical direct-response ad is a newspaper ad that provides information about the real estate agent and prompts the audience to contact the agent when the need arises.

Distributive negotiation   Distributive negotiation is the process of negotiation in which there is the perception that there exists a fixed "pie" and one needs to do one's best to obtain a bigger slice of the pie (zero-sum situation).

Dominance, Influence, Steadiness, and Compliance (DISC)   Real estate sales managers typically administer to potential recruits a personality inventory called the DISC that measures the extent to which applicants are high or low on dominance, influence, steadiness, and compliance.

Earnest money contract   An earnest money contract is essentially an agreement that states that the buyer has agreed to purchase a property for a specified amount and the buyer has paid a specified amount of money to demonstrate that intent.

Environmental impact statement   An environmental impact statement refers to a document that captures information about the effect of a proposed project on the total environment surrounding the planned structures. This should allow a neutral decision maker to judge the environmental benefits and costs of the project.

Equal Credit Opportunity Act (ECOA)   ECOA is a U.S. law enacted in 1974. In the context of real estate, this law applies mostly to mortgage lenders. The law states that it is unlawful for any mortgage lender to discriminate against any applicant who has the capacity to contract on the basis of race, color, religion, national origin, sex, marital status, or age.

Equal Employment Opportunity Commission (EEOC)   The EEOC is a federal agency in the U.S. designed to enforce other laws against workplace discrimination. The EEOC takes action against employers who are found to have engaged in discrimination based on an employee's race, color, national origin, religion, sex, age, disability, genetic information, or retaliation for reporting, participating in, or opposing a discriminatory practice at work. The EEOC also treats acts of discrimination against lesbians, gays, and bisexual individuals as illegal.

Equal Pay Act   The Equal Pay Act (of 1963) is a law aimed at prohibiting sex discrimination in the work place. The law prohibits employers from paying wages to women at a rate less than the rate at which the employer pays wages to men for equal work on jobs for which the performance requires equal skill, effort, and responsibility.

Escrow fees   Escrow fees are fees paid to an escrow agent—typically an attorney or a title company—for making sure that all the terms of the earnest money contract are implemented; earnest money is a deposit made by the buyer to the seller demonstrating the buyer's good faith in a transaction.

Exclusive agency listing   Exclusive agency listing is similar to the exclusive-right-to-sell listing, with the exception that the seller has the right to sell the property alone; and if he does, the brokerage firm of the seller agent does not receive the sales commission.

Exclusive-right-to-sell listing   This is listing agreement between the listing agent and a seller in which the brokerage firm of the listing agent is the firm that gets paid the commission (which in many cases is shared with the brokerage firm of the buyer representative, assuming that there was a buyer agent).

Expert power   Expert power is a form of social power commonly used in interpersonal communications (e.g., communication between a real estate agent and a prospective customer). It refers to social influence based on the special knowledge and professional experience of the communicator as perceived by the audience. The higher the perception of expertise, the greater the social influence of the communicator.

External analysis   External analysis is an assessment of the external environment of the real estate development firm. This assessment typically involves market analysis and competitive analysis.

Fair housing law   The U.S. Fair Housing Law of 1968 is designed to protect buyers (or renters) of a residential property from discrimination by property owners (or managers).

Farming   Farming is a prospecting method in which a real estate agent establishes expertise in real estate related to specific neighborhoods.

Filing fees per release   Filing fees per release are expenses related to filing and recording releases that clear the title of the property.

Floor plan   The floor plan of a residential structure reflects the design of the living area, the sleeping area, and the service area of the structure.

Floor time   Floor time is a passive prospecting method in which certain agents in a brokerage firm are assigned to handle incoming calls. Some of the incoming calls can be prospective clients.

Focus on Interests, Not Positions   This is a negotiation principle that advocates that real estate negotiators should not focus on positions (i.e., position on price alone) but address the interests underlying the positions.

Frequency   Frequency is a measurement criterion used by media planners to decide on the number of message insertions in one or more media vehicle in the context of a real estate marketing communications campaign. It refers to the number of times that the target audience is likely to be exposed to the focal ad in the context of a specific media vehicle and across a set of vehicles.

Goal selection   Goal selection is the second step in promotion planning in which the desired outcomes of the marketing communications campaign are spelled out (e.g., build brand awareness, create a positive image of the company, move the audience to take action by contacting the leasing office).

Gross rating points   Gross rating points is a measurement criterion used by media planners to select among various configurations of media vehicles with a set number of insertions in the context of a real estate marketing communications campaign. Gross rating points is a product of reach and frequency.

Housing expense ratio   Housing expense ratio is a formula used by real estate agents and mortgage lenders to qualify a prospective buyer. The formula is monthly housing expense divided by monthly gross income.

Income approach   This is a method of determining market value of rental property commonly used by property management companies. The method involves determining the market value of the property by determining the annual net income the rental property generates and dividing this figure by the market rate (current interest rate).

Insist on Using Objective Criteria   This is a negotiation principle that advocates that all price positions should be evaluated using criteria that are considered to be fair and credible.

Integrative negotiation   Integrative negotiation is a process of negotiation based on the assumption that the "pie" can be made bigger and that both parties can walk away with a slice of the pie that is fair and satisfactory (non-zero sum). That is, effective negotiation in real estate involves problem solving and making tradeoff decisions.

Internal analysis   Internal analysis is an assessment of the internal environment of the real estate development firm. This assessment typically involves a sales analysis and a customer analysis.

Interpersonal orientation   Interpersonal orientation refers to the extent to which real estate negotiators are sensitive to interpersonal aspects of their relationship with one another. A high interpersonal-orientation negotiator is one who has been sensitized, and is therefore likely to be especially reactive to variations in the other's behavior. A low interpersonal-orientation negotiator, on the other hand, is relatively insensitive to the interpersonal aspects of his relationship with the other, and is therefore likely to be less responsive to variations in the other's behavior.

Interstate Land Sales Full Disclosure Act (ILSFDA)   The ILSFDA is a U.S. law passed in 1968 to regulate interstate land sales—specifically to protect consumers from fraud and abuse in the sale or lease of land.

Invent Options for Mutual Gain   This is a negotiation principle that advocates the generation of multiple solutions that both negotiating parties may find satisfactory.

Job analysis   This is the process of breaking down a job description into activities and responsibilities.

Job Characteristics Model   The model argues that enriched or complex jobs are best composed of five core job characteristics: task significance, task variety, task identity, autonomy and feedback.

Job enrichment   Sales managers enrich jobs by providing the salesperson with greater autonomy and responsibility.

Laissez-faire leadership style   Laissez-faire leadership (also known as "delegative leadership") is a style in which sales managers take a passive approach to the management of the sales force.

Legitimate power   Legitimate power is a form of social power commonly used in interpersonal communications (e.g., communication between a real estate agent and a prospective customer). It refers to social influence based on the position of authority of the communicator as perceived by the audience. The higher the position of authority of the communicator, the greater the social influence of the communicator.

Loan discount fee   A loan discount fee is a fee charged by the mortgage lender to make yield on the loan competitive with other investment options; it is usually called "points," where 1 point equals 1 percent of the mortgage loan amount.

Low versus high context culture   A low context culture is one in which people communicate with one another through text or written materials. A high context culture means that communications among various parties engaged in business transactions tend to involve both oral and written communications, and that oral communications can substitute for written communications.

Management-by-exception   Management-by-exception refers to management designed to maintain the status quo. The status quo is the sales agents doing their job and meeting the sales quota. Deviations from the status quo require intervention by management.

Market analysis   Market analysis refers to collecting and analyzing data about issues directly related to target marketing, product strategy, place strategy, pricing strategy, and promotion strategy in the context of a real estate development firm.

Market approach   This is a method of determining market value of residential real estate. The method involves determining the price of a residential property as a direct function of the price of comparable homes ("comps") sold recently. Once these "comps" are identified, the real estate appraiser obtains information about these comparable properties through their sales record. The appraiser proceeds to make adjustments—adjustments to the price of each comp to further match the house being appraised.

Market segmentation   Market segmentation refers to the process of dividing the total market into several segments using criteria such as demographics, psychographics, and geographics.

Market selection   Market selection is the process by which the marketing manager in a real estate development firm selects the most viable market segments from a pool of segments (identified in a market segmentation procedure).

Media schedule   This is the fifth step in promotion planning, in which the real estate marketer examines the effectiveness of various configurations of media vehicles guided by reach, frequency, and cost considerations. Once a configuration is selected, it becomes "the media schedule."

Media selection   This is the fourth step in promotion planning, in which the various media categories (e.g., advertisements on television and in local/regional magazines) and media vehicles (e.g., the news program of television station XYZ, and local/regional magazines A, B, and C) are identified for possible inclusion in a media schedule.

Message selection   This is the third step in promotion planning, in which the core message of the marketing communications campaign of the real estate development firm is spelled out. That core message is traditionally captured in terms of a slogan.

Mobile advertising   Mobile advertising is a form of real estate advertising through mobile phones, with, for example, banner ads displayed on all mobile operating systems.

Motivational orientation   Motivational orientation refers most generally to one negotiator's attitudinal disposition toward another, which may be manifested in three styles: cooperative, competitive, and individualistic. A negotiator has a cooperative negotiation style to the extent that he has a positive interest in the other's welfare as well as his own. A competitive style denotes an interest in doing better than the other, while at the same time doing as well for oneself as possible. A negotiator with an individualistic style is simply interested in maximizing his own outcomes, regardless of how the other fares.

Multiple Listing Service (or MLS)   The MLS is a digital database controlled by the National Association of Realtors (NAR) in the U.S. that allows all "realtors" (real estate agents and their brokers, as members of the NAR) to list properties in the system, thus promoting the property to other realtors who have relationships with prospective buyers.

National Environmental Policy Act (NEPA)   NEPA is a U.S. law promoting the protection and enhancement of the environment.

National Flood Insurance Act   This piece of U.S. legislation was enacted in 1968. It led to the creation of the National Flood Insurance Program (NFIP). NFIP was designed to accomplish two goals: (1) to provide flood insurance for real estate property in floodplain communities, and (2) to identify areas of high and low flood hazard and, based on this assessment, establish flood insurance rates for real estate properties inside each flood hazard area.

Negotiation effectiveness   Negotiation effectiveness is a mutually satisfactory outcome in real estate negotiations.

Net listing   A net listing is a contract between a seller and a listing agent. This is an arrangement in which the broker of the seller representative agrees to pay the seller a set amount from any sale (plus closing costs). The remaining money goes to the brokerage firm of the seller agent.

Net worth   Net worth is a criterion used by real estate agents and mortgage lender to qualify prospective buyers. It is based on the balance sheet of the prospective buyer, on which liabilities are subtracted from assets.

Noise Pollution and Abatement Act   The Noise Pollution and Abatement Act of 1972 is a U.S. law designed to regulate noise pollution to protect human health and minimize noise annoyance to the general public.

One-time listing   A one-time listing refers to a listing that is limited to one buyer for one time, for one property.

Online classified advertising   Online classified advertising is a form of real estate advertising based in online classified media such as Zillow, Trulia, Realtor.com, Yahoo!, Yahoo!-Zillow Real Estate Network, Craigslist, and online classified ads in major national newspapers such as the *New York Times*.

Open house   This is a promotional event organized by the listing agent to promote a specific listing. The listing agent schedules a showing to the public at large. The agent advertises the location and time of the showing, and is present at the event to answer all pertinent questions about the property.

Open listing   An open listing is a contract between a seller and a listing agent. The contract allows anyone to sell the property, and only the broker of the agent that ends up selling the property gets the sales commission (which, in turn, is split with the brokerage firm of the buyer agent, assuming that the buyer has an agent). If the seller ends up selling the property himself, the seller does not have to pay a commission to any broker or agent.

Out-of-home advertising   Out-of-home advertising involves real estate advertising through bulletin boards, bus stops, inside and outside buses, train stations, taxis, airport terminals, banners, and building illumination.

People-oriented leadership style   People-oriented (or relationship-focused) leadership is an approach in which the sales manager focuses on the satisfaction, motivation, and general well-being of the sales agents.

Performance evaluation   Performance evaluation is the final step in promotion planning, in which the real estate marketer determines how the marketing communications campaign is to be evaluated.

Personal selling Personal selling is a form of marketing communications that real estate professionals use, which is based mostly on face-to-face communications.

Place strategy Place strategy refers mostly to site selection for real estate development.

Positional negotiation Positional negotiation is a style of negotiation that focuses on the real estate parties' positions on price (mostly) and how negotiations can effect a change in the position of the other party (hard positional) or one's own position (soft positional).

Positioning Positioning is a key message decision involved in the planning of a real estate promotion plan. This message translates into a slogan that is used in all promotional material.

Pregnancy Discrimination Act The Pregnancy Discrimination Act (of 1978) prohibits discrimination on the basis of pregnancy. Specifically, the law covers discrimination on the basis of pregnancy, childbirth, or related medical conditions.

Presenteeism Presenteeism refers to the situation in which employees (e.g., real estate agents) are physically present at work but do not use their time wisely.

Press release A press release is a form of public relations commonly used in real estate development firms that announces an important event (e.g., a real estate development firm announces the commencement of building a major shopping mall). Press releases are typically posted on the real estate development firm's website and disseminated to the local news media.

Principled negotiation The goal of principled negotiation is a wise outcome reached efficiently and amicably, not accommodating the other party (soft positional), nor beating them (hard positional). Principled negotiators try their best to separate the people from the problem, not being soft on the people and the problem (as in soft positional), nor hard on the people and the problem (as in hard positional).

Print advertising Print advertising involves real estate advertising through newspaper, direct mail, local real estate publications, yellow pages, and other local and regional consumer or trade magazines.

Product strategy Product strategy refers to decisions related to housing design, site orientation, energy conservation, and neighborhood features commonly made by architects and builders with the consultation of the marketing manager in real estate development firms.

Profile sheet A profile sheet is an information sheet that provides details regarding the "specs" of the property (information about the property, seller, size, amenities, etc.). These are typically left at the property (inside and outside the property) to be picked up by other agents and prospective buyers.

Promotion plan A promotion plan refers to a document that reflects many message and media decisions made by the real estate marketers—decisions such as audience selection, goal selection, message selection, media selection, media schedule, and performance evaluation.

Promotion strategy Promotion strategy refers to message and media decisions made by real estate marketers.

Prospecting Prospecting refers to methods used by real estate agents to identify and contact potential clients (e.g., direct mail campaign, referrals, canvassing).

Prospect strategy Prospect strategy refers to a market selection decision customarily made by the marketing professional in real estate development firms.

Publicity Publicity is a form of public relations commonly used by real estate development firms in which information is disseminated through the news media. The news media will publicize the information only if it is deemed unbiased and newsworthy.

Public relations   Public relations is a form of marketing communications performed at the corporate level, in which the focus is on corporate communications. Public relations in real estate development firms typically involves the use of press releases, corporate sponsorship of events, and publicity.

Reach   Reach is a measurement criterion used by media planners to select among various media vehicles in a real estate marketing communications campaign. It refers to the extent to which a focal media vehicle (e.g., local news program on a CBS affiliate television station) reaches a certain percent of a designated target market (e.g., 70% of high income households).

Real estate marketing   Real estate marketing involves anticipating, managing, and satisfying demand via the exchange process between buyer and seller of a property. As such, marketing encompasses all facets of real estate buyer/seller relationships. Specific marketing activities include strategic analysis, target marketing, property planning, site selection, pricing of the property, promotion planning, and marketing management.

Real Estate Settlement Procedures Act (RESPA)   RESPA is a U.S. law created in 1974. The purpose of the law is to crack down on real estate companies (e.g., lenders, real estate agents, real estate developers, and title insurance companies) that engage in providing undisclosed kickbacks to each other.

Referent power   Referent power is a form of social power commonly used in interpersonal communications (e.g., communication between a real estate agent and a prospective customer). It refers to social influence based on identification—the extent to which the audience can identify with the communicator. The higher the perception of identification, the greater the social influence of the communicator.

Reservation price   A reservation price is the bottom line (least desirable) price that real estate negotiators are willing to accept.

Resistance to yielding   Resistance to yielding refers most generally to one negotiator's attitude toward resistance to give in to the demands of the counterpart in negotiation.

Reward power   Reward power is a form of social power commonly used in interpersonal communications (e.g., communication between a real estate agent and a prospective customer). It refers to social influence based tangible resources—the extent to which the communicator can provide a tangible incentive for responding in certain ways. The higher the perception of reward value, the greater the social influence of the communicator.

Sales analysis   A sales analysis involves an assessment of the drivers of sales trends in a real estate development firm.

Separate the people from the problem   This is a negotiation principle commonly used in real estate negotiations in which the negotiator makes every attempt possible to divorce the bad feelings that may originate from the inherent conflict (two parties vying to get their way) from the problem at hand (finding common ground).

Showing   Showing is the process in which the buyer agent presents to prospective buyers available listings by taking the buyers on a tour to visit these listings.

Site orientation   The site orientation of a residential structure refers to the position of the structure in terms of North, South, East, and West coordinates (e.g., a house facing southeast).

Social contract   Social contract theory is a theory of ethics asserting that an action is judged as unethical if the action violates community norms and standards.

Social justice   Social justice is an ethics concept asserting that an action is judged to be unethical if it discriminates against any group of people using criteria such as religion, race, ethnicity, national origin, gender, sexual orientation, familial status, or age.

Social media   Social media is a form of real estate advertising based on digital social networks such as Facebook, Twitter, LinkedIn, and Pinterest.

Specialty advertising   Specialty advertising is a form of real estate advertising in which the message or slogan is imprinted on mugs, calendars, t-shirts, pens, letter openers, key chains, and other materials that are handed to prospective customers for personal use.

Staging   Staging refers to a task in which the listing agent makes arrangement with a staging company to make the property as presentable as possible, thus increasing the likelihood that the property will sell.

Stakeholder theory   Stakeholder theory is a theory of business ethics asserting that management should attend to the needs of all the firm's stakeholders (customers, employees, shareholders, the community, distributors, suppliers, and the environment), not only shareholders.

Strategic analysis   Strategic analysis involves an assessment of the internal and external environments. This is performed by the marketing executive in real estate development firms.

Survey fee   A survey fee is a fee charged by a licensed surveyor to determine the exact boundaries of the property.

Target price   Target price is the desired price that negotiators of real estate aim to achieve.

Task-oriented leadership style   Task-oriented leadership is an approach to sales management in which the manager focuses on the personal selling tasks to meet certain sales goals.

Teleological ethics   Teleological ethics is another term for consequentialism, which is a group of ethical theories asserting that the consequences of one's action should be the basis for any judgment about the rightness or wrongness of that action.

Termite inspection fees   Termite inspection fees are expenses related to the service provided by a licensed exterminator as to the condition of the property regarding termites and other wood-destroying insects.

Third-party relocation companies   These are firms that specialize in helping company executives and their families move to another location—the location of job transfer.

Title insurance   Title insurance is the fee for insurance against unknown liens, encumbrances, or defects to the property.

Tradeoffs   Tradeoffs are offers made by real estate negotiators that help achieve a mutually satisfactory agreement. Tradeoff offers are made by assembling a set of costs and benefits that go beyond price.

Transactional leadership style   Transactional leadership (also known as "managerial leadership") is an approach of sales management in which the sales manager focuses on the role of supervision, organization, and performance.

Transformational leadership style   Transformational leadership focuses on motivation and morale of the sales force to achieve sales goals.

Truth in Lending Act (TILA)   TILA is a law enacted in the U.S. in 1968 requiring lenders to fully disclose and explain the loan terms in understandable language. TILA also gives consumers the right to cancel a transaction involving a lien on a consumer's principal dwelling.

Utilitarianism   Utilitarianism is a school of thought in ethics asserting that the decision maker does not only consider the consequences of a given action but also evaluates the consequences of alternative actions. An action is judged as unethical if alternative actions could have been taken that were more likely to produce more good than harm in relation to all stakeholders.

Zone of Potential Agreement (ZOPA)   ZOPA is the range between the seller's reservation price and the buyer's reservation price in which an agreement can be reached.

Index

Page numbers in italic format indicate figures and tables.

promotion expenditures 7
promotion planning: audience selection for
59–60; description of 298; goal selection
for 60; media schedule for 61–2; media
selection for 60–1; message selection for 60;
performance evaluation for 62–3; promotion
campaign and 63
promotion-related services 85–7
promotion responsibilities 271–2
promotion strategy: cases/anecdotes 58;
description of 43; market analysis for 12;
media decisions and 53–9, 63; message
decisions and 51–3, 63
property see residential structure
property management 50, 152, 280, 283, 284
property taxes 35, 50, 51
prospective buyer: disclosing information to
86–7; follow-up and servicing 109–10; needs
assessment of 99–100; negotiating with 84;
positioning by an aspect of 52; qualifying
100–2; showing process for 102–4
prospective seller: analyzing needs of 74–5;
cases/anecdotes 81, 88–9; conclusion about
89–90; determining value of property of 73–4;
following with 88–9; making pitch to 75–7;
negotiating with 77–82, 104–5; servicing 82–8
prospect strategy: description of 13, 20; market
selection and 15–19; see also market segment
proximity factors 45
psychographic variables 17
public records 244
public relations 58–9
published reserve auction 76

quality standards: for commercial real estate 23;
for sleeping area design 26–7

rational negotiator 133–4
real estate: appraisal 156, 266, 280; auctions 76;
effect of physical attractiveness in 170; false
representation about 243; glossary 291–300;
high turnover in 163, 180; obtaining license
for 167; profit-sharing in 207; sales career in
202; work-life balance in 203–4
real estate agents: cases/anecdotes 198; Code of
Ethics for 276–87; failure to disclose issues
and 245–6; how to recruit 174–80; personal
features of 167–8; personality of 168–71;
public records and 244; see also commission;
ethical issues; negotiation
real estate development firm: advertising by
54–5, 61; corporate societal responsibility
of 269; energy conservation issues for 31–2;
marketing in 4, 5; site selection by 34–9;
zoning issues 35–8
real estate marketing: defined 5–6; media
decisions and 53–9; message decisions

in 51–3; pricing responsibilities and
270–1; product responsibility and 270;
professional designations in 187, 188;
promotion planning and 59–63; promotion
responsibilities and 271–2; recruitment
process in 164; strategic analysis and 6–13;
training process in 181; through video 55–6;
see also ethical issues; promotion strategy
real estate marketing laws: Clean Air Act
251; Clean Water Act 252; Coastal Zone
Management Act 253; conclusion about
253; Deceptive Trade Practices Act 243–4;
Equal Credit Opportunity Act 247–8;
Fair Housing Law 248–9; Interstate Land
Sales Full Disclosure Act 246–7; National
Environmental Policy Act 250–1; Noise
Pollution and Abatement Act 252; Real
Estate Settlement Procedures Act 247; Truth
in Lending Act 248
real estate pricing: approaches to 46–50, 51; cases/
anecdotes 48–9; factors influencing 43–6
Real Estate Settlement Procedures Act
(RESPA) 247
recognition programs 200–1
recruitment process: analyzing the job
164–6; conclusion about 190–1; job
qualifications assessment 166–8, 170–2; legal
considerations 171–2; methods for 172–3;
prospecting sources for candidates 172–3;
selection of prospects 177–80; validation
process 180–1; see also compensation plans
referent power 53, 230
referral fees 205, 206, 207
referrals and repeat business 73, 96
rental properties, income approach for 50, 51, 74
renters method of prospecting 97
reservation price 104, 105, 117
residential site 30–1
residential structure: closing process 106–8;
detecting deficiencies in 24; energy
conservation issues 31–3, 40; enhancing
curb appeal of 83; floor plan 23–7; judging
architecture of 27–30, 40; neighborhood
features 33–4, 40; presenting true picture of
87; price strategy for 47–50, 73–4; property
taxes for 35; quality standards for 23, 26–7;
sales trend for 7; showing process for 102–3;
staging of 79, 83
resistance to yielding 132, 299
retrofitting 32
reward power 54, 229
right to a healthy environment 264
right to be heard and to redress 266
right to be informed 264
right to choose and to fair trade 265
right to privacy 267
right to product safety 263–4